Agriculture Issues and Policies

nova
science publishers
www.novapublishers.com

Agriculture Issues and Policies

Strategies to Achieve Sustainable Development Goals (SDGs): A Road Map for Global Development
Rajani Srivastava, M.Sc., PhD (Editor)
2022. ISBN: 978-1-68507-836-2 (Hardcover)
2022. ISBN: 979-8-88697-027-2 (eBook)

Pistachios: Cultivation, Production and Consumption
Shaziya Haseeb Siddiqui, PhD,
Shoaib Alam Siddiqui, PhD (Editors)
2022. ISBN: 978-1-68507-949-9 (Hardcover)
2022. ISBN: 979-8-88697-063-0 (eBook)

Ecosystem Services: Types, Management and Benefits
Hanuman Singh Jatav, PhD,
Vishnu D. Rajput, PhD (Editors)
2022. ISBN: 978-1-68507-614-6 (Hardcover)
2022. ISBN: 978-1-68507-747-1 (eBook)

Underutilized Crops and Their Value Addition
Romesh Kumar Salgotra, PhD,
Monika Sood, PhD, Surbhi Jasrotia
2021. ISBN: 978-1-68507-443-2 (Hardcover)
2021. ISBN: 978-1-68507-481-4 (eBook)

Solanum melongena: Production, Cultivation and Nutrition
Abdul Majid Ansari, PhD,
Wajid Hasan, PhD,
M. Prakash, PhD (Editors)
2021. ISBN: 978-1-68507-311-4 (Hardcover)
2021. ISBN: 978-1-68507-323-7 (eBook)

More information about this series can be found at
https://novapublishers.com/product-category/series/agriculture-issues-and-policies/

Tarique Hassan Askary
Editor

Pest Management

Methods, Applications and Challenges

nova
science publishers
www.novapublishers.com

DOI: 10.52305/VCWF8745.

Library of Congress Cataloging-in-Publication Data

ISBN: 979-8-88697-268-9

Published by Nova Science Publishers, Inc. † New York

Contents

Preface

Insects constitute 70 percent of all animals in the world. A complex relationship exists between insects and plants but when they cause loss to crops and stored products, they are known as pests. These losses may vary depending upon the crop and population density of pests. Annual global losses due to pests run into billions of dollars. Farmers generally depend upon chemical pesticides to suppress the pests and get quick results, but they are normally not interested in their identification. This leads to indiscriminate use of toxic chemicals which causes an adverse effect on the environment and human health. Single factors or innovations alone cannot pave the way to the development of sustainable agriculture. Hence, the present situation demands a strategy that should be ecologically sound, easy to practice and cost-effective in a given ecosystem. Plenty of literature is available on chemical and microbial pesticides but the information is scattered. At present, there is no book that could provide detailed information including the remarkable developments and latest achievements about both chemical and microbial pesticides in a single volume. The current volume fulfills these objectives.

In this spirit, I targeted 58 experts from six countries who were authorities in their disciplines to compose the 13 chapters contained in this book. The volume is broadly divided into two parts. Part I contains six chapters that deal with the role of chemicals in pest management. The opening chapter begins with the challenges and opportunities in the application of insecticides. The five additional chapters provide information on the pests of temperate fruits, pulses, spices and rice and the recent advances in their management. The concluding chapter of Part I deals with the utilization of chemicals in plant parasitic nematode management and the challenges faced thereof.

In Part II, there are seven chapters devoted to microbes. The first chapter broadly explains the role of biopesticides in pest management, the new developments, challenges and future thrusts in this direction. The emphasis of the subsequent chapters shifts to different biocontrol agents viz., entomopathogenic nematodes, nematophagous fungi, bacteria and baculoviruses. One chapter deals exclusively with the integrated application of entomopathogenic nematodes and chemicals. The last chapter of the book deals extensively with the field application of biopesticides.

An attempt has been made to present all important aspects and necessary guidelines for pest management in as simple and lucid mannered as possible with some essential data and nice photographs. The present book would help agricultural scientists, research practitioners, extension workers, the farming community and especially those undergraduate and postgraduate students having entomology or nematology as their specialized subject.

I am thankful to all the contributors for their help in making this book possible. Finally, I express gratitude to my beloved wife Afrin, for her long-time support during the preparation of this book.

Tarique Hassan Askary
July 2022

Part I. Chemicals in Pest Management

Chapter 1

Application of Insecticides for the Management of Insect Pests: Challenges and Opportunities

Ishtiyaq Ahad*, **M. Parthiban, Danishta Aziz, Afreen Rehman and Jameela Rasool**
Division of Entomology, Faculty of Agriculture,
Sher-e-Kashmir University of Agricultural Sciences and Technology of Kashmir,
Jammu and Kashmir, India

Abstract

Insects are one of the dominant animals on the earth due to their tremendous adaptation to a wide variety of climatic conditions. In India, evidence indicates that pests cause 25 per cent loss in rice, 5-10 percent in wheat, 30 per cent in pulses, 35 per cent in oilseeds, 20 per cent in sugarcane and 50 per cent in cotton. Climate change, development of modern agriculture with intensified cropping and high yielding varieties have led to increase in the severity of pest incidence and as a result high economic loss. To check the economic loss caused by various insect pest populations, different groups of natural and synthetic pesticides, such as organochlorines, organophosphates, carbamates, and neonicotinoids were introduced in agriculture for successful management of insect pest. Despite insecticides having negative impact on environment and human health, application of insecticides play inevitable role in integrated pest management.

Keywords: insect pest, insecticide, environment, economic loss, management

---

* Corresponding Author's Email: drishtiyaqraina@gmail.com.

In: Pest Management: Methods, Applications and Challenges
Editor: Tarique Hassan Askary
ISBN: 979-8-88697-268-9

1. Introduction

The major challenge to food security is due to emerging outbreak of insect pest and invasive insect species. Globalization of agricultural trade led to more chances of introduction of invasive insect species which caused huge crop loss due to absence of natural enemies in the new areas. Intensive agricultural practice, introduction of high yielding varieties, increased agricultural trade and climate change resulted in minor pest to emerge as major pest and at the same time the invasion of exotic insect species (Rai et al., 2014). An average estimated yield loss due to insect pests is about 15-20 per cent. As far as India is concerned, the yield losses due to pests in rice, wheat, pulses, oilseed, sugarcane and cotton has been estimated 25, 5-10, 30, 35, 20 and 50 per cent, respectively. The cotton whitefly, *Bemisia tabaci* appeared in epidemic form in Punjab during 2015-16, destroying 2/3rd of cotton crop. The crop suffered huge losses worth Rs. 4200 crores despite the use of pesticides worth Rs. 150 crores (Kranthi, 2016).

As temperature and precipitation are key drivers of the abundance, distribution, and phenology of insects, studies suggest climate change is likely to modify the current distribution and impact of agricultural insect pests (Deutsch et al., 2018; Lehmann et al., 2020). The regional composition of crops, propensity to spray pesticides, and technology adoption such as seed treatments may vary irrespective of damage (Larsen and McComb 2021). Global warming increased the risk of plant hoppers outbreak in Bangladesh (Ali et al., 2014). One important and thus far poorly understood impact is the effect of land cover and climate change on agricultural insect pests and insecticides. However, both landscape and climate change is expected to increase future insecticide use (Larsen and McComb 2021). It is supposed that an altered pest population may change the cost of pest treatment (Chen and McCarl, 2001; Delcour et al., 2015). Insecticide usage varies with crop, region and year, whereas incidence of insect pest population varies according to prevailing weather pattern. Koleva et al., (2009) investigated the impact of climate change on the external cost of pesticide applications in US agriculture. The results were alarming as the average external cost of pesticide use in agriculture was US $42 per hectare.

In this chapter we aim to address the impact of insecticide application in insect pest management, besides, highlighting the opportunities to fit it as one of the components in integrated pest management programs.

2. Use of Insecticides in Agriculture

The research workers around the world are documenting the movement of agrochemical in the environment as well as residual poisoning to the human being (Goulson et al., 2015; Goswami et al., 2017; Hakme et al., 2018). It is well recorded that 0.1 per cent of pesticides reached the target pest population and the remaining 99.9 per cent of pesticides are leaked into surrounding environment causing water pollution, soil contamination, insect resistance towards insecticides, development of biotypes, loss of biodiversity and elimination of key species such as natural enemies and pollinators.

2.1. Earlier Evidence on the Negative Impact of Insecticide

The attention to pesticide poisoning was drawn among public only after the publication of silent spring by Rachel Carson. Initially, the contamination effect of organochlorine pesticides such as DDT (dichloro-diphenyl-trichloroethane) and its metabolites has been observed in wildlife including birds, mammals, reptiles and aquatic amphibians (Krief et al., 2017; Hatcher et al., 2008). Besides organochlorine, other group of pesticides such as organophosphates, carbamates, pyretheroids are the major causes of chronic toxicity in birds, rodents, fishes and amphibians. Theses pesticides disturb the endocrine system leading to harmful effect on wildlife (Bhandari et al., 2015). Among the organophosphate insecticides, Malathion and Chloripyriphos recorded more than 50 per cent of residual contamination in environment (Wang et al., 2016).

2.2. Impact of Insecticides on Pollinators

Acute and chronic toxicity of pesticide poisoning to global pollinators especially bees have been reported in terms of homing ability, reproductive process and foraging behaviour (Henry et al., 2012; Rundlöf et al., 2015). Honeybees are exposed to several pesticides that results in steep decline in bee population as well as adverse impact on pollinator dependent fruit crops. The exposure of thiomethaxam to bumble bee colonies offers low visitation rate and less pollen collection from apple tree (Stanley et al., 2015). Pesticide contamination impairs the ability of bees in terms of pollination capacities that ultimately have a negative impact on crop yield (Bommarco et al., 2013; Vanbergen, 2013).

2.3. Impact of Insecticides on Human Being

The overdose or incorrect application of insecticides shows several undesirable effects on human being. According to the market survey reports, humans are threatened by contamination caused by pesticides (Jones et al., 2014). Public exposure of pesticides via several pathways includes i. direct route such as oral, dermal, inhalation and ii. indirect route like food consumption and occupational exposure (Kennedy et al., 2015). Several human health problems related to wheezing, allergic, obesity, cancer, neurological disorders and endocrine problem have been reported due to excessive use of chemical pesticides (George and Shukla, 2011; Hoppin et al., 2017). The potential effect of pesticides can act as an agonist to modulate the estrogen receptor bindings (Hernández et al., 2013). Organochlorine pesticides cause cognitive impairment, which is approximately three times higher in the old aged people. Even after weight loss there are chances of pesticides reach to brain neurons (Lee et al., 2016). Drifting and volatile nature of pesticides are not only hazardous to the health of agricultural workers but also to the surrounding environment. The pesticide exposure during gestation cause adverse effects to the birth outcome (Kumar et al., 2019).

2.4. Residual Effects of Insecticides on Food Crop

Several factors that significantly affect the residual level of insecticides are i. properties of active ingredient and their formulation ii. Site condition iii. waiting period for harvest iv. type of crop v. pest infestation and vi. plant health. Higher residues are likely to occur in poorly developed crops. Some carcinogenic nature of pesticides are reported in apple such as captan, folpest, iprodione and procymaidone (Lozowicka, 2015). In order to overcome the toxic pesticide residues in food, several international standards are specified by codex Alimentarious, a joint body of Food and Agricultural Organization (FAO) and World Health Organization (WHO) to ensure the acceptable level of maximum residual limit (MRL) of the food consumption. Several attempts have been made to ban the toxic pesticides and reduce the MRL of pesticides.

Recent efforts are focused on target delivery of active ingredient and reduce the impact on non-target organism.

3. Opportunities

The discovery of synthetic insecticides between 1930 and 1950 brought revolution in modern agriculture with increased yield by protecting the crop from voracious insects. In integrated pest management (IPM), insecticide application plays a vital role in bringing the insect population below economic threshold level. The regulatory action taken by government bodies and policy makers restricted or banned the use of hazardous insecticides and monitored the impact of pesticides on environment (Perry et al., 2013).

Organophosphates are esters of phosphoric acid or thiophosphoric acid. Originally, the compounds were highly toxic to mammals but later on manufactured in such a way that it proved less toxic to mammals and toxic to target insects. Organophosphate pesticides such as malathion, parathion, diazinon and chlorpyrifos are among the most widely used pesticides in agriculture. Though, they comprise only approximately 10 per cent of the total pesticides used in the United States, however, they are applied to over 65 per cent of the total insecticide-treated agricultural crops. Organophosphate pesticides can be rapidly degraded by hydrolysis in sunlight, air, and soil and biodegraded by soil bacteria, although small amount may persist in food and drinking water.

3.1. Newer Class of Insecticides- Neonicotinoid

Nicotine is isolated from *Nicotiana tabacum*. It is used as systemic insecticides against sucking insect pests with its unique mode of action on nicotinic acetylcholine receptors of insect central nervous system (Thany, 2010). Sine nicotine is equally or more toxic to mammals than insects, scientists are working to synthesize new nicotine insecticide that would be only toxic to insects (David et al., 2007). Synthetic neonicotinoids are highly toxic to insects which target nicotinic acetylcholine receptor in post synaptic membrane (Yamamoto and Casida, 1999; Tomizawa and Casida, 2003).

Currently neonicotinoids and fibronil occupies third position in the world market with additional significance over conventional insecticides such as systemic action and unique mode of action targeting nicotinic acetylcholine receptor in post synaptic membrane (Pisa et al., 2021). During 1991, imidacloprid was introduced as the first neonicotinoids. Neonicotinoids are water soluble and plants absorb neonicotinoids through vascular system and spread it into all part, even in pollen and nectar. There are seven commercial compounds of neonicotinoids available in the market. N- nitroguanidines include imidacloprid, thiomethoxam, clothianidin and dinotefuran; N-nitromethylenes include nitenpyram and N- cyanoamidines include acetamiprid and thiocloprid (Jeschke et al., 2011, 2013). Neonicotinoids are applied in soil by different ways such as soil drenching, mixing in irrigation water and foliar spray. Moreover, they are most commonly applied as seed treatment which protects seedlings from any insect attack (Bonmatin et al., 2015). Although, there are reports of neonicotinoides creating another "pesticide treadmill" after DDT, developing insecticides resistance and depletion of natural enemies and pollinators (Parsons, 2017; Bakker et al., 2020). However, not much information is available regarding other pollinators such as butterflies, moth and flies (Wu et al., 2019).

3.2. Nano Pesticides

Globally researchers are focusing on reduction of pesticide use without posing any adverse effect on agricultural productivity and profitability. Application of nanotechnology is supposed to overcome several limitations faced from conventional pesticides. Nanopesticides are highly effective against insect pests and less harmful to non-target organisms (Kumar et al., 2019). The specified features of nano pesticides are:

i. Controlled release,
ii. Spatially- targeted release, and
iii. Self or remotely regulated release to successfully effect the target organism.

The main function of nanoencapsulation or nanocarrier is:

i. To protect the active ingredients before the target specific release,
ii. To improve the solubility and penetration of active ingredients into the target site and,

iii. To modify or control release of active ingredients into the surrounding environment (Kumar et al., 2019).

Different formulations of nanopesticides have been developed such as nanosuspension, nanoemulsion and nanocapsules (Kumar et al., 2019). However, few concerns of nanopesticide applications are the effect on non-target organisms as well as their environmental fate. Till now, the successes achieved by the application of nanopesticides are only at laboratory level and meagre information is available related to field studies (Li et al., 2006).

Conclusion

Globally, the current focus and the prime goal in all around research field are to manage natural resources and maintain environmental balance. The availability of new insecticide molecules with novel modes of action such as higher toxicity to target pests at very low doses, less toxicity to non-target organisms and low persistence in nature has further strengthened the role of insecticides in IPM. The site-specific application of nano formulated insecticides is a good approach in the management of crop insect pests but further researches are still needed particularly under field conditions to ascertain the technologies.

Acknowledgments

We are grateful to Head, Division of Entomology, Faculty of Agriculture, SKUAST-Kashmir, India for providing support in the preparation of the manuscript.

References

Ali, M. P., Huang, D., Nachman, G., Ahmed, N., Begum, M. A. and Rabbi, M. F. (2014). Will climate change affect outbreak patterns of plant hoppers in Bangladesh. *PLoS One*, 9(3), e91678. https://doi.org/10.1371/journal.pone.0091678.

Bakker, L., W. Van der Werf, P. Tittonell, K. A. G. Wyckhuys, and F. J. J. A. Bianchi. 2020. Neonicotinoids in global agriculture: evidence for a new pesticide treadmill? *Ecology and Society*, 25: 26. https://doi.org/10.5751/ES-11814-250326.

Bhandari, R. K., Deem, S. L., Holliday, D. K., Jandegian, C. M., Kassotis, C. D., Nagel, S. C., Tillitt, D. E., Vom Saal, F. S., and Rosenfeld, C. S. (2015). Effects of the environmental estrogenic contaminants bisphenol A and 17α-ethinyl estradiol on sexual development and adult behaviors in aquatic wildlife species. *General and Comparative Endocrinology*, 214: 195–219.

Bommarco, R., Kleijn, D. and Potts, S. G. (2013). Ecological intensification: Harnessing ecosystem services for food security. *Trends in Ecology and Evolution*, 28: 230–238.

Bonmatin, J.-M., Giorio, C., Girolami, V., Goulson, D., Kreutzweiser, D. P., Krupke, C., Liess, M., Long, E., Marzaro, M. and Mitchell, E. A. (2015). Environmental fate and exposure; neonicotinoids and fipronil. *Environmental Science and Pollution Research*, 22: 35–67.

Chen, C. C. and McCarl, B. A. (2001). An investigation of the relationship between pesticide usage and climate change. *Climatic Change*, 50: 475-487.

David, D., George, I. A. and Peter, J. V. (2007). Toxicology of the newer neonicotinoid insecticides: Imidacloprid poisoning in a human. *Clinical Toxicology*, 45: 485–486.

Delcour, I., Spanoghe, P. and Uyttendaele, M. (2015). Literature review: Impact of climate change on pesticide use. *Food Research International*, 68: 7-15.

Deutsch, C. A., Tewksbury, J. J., Tigchelaar, M., Battisti, D. S., Merrill, S. C., Huey, R. B. and Naylor, R. L. (2018). Increase in crop losses to insect pests in a warming climate. *Science,* 361: 916–919.

George, J. and Shukla, Y. (2011). Pesticides and cancer: Insights into toxicoproteomic-based findings. *Journal of Proteomics*, 74: 2713–2722.

Goswami, L., Kim, K. H., Deep, A., Das, P., Bhattacharya, S. S., Kumar, S. and Adelodun, A. A. (2017). Engineered nano particles: nature, behavior, and effect on the environment. *Journal of Environmental Management*, 196: 297–315.

Goulson, D., Nicholls, E., Botías, C. and Rotheray, E. L. (2015). Bee declines driven by combined stress from parasites, pesticides, and lack of flowers. *Science*, 347(6229). https://doi.org/10.1126/science.125595.

Hakme, E., Lozano, A., Ferrer, C., Díaz-Galiano, F. J. and Fernández-Alba, A. R. (2018). Analysis of pesticide residues in olive oil and other vegetable oils. *TrAC Trends in Analytical Chemistry*, 100: 167–179.

Hatcher, J. M., Delea, K. C., Richardson, J. R., Pennell, K. D. and Miller, G. W. (2008). Disruption of dopamine transport by DDT and its metabolites. *NeuroToxicology*, 29: 682–690.

Henry, M., Béguin, M., Requier, F., Rollin, O., Odoux, J. F., Aupinel, P., Aptel, J., Tchamitchian, S. and Decourtye, A. (2012). A common pesticide decreases foraging success and survival in honey bees. *Science*, 336 (6079): 348–350.

Hernández, A. F., Parrón, T., Tsatsakis, A. M., Requena, M., Alarcón, R. and López-Guarnido, O. (2013). Toxic effects of pesticide mixtures at a molecular level: their relevance to human health. *Toxicology*, 307: 136–145.

Hoppin, J. A., Umbach D. M., Long, S., London S. J., Henneberger, P. K., Blair, A., Alavanja, M., Freeman, L. E. B. and Sandler, D. P. (2017). Pesticides are associated with allergic and non-allergic wheeze among male farmers. *Environmental Health Perspectives*, 125: 535–543.

Jeschke, P., Nauen, R. and Beck, M. E. (2013). Nicotinic acetylcholine receptor agonists: a milestone for modern crop protection. *Angewandte Chemie International Edition*, 52: 9464–9485.

Jeschke, P., Nauen, R., Schindler, M. and Elbert, A. (2011). Overview of the status and global strategy for neonicotinoids. *Journal of Agricultural and Food Chemistry*, 59: 2897–2908.

Jones, K., Everard, M. and Harding, A. H. (2014). Investigation of gastrointestinal effects of organophosphate and carbamate pesticide residues on young children. *International Journal of Hygiene and Environmental Health*, 217: 392–398.

Kennedy, M. C., Glass, C. R., Bokkers, B., Hart, A. D. M., Hamey, P. Y., Kruisselbrink, J. W., de Boer, W. J., van der Voet, H., Garthwaite, D. G. and van Klaveren, J. D. (2015). A European model and case studies for aggregate exposure assessment of pesticides. *Food and Chemical Toxicology*, 79: 32–44.

Koleva, N. G. and Schneider, U. A. (2009). The impact of climate change on the external cost of pesticide applications in US agriculture. *International Journal of Agricultural Sustainability*, 7: 203-216.

Kranthi, K. R. (2016). Let not the whitefly see red. *Cotton Statistics and News*. 45: 1-6.

Krief, S., Berny, P., Gumisiriza, F., Gross, R., Demeneix, B., Fini, J. B., Chapman, C. A., Chapman, L. J., Seguya, A. and Wasswa, J. (2017). Agricultural expansion as risk to endangered wildlife: pesticide exposure in wild chimpanzees and baboons displaying facial dysplasia. *Science of The Total Environment*, 598: 647–656.

Kumar, S., Nehra, M., Dilbaghi, N., Marrazza, G., Hassan, A. A. and Kim, K. H. (2019). Nano-based smart pesticide formulations: emerging opportunities for agriculture. *Journal of Controlled Release*, 294: 131–153.

Larsen, A. E. and McComb, S. (2021). Land cover and climate changes drive regionally heterogeneous increases in US insecticide use. *Landscape Ecology*, 36: 159-177.

Lee, D. H., Lind, P. M., Jacobs, D. R., Salihovic, S., van Bavel, B. and Lind, L. (2016). Association between background exposure to organochlorine pesticides and the risk of cognitive impairment: a prospective study that accounts for weight change. *Environment International*, 89–90: 179–184.

Lehmann, P., Ammune´t, T., Barton, M., Battisti, A., Eigenbrode, S. D., Jepsen, J. U., Kalinkat, G., Neuvonen, S., Niemela, P., Terblanche, J. S., Okland, B. and Bjo¨rkman, C. (2020) Complex responses of global insect pests to climate warming. *Frontiers in Ecology and the Environment*, 18: 141-150.

Li, Z. Z., Xu, S. A., Wen, L. X., Liu, F., Liu, A. Q., Wang, Q., Sun, H. Y., Yu, W. and Chen, J. F. (2006). Controlled release of avermectin from porous hollow silica nano particles: Influence of shell thickness on loading efficiency, UV-shielding property and release. *Journal of Controlled Release*, 111: 81–88.

Lozowicka, B. (2015). Health risk for children and adults consuming apples with pesticide residue. *Science of The Total Environment*, 502: 184–198.

Mulliken, D. L., Zambone, J. D. and Rolph, C. G. (2004). DDT: A Persistent Lifesaver. *Natural Resources and Environment*, 19: 3-7.

Parsons, J. (2017). *Our Bumblebees are Disappearing: The Neonicotinoid Connection.* https://ir.library.oregonstate.edu/concern/defaults/ft848v51h.

Perry, A. S., Yamamoto, I., Ishaaya, I. and Perry, R. Y. (2013). *Insecticides in Agriculture and Environment: Retrospects and Prospects.* Springer, Heidelberg, Germany.

Pisa, L., Goulson, D., Yang, E. C., Gibbons, D., Sánchez-Bayo, F., Mitchell, E., Aebi, A., van der Sluijs, J., MacQuarrie, C. J. K., Giorio, C., Long, E. Y., McField, M., van Lexmond, M. B. and Bonmatin, J. M. (2021). An update of the Worldwide Integrated Assessment (WIA) on systemic insecticides. Part 2: impacts on organisms and ecosystems. *Environmental Science and Pollution Research International*, 28: 11749–11797.

Rai, A. B., Halder, J. and Kodandaram, M. H. (2014). Emerging insect pest problems in vegetable crops and their management in India: an appraisal. *Pest Management in Horticultural Ecosystems*, 20: 113-122.

Rundlöf, M., Andersson, G. K. S., Bommarco, R., Fries, I., Hederström, V., Herbertsson, L., Jonsson, O., Klatt, B. K., Pedersen, T. R., Yourstone, J. and Smith, H. G. (2015). Seed coating with a neonicotinoid insecticide negatively affects wild bees. *Nature*, 521: 77–80.

Stanley, D. A., Garratt, M. P. D., Wickens, J. B., Wickens, V. J., Potts, S. G. and Raine, N. E. (2015). Neonicotinoid pesticide exposure impairs crop pollination services provided by bumble bees. *Nature*, 528: 548–550.

Thany, S. H. (2010). Neonicotinoid Insecticides. In: *Insect Nicotinic Acetylcholine Receptors* (Ed. Thany, S. H.). Springer, New York, USA, pp. 75–83.

Tomizawa, M. and Casida, J. E. (2003). Selective toxicity of neonicotinoids attributable to specificity of insect and mammalian nicotinic receptors. *Annual Review of Entomology*, 48: 339–364.

Vanbergen, A. J. (2013). Threats to an ecosystem service: pressures on pollinators. *Frontiers in Ecology and the Environment*, 11: 251-259.

Wang, L., Espinoza, H. M., MacDonald, J. W., Bammler, T. K., Williams, C. R., Yeh, A., Louie, K. W., Marcinek, D. J. and Gallagher, E. P. (2016). Olfactory transcriptional analysis of salmon exposed to mixtures of chlorpyrifos and malathion reveal novel molecular pathways of neurobehavioral injury. *Toxicological Sciences*, 149: 145–157.

Wu, C., Dong, F., Mei, X., Ning, J. and She, D. (2019). Distribution, dissipation, and metabolism of neonicotinoid insecticides in the cotton ecosystem under foliar spray and root irrigation. *Journal of Agricultural and Food Chemistry*, 67: 12374–12381.

Yamamoto, I. and Casida, J. E. (1999). *Nicotinoid Insecticides and the Nicotinic Acetylcholine Receptor.* Springer, Tokyo, Japan.

Chapter 2

Recent Advances in the Pest Management of Temperate Fruit Crops

Wasim Muzamil Dass[1,*], Hidayatullah Tak[1], Ishfaq Majeed[1], Aroosa Khalil[2], Maheen Khurshid[3] and A. R. Ramya[3]

[1]Department of Zoology, University of Kashmir,
Srinagar, Jammu and Kashmir, India
[2]Division of Fruit Sciences, Faculty of Horticulture,
Sher-e-Kashmir University of Agricultural Sciences and Technology of Kashmir,
Srinagar, Jammu and Kashmir, India
[3]ICAR-Central Institute of Temperate Horticulture,
Srinagar, Jammu and Kashmir, India

Abstract

Fruit crops *viz.*, apple, pear, peach, plum, cherry, strawberry, apricot, almond and walnut are grown in the temperate zones of the world because high altitude and chilling temperature that breaks bud dormancy and initiate bud breaking are sine qua non for their top-notch quality and high production. These temperate fruit crops are very momentous part of diet as they are major source of phytonutrients, dietary fibers and antioxidants. The global production of temperate fruits has been estimated 301,599 x 10^6 tons, hence chips into world's economy by a whacking amount but due to insect pests this portion of economy is dwindled to a great extent. In order to fetter this loss, different chemical pesticides are applied for their management albeit they have tremendous contribution to cope up the damage, yet pose serious threats to

* Corresponding Author's Email: wasimmuzamil913@gmail.com.

In: Pest Management: Methods, Applications and Challenges
Editor: Tarique Hassan Askary
ISBN: 979-8-88697-268-9

environment. Under these circumstances, integrated pest management (IPM) is the strategy to manage the pests and mitigate the negative effects of chemicals in the environment. The pest control industries are relentlessly examining novel technologies such as forecasting models, innovative spray equipment and products that will mitigate the damage caused by insect pests and ultimately help in increasing the quality and quantity of fruit crops.

Keywords: temperate, fruits, chemical, pesticides, insect, integrated pest management

1. Introduction

Temperate fruits also known as 'Hill fruits' or 'Cold region fruits' generally grow at higher altitudes. Apple, pear, plum, peach, strawberry, walnut, almond and apricot are the temperate fruits which bear an intensive cold as chilling is necessary for their optimal growth and top-notch quality and quantity. More importantly, chilling is a prerequisite for breaking bud dormancy and initiation of bud growth (Pereira et al., 2002). Temperate fruit crops are grown and adapted naturally in the areas having well defined seasons where maximium growth occurs during spring and summer and least or zero growth in autumn and winter where temperature is close to zero or below zero degrees celsius (Dalastra et al., 2009). Annual production of temperate fruits in the world was recorded 30, 1599 x 10^6 tons (Hardenburg et al., 1986). Around 50 per cent of the total production is produced by China, USA, Brazil, India and European Union. The area under temperate fruit crops in world reached 23.5 million hectares in 2009 (FAO Agricultural statistics, 2009). However, the crops are attacked by several pests which damage and dwindles their production and estimated yield loss vary from 10-30 per cent. The synthetic chemicals used as plant protection measure have undoubtedly fetched good results yet pose many drastic effects such as pest resistance, chemical residues and adverse effect on non-target organisms (Jallow et al., 2017). Hence there is a need of such technology wherein minimum or no chemical is used and pest is suppressed significantly in a tactful manner.

IPM includes a spectrum of management strategies wherein pest population is suppressed by minimum use of chemicals (Bale et al., 2008; Chandler et al., 2011). It involves several methods including manipulation of environment for discouraging pests and encouraging beneficial organisms, mass trapping of pests (El-Sayed et al., 2006; Witzgall et al., 2010), use of biological control agents and several other physical and mechanical tactics of pest control.

The present chapter is focused on the recent techniques involved in the management of insect pests associated with different temperate fruits crops.

2. Pesticide Application

Table 1. Synthetic pesticides in the management of insect pests

Pesticide	Description
Acephate	Generally used as horticultural insecticide for protecting the crops from biting and sucking insects, but it is toxic to bees (Kidd et al., 1991).
Bifenthrin	It is used in orchards against various insect pests such as aphids, moths, beetles, mites, thrips and caterpillars. It has long residual time in soil and toxic to aquatic organisms (Dai et al., 2010).
Imidacloprid	Imidacloprid is the most widely used insecticide in the world. It is generally applied for the management of beetles, fleas, aphids, stink bugs, termites, locusts, thrips, carpenter ants and cockroaches. However, it is considered most lethal to honey bees.
Methamidophos	A highly active, systemic, residual organophosphate insecticide used against the pests of temperate fruits. It is harmful to beneficial insects (Gary et al., 1989).
Phosmet	It is mainly applied on apple trees to control codling moth. It is however used on other fruit crops like ornamentals and vines for managing aphids, suckers, mites and fruit flies.
Pyridaben	Applied against several insect pests of fruit and nut crops. It is however toxic to honey bees, therefore it should not be sprayed when honey bees are in close proximity.

Pesticide application is the practical way of applying pesticides to their target insect pests. Several pesticides are being used by the farmers to gain quick results in pest management (Table 1). The exercise of pest management by applying pesticides rationally is highly multidisciplinary, collaborating many aspects of biology and chemistry. The application of pesticides on fruit trees to control pests are applied as spray applications by different means and one of the best pesticide application is by mechanical sprayers. Hydraulic sprayers consist of a tank, a pump and a nozzle or many nozzles. These sprayers change the formulation that contains water and chemical pesticide into droplets such as tiny particles or large rain type droplets. The conversion is accomplished under pressure by forcing the spray mixture via a spray nozzle. The droplet size can be altered by using different nozzles of different sizes.

Pesticide application is also executed by aerial ways, also called 'crop dusting.' In this method spraying is performed from an agricultural aircraft. It is an aircraft for aerial application of pesticides. Most of the agricultural aircrafts have fix wing such as Air tractor, Cessna Ag-Wagon, Gippsland GA200, Rockwell thrush commander but helicopters are also used. Aerial application of pesticides is severely limited in many countries because of public health hazards and environmental problems like spray drift and this is mostly banned with few restricted exceptions in 2009 (Directive, 2009). However, aerial sprays were controversial for the past few years because of environmental concerns about pesticide drifts. Pesticide drift refers to diffusion of pesticides and contamination to run off water and soil, which results in human health problems and environmental contamination (Spanoghe et al., 2004). Now it is facing restrictions and banned in many countries.

All these types of chemical treatments pose drastic negative impacts on human health and environment hence this glitch needs some alternatives with least effects.

3. Integrated Pest Management (IPM)

The widespread application of synthetic chemical pesticides has undoubtedly proved a boon in managing insect pests nevertheless insect pests still cause huge crop loss to horticultural and agricultural crops (Casida and Quistad, 1998; Culliney, 2014). Moreover unabated application of synthetic pesticides pose negative effects on environment, human health, development of resistance to pesticides in insects (Hawkins et al., 2019; Basit, 2019), soil and water pollution (Rodríguez et al., 2018), chemical residues on fruits (Jallow et al., 2017) and drop in the number of beneficial insect species (Jacob et al., 2019). In order to mitigate these effects, IPM acts as an umbrella which incorporates a spectrum of strategies to control insect pests. IPM is economically affordable, eco-friendly, having least effects on environment, non-target organisms and is very effective against insect pests. Besides mechanical and cultural techniques, IPM involves the use of biocontrol agents, semio-chemicals, plant extracts, manipulation of environment to discourage pests or encourage beneficial organisms (Boiteau et al., 2001) and mass trapping of pests (Witzgall et al., 2010).

3.1. Biorational Control

Biorational (biological + rational) approach is the usage of selective means based on chemical diversity, physical and biological resources for curbing insect pests that are compatible with environment and natural enemies of insect pests. Biorational approach is a lift for elevating IPM strategies. During last three decades formulation of new compounds shackle the hormonal regulation of developmental processes e.g., ecdysone agonists, juvenile hormone and inhibitors of chitin synthesis (Ishaaya, 2001). Juvenile hormone (JH) is the major target of potential insecticides. JH is produced by corpora allata which is a major regulator of insect metamorphosis and reproduction. JH biosynthetic enzyme inhibitors block the pathways of formation of JH. Thus it is lethal for insect pests that manages the insect population without causing negative effects on environment.

3.2. Semiochemicals

Semiochemicals have good potential for managing insect pests as these are organic compounds being used for communication of specific chemical messages that could change physiology or behaviour (El-Sayed, 2015). It is derived from Greek word "Semeon" referring to signal or sign. Insects use these organic compounds to search for mate, food source, avoiding competition, escaping from natural enemies and subsiding host defenses. They are also used for communication of messages over long distances and are very active biologically even at meager concentrations in environment. For insect pest management expensive equipments are necessary for their extraction and chemical characterization. These include solid phase microextraction (SPME), gas chromatography electroantenography (GCEAS), gas chromatography mass spectrometry (GC-MS), and nuclear magnetic resonance (NMR) (Soroker et al., 2015). Semiochemicals are classified into two groups, *viz.*, pheromones: mediating intraspecific interactions and allelochemicals: mediating interspecific interactions.

Pheromone term was coined by Carlson and Luscher (Karlson and Luscher, 1959), to describe signals of chemical nature that execute intraspecific interactions. Pheromones are released to attract insect pests in a particular area and then killed with an insecticide. Insect sex pheromones are the semiochemicals used for managing pests. They are ecofriendly, safe to use, species specific, do not persist in environment for long time and do not have any negative effect on non-target organisms (Horowitz et al., 2009).

Pests are living in an environment full of many volatile compounds. These compounds ultimately change the behaviour and physiology of insect pest by interacting with each other. Many insect pests acquire host plant compounds and use them as sex pheromone precursors (Reddy and Guerrero, 2004). Baker (2011) used sex pheromone of moths to manage insect pests. Sex pheromones are applied for gauging the density of pest population. Pheromones make the monitoring procedure quick, decrease the load of pesticides in environment and limit the overall plant protection cost. Insect population can be dwindled in apple orchard up to 50 per cent by successful application of pheromones (Nordlund et al., 1981).

Allelochemicals are secondary metabolites or primary metabolites of non-nutritional value which has the ability to hamper growth, reproduction and behaviour of target individual. They are applied preferably to kill and mitigate insect pest population. Though, allelochemicals execute a wide range of complex interactions, one way is these volatile materials are released into environment that poses their effect over some distance from emitter. These include short chain alcohols, aldehydes, ketones, esters and aromatic phenols. The second way includes compounds produced for defense which reside in the body of producer. This group contains toxins and phytochemicals. However problems arise regarding their application in fields and due to these challenges their utilization for pest control are yet at the beginning (Soroker et al., 2015).

3.3. Plant Extracts

During past few decades worldwide concern for protecting environment and plummeting the use of synthetic pesticides have laid stress on use of plant products as one way to kill crop insect pests (Wijewardane and Guleria, 2009). A lot of research has been done globally for the effectiveness of plant extracts in controlling insect pests in temperate fruits and enhancing their quality and quantity (MacHardy, 2000). Production of apple in presence of insect pests is formidable challenge to have alternative management because synthetic insecticides are most effective against pests but public concern regarding adverse effects on human health, wildlife, soil and water have led to look at low risk control measures (Altieri and Nicholls, 2005). Biodiversity of plants has fetched a tremendous source of biologically potent materials for controlling insect pests. In the past few decades, studies have been conducted to test the efficacy of plant extracts against a variety of insect pests (Akhtar and Isman, 2004; Pascual-Villalobos and Robledo, 1999). In an experiment toxicity and lethality of more than 30 plant extracts were assessed on larvae of colorado bettle (Gokce et al., 2006). It was observed that these extracts were lethal to the pest and very promising under field conditions. These plant extracts have also been reported to interfere with the oviposition, therefore proved to be highly gruesome to aphids (Das et al., 2008) and lepidopterans. Larvicidal action of acetone, chloroform, ethyl acetate, hexane, methanol, peel, leaf and flower extracts of *Citrus sinensis, Cimum canum, Ocimum sanctum and Rhinocanthus nasutus* against lepidopteran insects are pretty promising and ideally ecofriendly (Kamaraj et al., 2008). Essential oils are extracted from aromatic plants and 49 per cent of these plants belong to Lamiaceae and Compositae families found in mediterrian type of climatic regions (Ross and Sombrero, 1991). Aromatic plants harbour mixture of low molecular weight isoprenoid volatile compounds which are present in specialized tissues e.g., trichomes, canals, cavities or ducts. Essential oils are able to deter oviposition of various insect pests such as *Drosophilla suzukii* which is a small fruitfly, native to South East Asia causing huge destruction to many fruit crops. The main loss to fruit industry is primarily due to its oviposition behaviour, as the adult female oviposits in ripened fruits (Karageorgi et al., 2017). Management of *D. suzukii* by synthetic insecticides is unsustainable so to overcome the limitations caused by these insecticides, various products from plants especially essential oils are used to manage these insects. These oils are bioactive, biodegradable and ecologically safer (Mossa, 2016). Essentials oils are being tested and applied as fumigants, contact insecticides and as deterrents to oviposition, without causing any negative effects on the environment (Erland et al., 2015).

Neem (*Azadirachta indica*) derived insecticides are being applied since long time in organic agriculture. Neem extracts affect the behaviour, sterility of adults and growth regulations, hence check the population of insect pests. Neem derived insecticides are effective against hundreds of insect pests belonging to orders Diptera, Lepidoptera and Hemiptera. Azadirachtin is potent component of neem derived insecticide that negatively affects the insect pests. Also there are many more compounds that show insecticidal properties yet pose no adverse effects on environment or human health. Hence they are preferred against insect pests for better crop production.

3.4. Biocontrol Agents

Biocontrol agents such as entomopathogenic fungi (EPF), bacteria, entomopathogenic nematodes (EPN) and viruses are detrimental against target insect pests but safe to other organisms or beneficial insects. *Metarhizium anisopliae*, *M. acridum* and *Beauveria bassiana* are EPF that are being used to manage crop insect pests in several countries of the world. EPF form appressoria (Bidochka et al., 2000) which firmly link it to the epicuticle of the insect. Enzymes such as proteases are secreted by EPF which provide help to penetrate the host insect cuticle (Charnley, 1997; Butt, 2002). Haemocoel of insect get invaded by fungus (Clarkson et al., 1998) that ultimately result in killing of the host (Butt, 2002). *M. anisopliae* has a wide host range but *M. acridum* is generally virulent against grass hoppers and locusts. EPF are being mass produced, formulated and commercialized by different countries under various trade names. However high price and inconsistent field results of EPF are still a problem for the crop growers.

Bacillus thuringiensis (Bt), a bacterium is very lethal to insect larvae. Endospores cause infection which consists of one or more proteins that have insecticidal properties. Being highly selective, Bt. is preferably applied in IPM programs of apple for limiting the population of different moth species such as Leafrollers (Lacey and Shapiro-Ilan, 2008; Andermatt et al., 1988; Ioriatti et al., 1996). Another spore forming bacteria is *Bacillus firmus* which is marketed by the trade names Nortica® and Poncho® VOTiVO® is also leathal against insect pests.

EPN belonging to two families steinernematidae and heterorhabditidae has been found very successful in controlling insect pests of various orders *viz.*, Lepidoptera, Coleoptera, Diptera, Hymenoptera (Abd-Elgawad et al., 2017; Askary and Abd-Elgawad, 2017; Sankaranarayanan and Askary, 2017; Askary and Ahmad, 2020; Askary and Ahmad, 2021a). The third stage juveniles also called as infective juveniles (IJs) of *Steinernema* and *Heterorhabditis* enter into insect body via mouth, anus, spiracles or abrading the cuticle ultimately reach the haemocoel where they release their symbiotic bacteria of genus *Xenorhabdus* and *Photorhabdus* respectively present in their gut. The bacteria find a suitable medium in the form of haemolymph where they multiply exponentially, release toxins that cause septicaemia resulting in the death of the insect within 24-72 hours (Askary, 2010; Askary et al., 2018; Askary and Abd-Elgawad, 2021). In the past few decades EPN species has been isolated (Bhat et al., 2019; Bhat et al., 2020; Askary et al., 2020), mass multiplied (Askary and Ahmad, 2017), formulated (Askary and Ahmad,

2021b) and commercialized in different countries of the world (Askary et al., 2017).

Moreover, viral particles are also detrimental to regulate population of insect pests. Most promising virus for managing apple insect pests is granulovirus. Firstly it was isolated from dead larvae present in orchards of Maxico (Tanada, 1964). Granulovirus of coding moth, *Cydia pomonella* offer novel ways of highly selective means for managing insect pests of temperate fruits (Falcon and Huber, 1991). In IPM program the commercialized products that are applied contain virus in aqueous suspension which is sprayed while eggs are hatching. In Europe and North America commercialized products of granulovirus are now registered and available in the market hence used by orchardists globally (Lacey et al., 2008).

Conclusion

In temperate region fruit cultivation is part and parcel of diet and economy but this very asset is degraded by various insect pests both qualitatively and quantitatively. Albeit these pests are managed by different insecticides easily and quickly but on the other hand insecticides pose serious adverse effects on environment, beneficial organisms and human health. Therefore, correct and reliable knowledge related to crops and their pests is very essential for growers engaged in cultivation of temperate fruit crops. The role of electronic media is also very important for keeping farmers updated about crops and their concerned pests. Awareness programmes should be conducted among farmers regarding indiscriminate application of pesticides and its consequences, simultaneously growers should be trained for managing the pests by eco-friendly tactics such as use of beneficial organisms, semiochemicals, plant extracts, pest resistant plant cultivars, cultural and mechanical practices and when necessary judicious use of chemical pesticides.

Acknowledgments

We are grateful to Head, Department of Zoology, University of Kashmir, India for providing support in the preparation of the manuscript.

References

Abd-Elgawad, M. M. M., Askary, T. H. and Coupland, J. (2017). *Biocontrol Agents: Entomopathogenic and Slug Parasitic Nematodes*. CAB International, Wallingford, UK.

Akhtar, Y. and Isman, M. B. (2004). Comparative growth inhibitory and antifeedant effects of plant extracts and pure allelochemicals on four phytophagous insect species. *Journal of Applied Entomology*, 128: 32–38.

Altieri, M. A. and Nicholls, C. I. (2005). Agroecology and the search for a truly sustainable agriculture. United Nations Environment Programme Environmental Training Network for Latin America and the Caribbean: Mexico City. *Mexico*, pp. 13–27.

Andermatt, M., Mani, E., Wildbolz, T. H. and Lüthy, P. (1998). Susceptibility of *Cydia pomonella* to *Bacillus thuringiensis* under laboratory and field conditions. *Entomologia Experimentalis et Applicata*, 49: 291–295.

Askary, T. H. (2010). Nematodes as biocontrol agents. In: *Sociology, Organic Farming, Climate Change and Soil Science* (Ed. Lichtfouse, E.). Springer, The Netherlands, pp. 347-378.

Askary, T. H. and Abd-Elgawad, M. M. M. (2021). Opportunities and challenges of entomopathogenic nematodes as biocontrol agents in their tripartite interactions. *Egyptian Journal of Biological Pest Control*, 31: 42. https://doi.org/10.1186/s41938-021-00391-9.

Askary, T. H. and Abd-Elgawad, M. M. M. (2017). Beneficial nematodes in agroecosystem: A global perspective. In: *Biocontrol Agents: Entomopathogenic and Slug Parasitic Nematodes* (Eds. Abd-Elgawad, M. M. M., Askary, T. H. and Coupland, J.). CAB International, Wallingford, UK, pp. 3-25.

Askary, T. H. and Ahmad, M. J. (2021a). Biocidal efficacy of some native isolates of entomopathogenic nematodes against oriental armyworm, *Mythimna separata* Walker (Lepidoptera:Noctuidae). *Indian Journal of Nematology*, 51: 67-73.

Askary, T. H. and Ahmad, M. J. (2021b). Survival and virulence capacity of native strain of entomopathogenic nematode, *Steinernema cholashanense* in different formulations. *Agricultural Research*, 10: 390-397.

Askary, T. H. and Ahmad, M. J. (2020). Efficacy of entomopathogenic nematodes against the cabbage butterfly (*Pieris brassicae* (L.) (Lepidoptera: Pieridae) infesting cabbage under field conditions. *Egyptian Journal of Biological Pest Control*, 30: 39. https://doi.org/10.1186/s41938-020-00243-y.

Askary, T. H. and Ahmad, M. J. (2017). Entomopathogenic nematodes: mass production, formulation and application. In: *Biocontrol Agents: Entomopathogenic and Slug Parasitic Nematodes* (Eds. Abd-Elgawad, M. M. M., Askary, T. H. and Coupland, J.). CAB International, Wallingford, UK, pp. 261-286.

Askary, T. H., Ahmad, M. J., Wani, A. R., Mohiddin, S. and Sofi, M. A. (2018). Behavioural ecology of entomopathogenic nematodes, *Steinernema* and *Heterorhabditis* for insect biocontrol. In: *Sustainable Agriculture Reviews 31* (Ed. Lichtfouse, E.). Springer, The Netherlands, pp. 425-441.

Askary, T. H., Bhat, A. H., Ahmad, M. J., Chaubey, A. K. and Spiridonov, S. E. (2020). *Steinernema feltiae* (Rhabditida: Steinernematidae) from hilly areas of Kashmir valley, India with a note on its geographical distribution. *Russian Journal of Nematology*, 28: 99 – 106.

Askary, T. H., Nermut, J., Ahmad, M. J. and Ganai, M. A. (2017). Future thrusts in expanding the use of entomopathogenic and slug parasitic nematodes in agriculture. In: *Biocontrol Agents: Entomopathogenic and Slug Parasitic Nematodes* (Eds. Abd-Elgawad, M. M. M., Askary, T. H. and Coupland, J.). CAB International, Wallingford, UK, pp. 620-627.

Baker, T. C. (2011). Insect pheromones: useful lessons for Crustacean pheromone programs? In: *Chemical Communication in Crustacean* (Eds. Breithaupt, T. and Thiel, M.). Springer, New York, pp. 531-550.

Bale, J. S., van Lenteren, J. C. and Bigler, F. (2008). Biological control and sustainable food production. *Philosophical Transactions of the Royal Society B: Biological Sciences*, 363: 761–76.

Bhat, A. H., Chaubey, A. K. and Askary, T. H. (2020). Global distribution of entomopathogenic nematodes, *Steinernema* and *Heterorhabditis*. *Egyptian Journal of Biological Pest Control*, 30: 31. https://doi.org/10.1186/s41938-020-0212-y.

Bhat, A. H., Askary, T. H., Ahmad, M. J., Suman, Aasha and Chaubey, A. K. (2019). Description of *Heterorhabditis bacteriophora* (Nematoda: Heterorhabditidae) isolated from hilly areas of Kashmir valley. *Egyptian Journal of Biological Pest Control*, 29: 96. https://doi.org/10.1186/s41938-019-0197-6.

Bidochka, M. J., Kamp, A. M. and de Croos, A. (2000). Insect pathogenic fungi: from genes to populations. In: *Fungal Pathology* (Ed. Kronstad, J. W.), Kluwer Academic Publishers, The Netherlands, pp. 171–93.

Butt, T. M. (2002). Use of entomogenous fungi for the control of insect pests. In: *The Mycota XI: Agricultural Applications* (Ed. Kempken, F.). Springer, Germany, pp. 111–34.

Casida, J. E. and Quistad, G. B. (1998). Golden age of insecticide research: past, present, or future? *Annual Review of Entomology*, 43: 1–16.

Chandler, D., Bailey, A. S., Tatchell, G. M., Davidson, G., Greaves, J. and Grant, W. P. (2011). The development, regulation and use of biopesticides for integrated pest management. *Philosophical Transactions of the Royal Society of London B: Biological Sciences*, 366: 1987–1998.

Charnley, A. K. (1997). Entomopathogenic fungi and their role in pest control. In: *The Mycota IV: Environmental and Microbial Relationships* (Eds. Wicklow, T. D. and Soderstrom, B.). Springer, Germany, pp. 185–201.

Clarkson, J. M., Scree, S., Bailey, A., Cob, B. and Charnley, K. (1998). Fungal pathogenesis in insects. In: *Molecular Varability of Fungal Pathogens* (Eds. Bridge, P. D., Couteaudier, Y. and Clarkson, J. M.). CAB International, Wallingford, UK, pp. 83–94.

Culliney, T. W. (2014). Crop losses to arthropods. In: *Integrated Pest Management* (Eds. Pimentel, D. and Peshin, R.). Springer, Netherlands, pp. 201–225.

Dai, P. L., Wang, Q., Sun, J. H., Feng, L., Wang, X., Wu, Y. Y. and Zhou, T. (2010). Effects of sub lethal concentrations of bifenthrin and deltamethrin on fecundity, growth, and development of the honeybee Apis mellifera logistica. *Environmental Toxicology and Chemistry,* 29: 644–649.

Dalastra, I. M., Pio, R., Campagnolo, M. A., Dalastra, G. M., Chagas, E. A. and Guimaraes, V. F. (2009). Pruning time in the production of 'Roxo de Valinhos' green fig in organic system in the west region of paraná state. *Revista Brasileira de Fruticultura*, 31: 447-453.

Das, B., Sarker, P. and Rahman, M. (2008). Aphicidal activity of some indigenous plant extracts against bean aphid *Aphis craccivora* Koch (Homoptera: Aphididae). *Journal of Pest Science*, 81: 153–159.

Directive (2009). Directive 2009/128/EC of the European Parliament and of the Council of 21 October 2009 establishing a framework for Community action to achieve the sustainable use of pesticides (Text with EEA relevance). *Official Journal of the European Union, 24(11).*

El-Sayed, A. M. (2015). *The Pherobase: Database of Insect Pheromones and Semiochemicals*, 2008.9856

El-Sayed, A. M., Suckling, D. M., Wearing, C. H. and Byers, J. A. (2006). Potential of mass trapping for long-term pest management and eradication of invasive species. *Journal of Economic Entomology*, 99: 1550–1564.

Erland, L. A. E., Rheault, M. R. and Mahmoud, S. S. (2015). Insecticidal and oviposition deterrent effects of essential oils and their constituents against the invasive pest *Drosophila suzukii* (Matsumura) (Diptera: Drosophilidae). *Crop Protection*, 78: 20–26.

Falcon, L. A. and Huber, J. (1991). Biological control of the codling moth. In: *Tortricid Pests, Their Biology, Natural Enemies and Control* (Eds. van Der Geest, L. P. S. and Evenhuis, H. H.). Elsevier Science Publishers, The Netherlands, pp. 355–369.

FAO Agricultural Statistics. (2009). Global strategy to improve agricultural and rural statistics. *The World Bank, Report No. 56719-GLB.* https://www.ibge.gov.br/arquivo/projetos/prpa/segundo_texto.pdf.

Gary, N. E. and Lorenzen, K. (1989). Effect of methamidophos on honey bees (Hymenoptera: Apidae) during alfalfa pollination. *Journal of Economic Entomology,* 82: 1067-1072.

Gokce, A., Whalon, M. E., Cam, H. T., Yanar, Y., Demirtas, I. M. and Goren, N. (2006). Plant extract contact toxicities to various develotnh6y6pmental stages of Colorado potato beetles (Coleoptera: Chrysomelidae). *Annals of Applied Biology*, 148: 197–202.

Gottar, M., Gobert, V., Matskevich, A. A., Reichhart, J. M., Wang, C., Butt, T. M., Belvin, M., Hoffmann, J. A. and Ferrandon, D. (2006). Dual detection of fungal infections in drosophila via recognition of glucans and sensing of virulence factors. *Cell*, 127: 1425–1437.

Hardenburg, R. E., Watada, E. E., and Wang C. Y. (1986). *The Commercial Storage of Fruits, Vegetables, and Florist and Nursery Stocks. USDA Handbook No. 96.* Washington, USA.

Hawkins, N. J., Bass, C., Dixon, A. and Neve, P. (2019). The evolutionary origins of pesticide resistance. *Biological Reviews*, 94: 135–155.

Horowitz, R., Ellsworth, P. C. and Ishaaya, I. (2009). Biorational pest control – an overview. In: *Biorational Control of Arthropod Pests* (Eds. Ishaaya, I. and Orowitz, A. R.). Springer, The Netherlands, pp. 1–20.

Ioriatti, C., Pasqualini, E. and Delaiti, M. (1996). Effectiveness of *Bacillus thuringiensis* Berliner on three species of apple leafrollers. *Boll Ist. Entomol. Guido Grandi' Stud Bologna*, 50: 73–93.

Ishaaya, I. (2001). Biochemical processes related to insecticide action: an overview. In: *Biochemical Sites of Insecticide Action and Resistance* (Ed. Ishaaya, I.). Springer, Germany, pp. 1–16. https://doi.org/10.1007/978-3-642-59549-3_1.

Jacob, C. R., Zanardi, O. Z., Malaquias, J. B., Souza, C. A. and Yamamoto, P. T. (2019). The impact of four widely used neonicotinoid insecticides on *Tetragonisca angustula* (Latreille) (Hymenoptera: Apidae). *Chemosphere*, 224: 65–70.

Jallow, M. F. A., Awadh, D. G., Albaho, M. S., Devi, V. Y. and Ahmad, N. (2017). Monitoring of pesticide residues in commonly used fruits and vegetables in Kuwait. *International Journal of Environmental Research and Public Health,* 14: 833. https://doi.org/10.3390/ijerph14080833.

Kamaraj, C., Rahuman, A. and Bagavan, A. (2008). Antifeedant and larvicidal effects of plant extracts against Spodoptera litura (F.), *Aedes aegypti* L. and *Culex quinquefasciatus* Say. *Parasitology Research*, 103: 325–331.

Karageorgi, M., Bräcker, L. B., Lebreton, S., Minervino, C., Cavey, M., Siju, K. P., Grunwald Kadow, I. C., Gompel, N. and Prud'homme, B. (2017). Evolution of multiple sensory systems drives novel egg-laying behavior in the fruit pest *Drosophila suzukii*. *Current Biology*, 27: 847–853.

Karlson, P. and Luscher, M. (1959). Pheromones: a new term for a class of biologically active substances. *Nature*, 183: 55–56.

Kidd, H. and James, D. R. (1991). *The Agrochemicals Handbook, Third Edition.* Royal Society of Chemistry Information Services, Cambridge, UK.

Lacey, L. A. and Shapiro-Ilan, D. I. (2008). Microbial control of insect pests in temperate orchard systems: potential for incorporation into IPM. *Annual Review of Entomology*, 53: 121–144.

Lacey, L. A., Thomson, D., Vincent, C. and Arhturs, S. P. (2008). Codling moth granulovirus: A comprehensive review. *Biocontrol Science and Technology*, 18: 639. https://doi.org/10.1080/09583150802267046.

MacHardy, W. E. (2000). Current Status of IPM in Apple orchards. *Crop Protection*, 19: 801–806.

Mossa, A. T. H. (2016). Green Pesticides: essential oils as biopesticides in insect-pest management. *Journal of Environmental Science and Technology*, 9: 354–378.

Nordlund, D. A., Jones, R. L. and Lewis, W. J. (1981). *Semiochemicals: Their Role in Pest Control*. John Wiley and Sons, New York, USA.

Pascual-Villalobos, M. J. and Robledo, A. (1999). Anti-insect activity of plant extracts from the wild flora in southeastern Spain. *Biochemical Systematics and Ecology*, 27: 1–10.

Pereira, A. R., Angelocci, L. R. and Sentelhas, P. C. (2002). Agrometeorologia: fundamentos e aplicações práticas [*Agrometeorology: fundamentals and practical applications*]. Guaíba: Agropecuaria. Accessed: 13 July, 2022.

Reddy, G. V. P. and Guerrero, A. (2004). Interactions of insect pheromones and plant semiochemicals. *Trends in Plant Science*, 9: 253–261.

Ross, J. D. and Sombrero, D. (1991). Environmental control of essential oil production in Mediterranean plants. In: *Ecological Chemistry and Biochemistry of Plant Terpenoids*

(Eds. Harborne, J. B. and Tomes-Barberan, F. A.). Clarendon Press, New York, USA, pp. 64–94.

Sankaranarayanan, C. and Askary, T. H. (2017). Status of Entomopathogenic Nematodes in Integrated Pest Management Strategies in India. In: *Biocontrol Agents: Entomopathogenic and Slug Parasitic Nematodes* (Eds. Abd-Elgawad, M. M. M., Askary, T. H. and Coupland, J.). CAB International, Wallingford, UK, pp. 362-382.

Soroker, V., Harari, A. and Faleiro, J. R. (2015). The role of semiochemicals in date pest management. In: *Sustainable Pest Management in Date Palm: Current Status and Emerging Challenges. Sustainability in Plant and Crop Protection* (Eds. Wakil, W., Faleiro, J. R. and Miller, T.). Springer, Switzerland, pp. 315-346.

Spanoghe, P., Maes, A. and Steurbaut, W. (2004). Limitation of point source pesticide pollution: Results of bioremediation system. *Communications in Agricultural and Applied Biological Sciences*, 69: 719–732.

Tanada, J. (1964). A granulosis-virus of the codling moth, *Carpocapsa pomonella* L. (Oleuthreutidae, Lepidoptera). *Journal of Insect Pathology*, 6: 378–380.

Wijewardane, R. M. N. A. and Guleria, S. P. S. (2009). Combined effect of Pre-cooling, application of natural extracts and packaging on the storage quality of Apple (*Malus domestica*) cv. Royal Delicious. *Tropical Agricultural Research*, 21: 10- 20.

Witzgall, P., Kirsch, P. and Cork, A. (2010). Sex pheromones and their impact on pest management. *Journal of Chemical Ecology*, 36: 80–100.

Chapter 3

Recent Advances in the Pest Management of Pulse Crops

A. G. Sreenivas[1,*], Shobarani[1], Rachappa V. Haveri[1], G. Poornima[1], Vijayalakshmi[1], N. M. Sunil Kumar[1], Raju G. Tegelli[1] and Ishtiyaq Ahad[2]
[1]Department of Agricultural Entomology,
University of Agricultural Sciences,
Raichur, Karnataka, India
[2]Division of Entomology, Faculty of Agriculture,
Sher-e-Kashmir University of Agricultural Sciences and Technology of Kashmir,
Jammu and Kashmir, India

Abstract

Pulses are an excellent source of dietary protein for millions of people around the world. However, they are attacked by a variety of insect pests that cause an annual loss varying between 2.0 to 2.4 million tons. The major insect pests of pulses are pod borer, *Helicoverpa armigera* (Hubner), spotted pod borer, *Maruca vitrata* (Fabricius) and pod fly, *Melanagromyza obtusa* (Malloch). Besides, bruchids, *Callosobruchus* spp. is also a threat to pulses during storage. In order to regulate the population of insect pests, different management strategies are adopted such as cultural, biological, chemical and use of resistant plant cultivars. Screening of genetic resources for insect resistance has been found promising in some pulse crops. Forecasting models and pheromone-based monitoring/ mass trapping methods also help in managing pest population. A single method has certain limitations. Hence, different

* Corresponding Author's Email: agsreenivas@gmail.com.

In: Pest Management: Methods, Applications and Challenges
Editor: Tarique Hassan Askary
ISBN: 979-8-88697-268-9

available tactics available at farming level may be integrated to manage the pests in economically feasible and eco-friendly manner.

Keywords: pulses, insect pest, resistance, cultural, biological, chemical

1. Introduction

Pulses are a major source of protein. They contribute significantly to the nutritional security for the majority of people globally. Pulses complement to cereals and are the best solution to protein-calorie malnutrition. India is the largest producer of pulses in the world with 28-30 per cent share in the global production. Chickpea occupies 38 per cent of total pulses area followed by pigeon pea which occupies 22 per cent. The other pulse crops grown in the country are urdbean, mungbean, lentil, French bean, horse gram and field pea. In India, the major pulses growing states are Madhya Pradesh, Maharashtra, Karnataka, Uttar Pradesh, Gujarat, Tamil Nadu and Andhra Pradesh, together accounting for 74.8 per cent of the total pulse production of 24.42 million tons from an area of 28.60 million hectare (Anonymous, 2021).

Insect pests are major constraints in pulse production. Annually about 2.0 to 2.4 million tons of pulses are lost due to the damage caused by insect pests (Reddy, 2009). As a management tactic, over-reliance on chemical pesticides has impaired the natural crop ecosystem balance by disrupting parasitoid and predator populations, thereby causing outbreaks of secondary pests. The use of chemicals also contributes to a vicious cycle of resistance in pests, leading to further investment in pesticide development but little change in crop losses to pests, which are currently estimated at nearly the same per cent, what it was 50 years ago. As a result, induced pest outbreaks, caused by inappropriate pesticide use, have increased. Excessive use of pesticides also exposes farmers to serious health risks and has negative consequences on the environment, and sometimes on crop yield. Although pulse crops are prone to as many as 200 pests (Lateef and Reed 1990; Shanower et al., 1999) and seed borne diseases, however, pod-borer (*Helicoverpa armigera* Hubner) in chickpea and pigeon pea has been a major concern as it has developed resistance against many chemical insecticides (Kranti et al., 2002). Incidence of pod borer, if not controlled, can cause huge yield losses alone and discourage farmers to grow pulse crops. Of late, spotted pod-borer (*Maruca vitrata* Fabricius) and pod-fly (*Melanagrozama abtusa* Malloch) have emerged as potential pests in key pulse growing areas, causing yield losses especially in long duration crops

(Gopali et al., 2010, 2013; Sharma et al., 2011). Pod fly incidence is increasing as a result of cultivation of long duration varieties. Pod fly infestation is hidden and difficult to detect before it has already damaged developing seeds (Sharma et al., 2011). Undoubtedly, availability of new pesticides has helped in management of lepidopteran pests more effectively (Sharma et al., 2012) but replacing of generic contact pesticides has resulted in upsurge of sucking insects such as jassids, stem borer as well as pod bugs, which often sucks away saps from developing grains rendering them to shriveled condition. In addition, heavy damage to pulses grain is caused by pests during storage, which often comes from field.

Our focus in this chapter is to describe the major insect pests of pulse crops, besides throwing light on the recent advances in the management of these pests in field as well as during storage condition.

2. Insect Pests of Pigeon Pea

Several insect pests are known to ravage the pigeon pea crop. Some key pests are mentioned below (Table 1).

Table 1. Insect pests of pigeon pea showing their incidence in different months of a calendar year

Insect Pest	Month											
	Jan	Feb	Mar	Apr	May	Jun	Jul	Aug	Sep	Oct	Nov	Dec
Aphids												
Pod borer												
Pod fly												
Mites												
Scale												
Pod bugs												
Cow bug												
Blister beetle												

2.1. Pod Borer: *Helicoverpa armigera* (Hubner) (Lepidoptera: Noctuidae)

2.1.1. Identification

Helicoverpa is one of the major polyphagous pests distributed throughout the country. The adult female moths have yellowish to orange forewings whereas in male moths it is greenish grey in colour, with a light dark transverse band in the distal third. Hind wings are pale grey in colour with a darker broad marginal band having small brown marking at base. Adult moths have a wingspan of 35-40 mm (Figure 1).

Figure 1. Adult pod borer, *Helicoverpa armigera.*

2.1.2. Biology

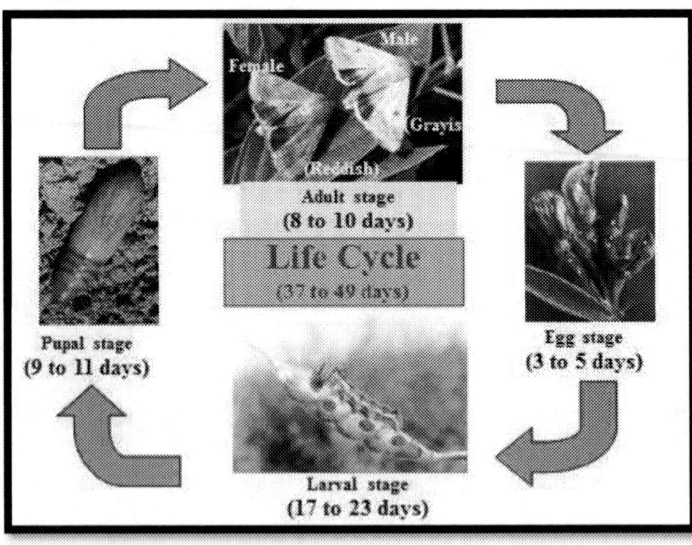

Figure 2. Life cycle of pod borer, *Helicoverpa armigera.*

Females lay several hundred eggs on all parts of plant, but preferably on young shoots and florets. The eggs hatch in less than three days at optimum temperature of 27-28°C. The newly hatched larvae feed for a short time on the tender leaflets by scrapping green tissue and then shift to flower buds and tender shoots. Under normal conditions larval and pupal period last for 12-16 and 6-10 days, respectively. The colour of the larvae varies from yellow to green, pink, orange, brown or black but all have characteristic light and dark stripes along either side of the body. The pupal period lasts for 8-10 days (Figure 2).

2.1.3. Nature of Damage

Larva enters slowly into the pod and feeds on the developing seeds. While feeding, the half portion of larva remains inside the pod (Figure 3). They cut hole on one to another locule and one larva can feed upon 20-25 pods in its lifetime.

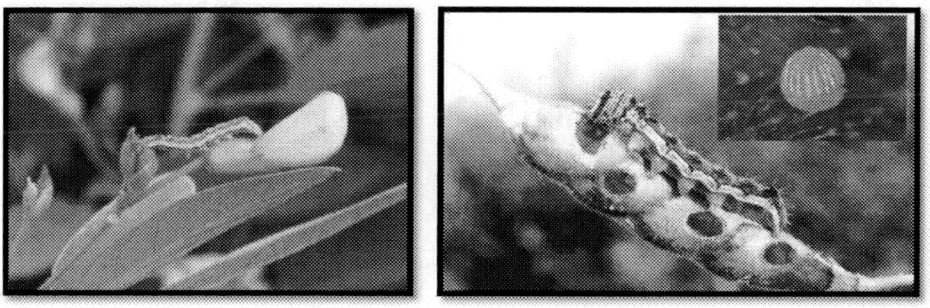

Figure 3. Pod damaged by pod borer, *Helicoverpa armigera.*

2.2. Spotted Pod Borer: *Maruca vitrata* (Geyer) (Lepidoptera: Pyralidae)

It is a sporadic pest of pigeon pea and commonly damages the short duration crop. Pigeon pea varieties with indeterminate growth and spreading types are more susceptible to *M. vitrata*.

2.2.1. Identification

The adult moth has three white spotted brown forewings and grayish–white hind wing with distal brown markings (Figure 4). The eggs are laid singly in the sepals or petals of the flower or buds or on the pods. The immature larvae are dull to yellow-white in colour and often reach a length of 18 mm. Each segment has dark spots which form a distinct series along the length of the body. This pattern is quite obvious on the upper surface. The larvae head is dark brown to black in colour (Figure 5). The newly hatched larvae feed on the tender leaves, flower buds and developing seeds in the pods for about 3 weeks. Thereafter, it emerges from the pods during night and crawls to descend to the soil and pupates beneath leaf debris. Pupation takes place in silken cocoon in pod or rarely in the soil. The hidden nature of larval (the damaging stage) and pupal stage makes it difficult to manage them by chemicals or other conventional means. The adults can be monitored through light traps though there are variations in the catches in different months in different regions of the country.

Figure 4. Adult of *Maruca vitrata.*

Figure 5. Larval instars of *Maruca vitrata.*

2.2.2. Nature of Damage

M. vitrata is a serious pest of early maturing varieties of pigeon pea. Larvae of *M. vitrata* cause extensive damage to floral buds and flowers. They damage flowers causing discoloration and shedding. The characteristic symptoms observed on pigeon pea plant is webbing together of flowers, pods, and leaves with frass often on pods and shoot tips. Damaged pods have small, darkened entry holes on the surface and larva is found inside. Many a times leaves and pods are stuck together by webbing with signs of surface feeding (Figure 6).

Figure 6. Damage symptoms on flowers and pods caused by *Maruca vitrata.*

2.3. Pod Fly: *Melanagromyza obtusa* (Malloch) (Diptera: Agromyzidae)

M. obtusa is widely distributed in India. The pest causes major economic losses in northern part of India, growing pigeon pea especially long duration varieties. The infected seeds do not germinate. There are no obvious external symptoms of pod fly attack till the fully grown larvae chew holes in the pod walls leaving a "window" through which the flies emerge after pupation in the pod. The damaged seeds are of no value.

The small black fly lays eggs through the wall of the developing pod and the size of white legless larva is approximately 3 mm long. They feed inside the green seed. Generally, only one larva of *M. obtusa* develops in one seed. The brown puparium is formed inside the pod but outside the seed. One generation takes about 3 weeks under optimum conditions. They prefer egg laying on partially matured pods than on tender or fully matured pods.

A single fly lays 80-100 eggs which incubate for 3 days and hatching maggot bores into the pod and feeds on developing seed. Usually, one seed is sufficient for their development. The larva pupates inside the pod after making a small hole on the pod, covering it with a thin membranous stricture. The pupal period may lasts for 7-15 days leading to emergence of adult through the exit hole made by the larva. All the immature stages remain within the developing pod and very difficult to monitor without damaging the pod (Figure 7).

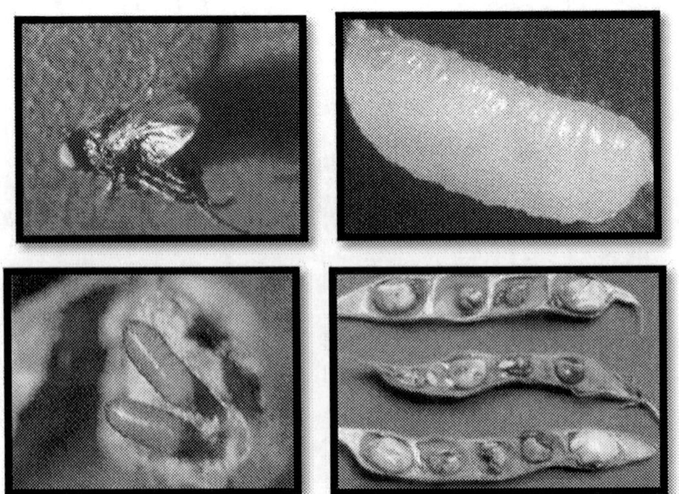

Figure 7. Different stages of *Melanagromyza obtusa.*

2.4. Pod Sucking Bugs: *Clavigralla gibbosa, C. scutellaris, C. tomentosicollis, Riptortus dentipes, Anoplocnemis curvipes* (Hemiptera: Coreidae) *Nezara viridula* (Hemiptera: Pentatomidae)

The most common pod-sucking bugs are giant coreid bug (*Anoplocnemis* spp.), spiny brown bug (*Clavigralla* spp.), Riptortus bug (*Riptortus* spp.) and green stink bug (*Nezara viridula*). These bugs suck sap from developing seeds through the pod wall, making them shrivelled with dark patches often associated with saprophytic fungi. Attacked seeds do not germinate and are also not acceptable as food. Sometimes during feeding, fungal spores are transmitted through mouth parts, resulting in rotting of the seeds. *Clavigralla* spp. is one of the most important pests of pigeon pea in Maharashtra and Karnataka state of India. The adults of *Clavigralla gibbosa* feed on pods and lay eggs in clusters of 3-15 which are deep brown in colour (Figure 8 & 9). Incubation period lasts for 6-8 days, nymphal period 17-20 days and adult longevity 25-62 days. Thus, total life cycle takes place between 31-88 days (Figure 10).

Figure 8. *Clavigralla* eggs on leaves.

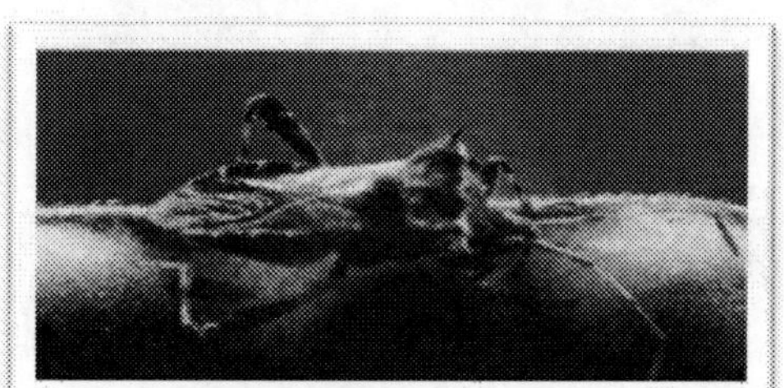

Figure 9. *Clavigralla* adult feeding on pod.

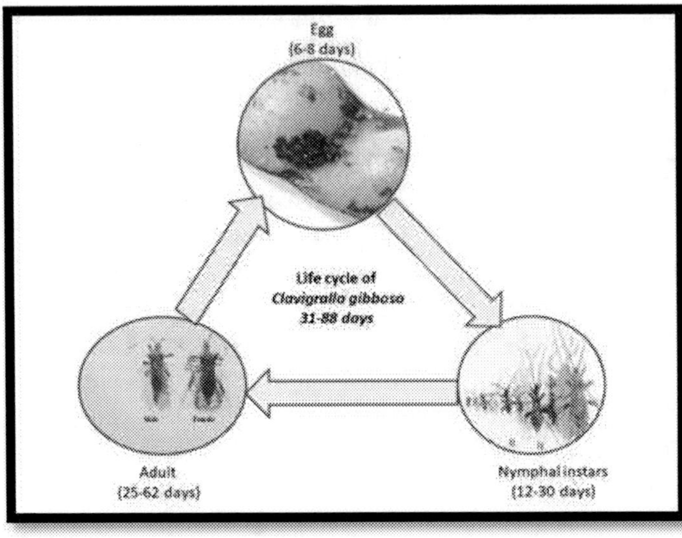

Figure 10. Life cycle of *Clavigralla gibbosa.*

2.5. Plume Moth: *Exelastis atomosa* (Pterophoridae: Lepidoptera)

2.5.1. Identification

Adults are light brown in colour, wings deeply fissured, the forewings elongate and longitudinally cleft into two plumes and hind wings into three plumes. Legs are long and slender. Abdomen is dark brown in colour (Figure 11).

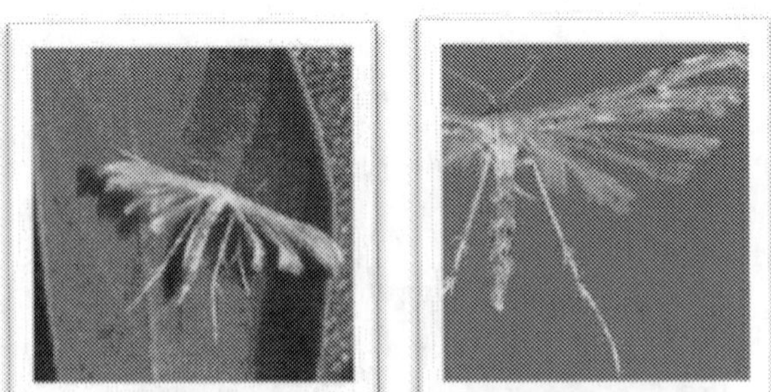

Figure 11. Adults of plume moth, *Exelastis atomosa.*

2.5.2. Biology

The moth is slender delicate with plumose light brown wings. Eggs are greenish, oval, laid singly on flower buds or pods. Incubation period lasts for 4 days. Larva is greenish brown with spines radiating all over the body from tubercles. Larval period is completed in 14- 30 days, pupation occurs outside the pod, lasting for 4-8 days.

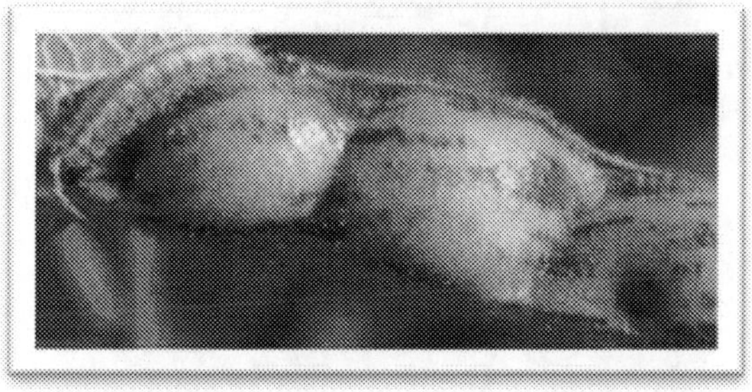

Figure 12. Pigeon pea pod damaged by plume moth.

2.5.3. Nature of Damage

The larvae feed on flower buds and seeds of pod by remaining outside (Figure 12).

2.6. Pod Borer: *Etiella zincknella* (Phycitidae: Lepidoptera)

Figure 13. Adult pod borer, *Etiella zincknella.*

2.6.1. Biology

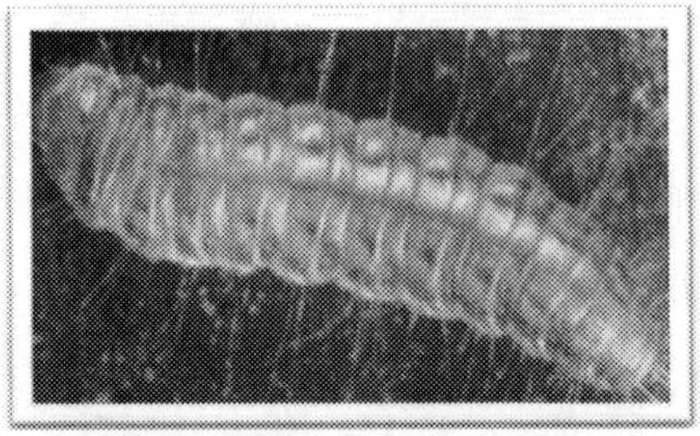
A

B

Figure 14. A. Larva of pod borer, *Etiella zincknella;* B. Pod damaged by larva.

The moth is small with white streak on the greyish brown forewing and pale hind wing (Figure 13). Adult females lay eggs singly on tender pods. Larva is greenish with five black spots on the prothorax. The grown-up larvae become pinkish and pupate in earthen cocoon. Egg, larval and pupal period lasts for 10, 10-17 and 9-12 days, respectively. The larvae bore into pods and cause damage (Figure 14).

2.7. Blue Butterflies: (Lycaenidae: Lepidoptera)

1. *Euchrysop cnejus*
2. *Lampides boetics*

2.7.1. Identification

The larvae are flat pale, greenish or yellowish with a red line and short black hairs on the body. Butterfly is active bluish, a tail like process at the base of hind wing possessing a dot above. Below the front margin of the forewing there are half-moon shaped marking (Figure 15).

Figure 15. Adult of blue butterflies.

2.7.2. Nature of Damage

Larvae bore into buds, pods, flowers and seeds of pigeon pea. Adult female lays sculptured round eggs singly on unopened buds.

2.8. Minor Pests of Pigeon Peas

1. Aphids
    - *Aphis craccivora*: Infest tender shoots and inflorescence of pigeon pea.
2. Blister beetle
    - *Mylabris pustulata:* Feed on flowers.
3. Stink bug
    - *Coptosoma cribraria*: Greenish brown stink bug infests tender shoots.
4. Stem fly
    - *Melanagromyza phaseoli*: Causes withering and drying of shoots in seedling stage.
5. Gall weevil
    - *Alcidodes collaris, A. Pietues, Colobodes dolichotis*: Causes gall at the base of the stem, the plants fall down in case of heavy wind.
6. Mite
    - *Aceria cajani:* Microscopic mite which is responsible for transmitting the sterility mosaic. This results in sterility.

Table 2. Key insect pests of pigeon pea and their economic threshold values

Common name	Scientific name	ETL value
Pod borer	*Helioverpa armigera*	2 eggs/plant or 1 larvae/ plant at flowering stage; 1 larvae/ plant at pod stage. 4 - 5 moths/ trap/ day >5% pod damage
Pod fly	*Melanagromyza obtusa*	2.5% pod damage
Leaf webber	*Maruca vitrata*	1 web/ plant
Pod sucking bugs	Clavigralla gibbosa C. scutellaris C. tomentosicollis Riptortus dentipes Anoplocnemis curvipes Nezara viridula	2 bugs/ plant
Plume moth	*Exelastis atomosa*	larva/ plant

Economic threshold level (ETL) values of key insect pests of pigeon pea are mentioned in Table 2.

2.9. Storage Pests

2.9.1. Bruchids: Callosobruchus maculatus (F.), Callosobruchus analis (F.), Callosobruchus chinensis (L.) (Coleoptera: Bruchidae)

Losses due to storage insects are considerably higher in pulses than in cereals. Bruchids especially *Callosobruchus maculatus* (F.) damage crops and other pulses in the field as well as in the storage. These insects begin their infestation when the pods are in ripening stage in the field and are subsequently carried with grain into the stores resulting in considerable losses. The pest generates exceedingly high levels of infestation even when they pass only one or two generations.

The adult beetles are mottled brown, which lay eggs on the pods when they are in the field or when the seeds are kept in storage. The common symptoms are round exit hole occasionally with attached white eggs on the pod wall (Figure 16). The larvae of the bruchids feed on the pulse seed contents reducing their degree of usefulness, making them unfit either for planting or for human consumption. None of the pigeon pea cultivars are reported resistant to bruchids. The susceptibility varied significantly among pigeon pea cultivars with variable seed coat thickness.

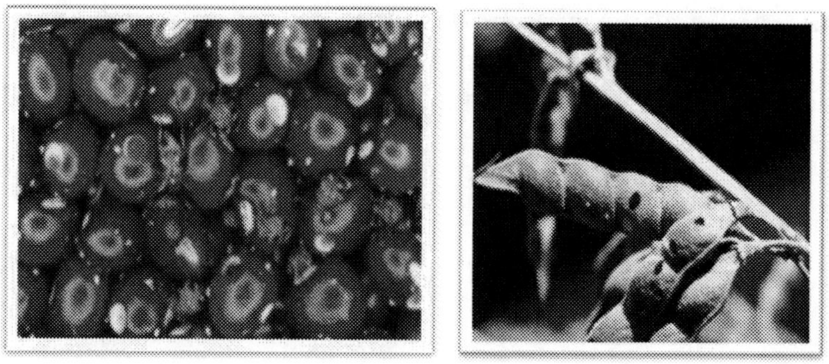

A B

Figure 16. A. Eggs and adult of bruchids on pigeon pea grain;
B. Infected pods showing exit holes.

2.10. Integrated Pest Management (IPM)

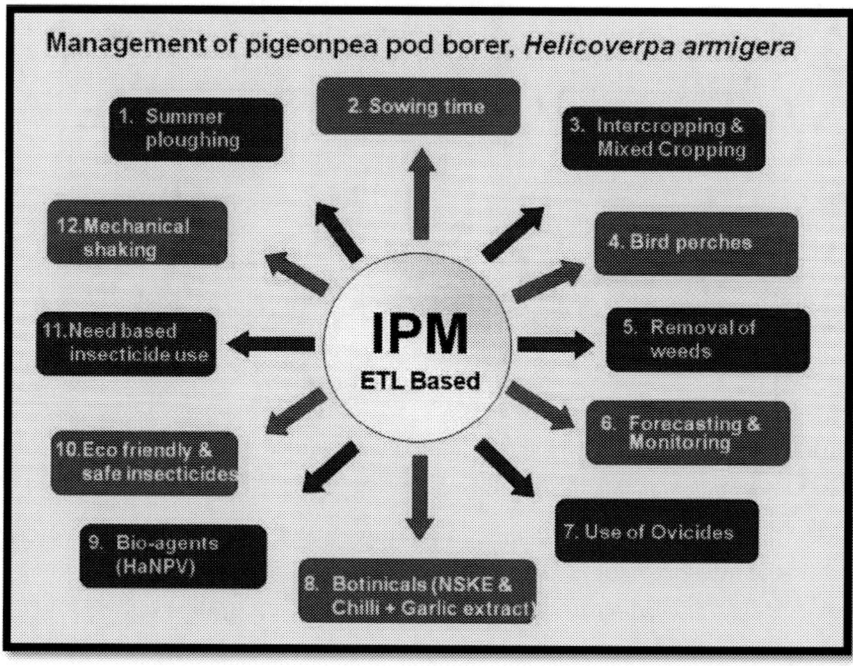

Figure 17. Diagram showing various components of Integrated Pest Management (IPM).

IPM is the core of sustainable agriculture that includes the pest management methods such as monitoring, cultural, physical or mechanical, biological and chemical (Figure 17). These methods when used in integration should be compatible with each other.

2.10.1. Cultural Control

1. Deep summer ploughing helps in killing the resting stages of insect pests. The insects exposed are also eaten by perching birds (Figure 18).
2. Synchronized sowing of tolerant/ resistant/ multiple pest resistant varieties *viz*., Sharad, Asha, Maruthi, Bahar, Abahya, BSMR-736, BDN 1, ICPL 87 (Pragati), ICP 8863, Birsa Arhar, Jawahar, Pusa-991, Pusa-9 and Jagrati.

A B

Figure 18. A. Deep summer ploughing exposed the resting stage of insects on soil surface; B. Perching birds making its prey to insects, exposed after deep ploughing.

3. Early planting by mid-June in Northwest Plain Zone of India avoids the infestation of *H. armigera.* Timely planting of pigeon pea cultivar Bahar in Northeast Plain Zone protects the crop from the attack of *H. armigera.*

A B

Figure 19. A. Tall maize variety planted around the pigeon pea growing field; B. Perching of insect by a black drongo.

4. Early sowing with early to mid-maturing varieties of pigeon pea *viz.*, TS3R, GRG 811, GRG 152 avoids the crop from the infestation of *M. obtusa*.
5. Planting of castor or tall sorghum/ maize varieties on borders conserves natural enemies and act as live perches for predatory birds *viz.*, black drongo, blue jay and black myna (Figure 19).
6. To facilitate spraying operation, short statured crops *viz.*, cowpea, moong bean, urd bean or fodder soya bean may be grown in one meter wide band after every 8-10 rows of pigeon pea (Figure 20). These crops can be harvested before flowering of pigeon pea and the space created will be used for free movement of operator/ machine.

Figure 20. Short statured crop grown with the main crop (pigeon pea) to facilitate spraying operation.

7. Growing of pigeon pea cultivars having different maturity periods should be avoided in endemic areas because this will provide pods to remain in field over a long period and thus allow the insect pests, particularly pod fly to complete several generations.
8. Intercropping of pigeon pea with sorghum is useful to reduce wilt, conserve beneficial insects and serve as perches for insectivorous birds (Figure 21).
9. Growing of marigold as trap crop on border for pod borer management (Figure 22).

Figure 21. Intercropping of pigeon pea with sorghum.

Figure 22. Marigold grown on border of pigeon pea growing field.

10. Seed treatment with chlorpyriphos 20 EC @ 8 ml/ kg pigeon pea seed is recommended to avoid the infestation of soil pests such as termite, root grub and cutworm.

11. During storage care should be taken that grains do not harbour infested grains of old produce and bruchids surviving on them.

2.10.2. Mechanical Control

1. Pheromone traps @ 5-10/ ha should be used for monitoring of adult male population of *H. armigera* (Figure 23). A threshold level of 4-5 moths/ trap/ day is indicative of its peak activity period and indicator of initiation of management practices.

Figure 23. A pheromone trap.

2. In case perching birds are not there, branched sticks, 1.5-2.0 feet taller than crop in field at the distance of 25-30 feet should be placed (Figure 24).

Figure 24. Perching birds on branched stick.

3. For insect pests like blister beetle, hairy caterpillars and Lepidopteran larvae, hand picking, jarring on cloth sheets or in bags is effective in a limited way in small areas (Figure 25).

Figure 25. Plant shaking for hand picking of insect pests.

4. In case of severe pod borer infestations, shaking of plants and collection of larvae for preparation of HaNPV (*Helicoverpa armigera* nuclear polyhedrosis virus) locally, cuts down the management cost.
5. Remove and destroy sterility mosaic affected plants.

2.10.3. Biological Control

1. Seed treatment with *Trichoderma* @ 5 gm/ kg seed.

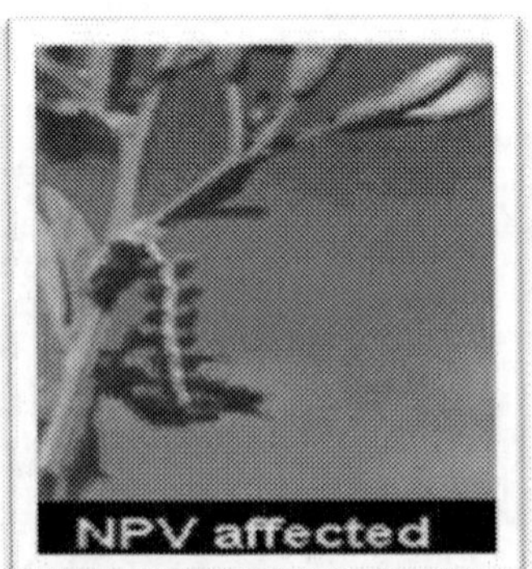

Figure 26. A dead larva after spray with HaNPV.

2. Spiders, *Campoletis*, *Chrysoperla*, Reduvid bugs, predatory wasps and birds play significant role in bringing down pest population. So, in order to conserve them indiscriminate use of pesticides should be avoided, instead the IPM must be followed.
3. Spray HaNPV @ 250 LE (larval equivalent)/ ha (9 x 10^{11}) admixed with UV (ultra violet) retardant Tinopol/ Ranipal @ 0.1 per cent (Figure 26).
4. Spray with commercial Bt (*Bacillus thuringiensis*) products @ 0.75-1.0 kg/ ha in case of moderate incidence of the pest.

2.10.4. Botanicals

1. Application of water extract of crude neem seeds i.e., neem seed kernel extract (NSKE 5%) or commercial neem-based formulations (1500 ppm) @ 3 ml/ litre of water against *H. armigera* and other lepidopterous pests, as and when ETL is attained.
2. Application of plant products such as neem leaf and bark, turmeric, henna, marigold, dhatura, tulsi, eucalyptus, lantana, lemon grass, cinnamon or nutmeg @ 1-12 gm/ kg seed.

2.10.5. Physical Control

1. Harvested grains should be dried thoroughly in scorching sunlight on cemented floor to prevent excessive moisture. This method will prevent bruchids to survive during storage.
2. It is better to use black plastic or tarpaulin sheet to reduce moisture content of grains and expose different stages of bruchids to direct sunlight.
3. Avoid exposing grains for longer duration in humidity prone areas.

2.10.6. Chemical Control

1. Spray ovicidal insecticides *viz.*, Profenophos 50 EC @ 2.0 ml/ litre of water or Thiodicarb 75 WP @ 1 gm/ litre of water or Methomyl 40 SP @ 1.0 gm per litre of water.
2. Systemic insecticides *viz.*, Thiamethoxam 0.3 gm/ litre of water or Acephate @ 1.0 gm/ litre of water or Imidacloprid 0.4 ml/ litre of water are effective in controlling pod bugs.

3. Gunny bags may be disinfected by dipping or soaking them in above mentioned insecticidal solution and should be dried thoroughly before their use.

3. Insect Pests of Chickpeas

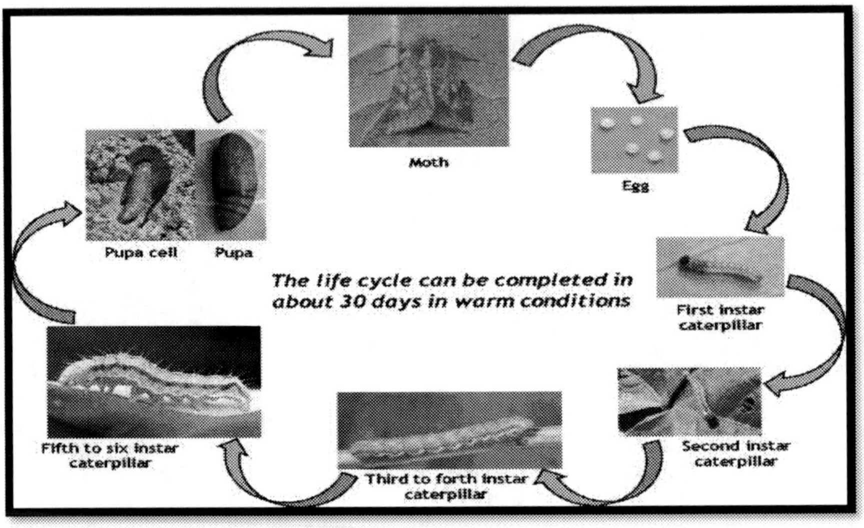

Figure 27. Life cycle of *Helicoverpa armigera.*

3.1. Pod Borer: *Helicoverpa armigera* (Hubner) (Lepidoptera: Noctuidae)

3.1.1. Biology

Eggs are spherical in shape and creamy white in colour, laid on the host plant singly. The egg period is 7 days. Full grown larva is greenish with dark brown, gray lines and dark and pale bands. It shows colour variation from greenish to brown. The larval duration is 14 days. Pupa is brown in color, pupation occur in soil, leaf, pod and crop debris. Adults are light pale brownish, yellow stout moth, forewings are grey to pale brown with V-shaped speck and hind wings are pale smoky white with a broad blackish outer margin. One generation is completed in 28 days under favorable conditions (Figure 27).

3.1.2. Nature of Damage

H. armigera is a perennial and persistent pest of chickpea and causes huge loss to the crop, particularly during podding stage. Its damage on flowers and green pods are more severe than the other plant parts. The young pod borer larvae (1st, 2nd and 3rd instar) occasionally enter the pod and feed upon the developing chickpea grains, but more often they feed from outside the pod with only the anterior part of their body into the pod (Saxena 1978; Singh and Singh 1987). Patel et al. (2010) reported that pod borer caused maximum entry holes at the basal region of chickpea pods irrespective of the genotypes.

3.2. Semi Looper, *Autographa nigrisigna*

3.2.1. Biology

The adult moths of *A. nigrisina* have typically patterned forewing. The larvae measure about 25 mm long.

3.2.2. Nature of Damage

Larvae feed on the leaves, flowers, buds, tender pods and developing seed. Larvae makeragged and irregular holes on the pod (contrasted with the damage caused by pod borer).

3.3. Tobacco Caterpillar, *Spodoptera exigua*

3.3.1. Biology

Moths lay eggs on the underside of leaves in masses of 100 to 300. The incubation period lasts for 2-3 days. Larval period lasts for 13-15 days and pupal period 7-9 days. Pupation occurs in the soil. Total life cycle takes place between 28-32 days; 6-8 generations are completed in one year (Figure 28).

3.3.2. Nature of Damage

Spodoptera exigua is an early season pest. It causes defoliation at early stage leads to sparse plant population that looks similar to cattle grazed field. Larvae feed on the tender leaves. They scrape out the green matter in the leaf leaving the epidermis alone. The second and third instar larvae feed by making small holes, late instars feed on entire lamina, petiole and even the twigs on the terminal shoots of plant. Later, the larvae spread in the field causing defoliation. The larvae also feed on the developing pods.

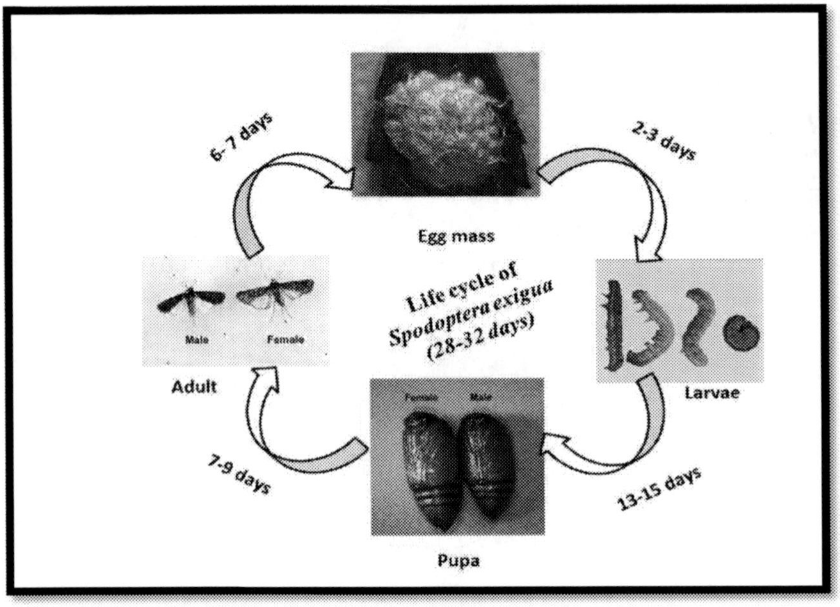

Figure 28. Life cycle of *Spodoptera exigua.*

3.4. Pod Bug: *Riptortus pedestris* (Coreidae: Hemiptera)

3.4.1. Biology

The female bug lays 115 eggs (singly) on an average. The egg period remains for 3-4 days. The nymphs are brownish black and hemispherical which resembles brown ants. The nymphal stage undergoes 5 instars in 16 days.

3.4.2. Nature of Damage

The nymphs and adults suck the juice of unripe seeds from the green pods. In case of severe infestation, the tender parts get shriveled and later dry up. The bugs are seen clustered around the pods.

3.5. Leaf Webber: *Eucosma critica* (Eucosmidae: Lepidoptera)

3.5.1. Biology

The small and dark brown moth lays eggs singly or in rows on leaves, petioles or stems. The grooves or depressions are preferred for egg laying. Around 80-

100 eggs are laid by one female. The incubation period lasts for 3-4 days. Larval period remains for 14-21 days and pupal period for 4-6 days. Pupation occurs within the folded buds, flowers or pods in a silken cocoon.

3.5.2. Nature of Damage

The damage is caused by the larvae, which bores into the tender shoots of folded leaves and feed from within. Since the terminal leaves get spun together, the growing tip gets damaged. The growth of the main shoot is affected.

3.6. Integrated Pest Management

3.6.1. Cultural Control

1. Deep summer ploughing is effective in destroying the eggs, larvae and pupae of various pests.
2. Use certified seeds of recommended variety having resistance/ tolerance to key insect pests.
3. Intercropping of chickpea with linseed (4 rows of chickpea + 2 rows of linseed) or coriander (9 rows of chickpea + 1 row of coriander).
4. Broadcasting of 2.0 kg puffed rice per hectare at pod formation stage will attract more insectivorous birds.
5. At the time of chickpea sowing, sunflower or sorghum seeds @ 250 gm/ ha will act as trap crop.
6. Host plant resistance is one of the most effective and economically viable methods for effective pest management. However, only low to moderate levels of resistance have been identified against *H. armigera* in the cultivated chickpea germplasm.
7. Plant density also affects the extent of damage caused by pod borer. In general, a denser plant population favours increased pod damage. However, higher plant densities may not necessarily result in yield loss due to compensation in the total pod number per unit area.
8. Low to moderate doses of NPK fertilizer will lead chickpea for lower pod borer infestation, but higher doses of NPK will lead to higher pod borer infestation (Hossain *et al.* 2009).
9. Applying organic manures like farmyard manure, neem cake and vermicompost will result in the lowest pest population as compared

to the application of inorganic fertilizers (Rao, 2003; Muddukumar, 2007; Singh and Singh, 2007).

Figure 29. Intercropping with chickpea plants.

3.6.2. Mechanical Control

1. Monitoring of pod borer through pheromone traps (5 traps/ acre) installed in field and replacing the rubber lure once in 15 days. The pest control measures should be taken when 5 adult insects are found/ trap or one larva or two eggs per 10 plants.
2. Once the crop attains bud forming stage, pheromone traps @ 5/ ha should be installed for monitoring of *H. armigera* adults in the field. Adult catches will help in getting ready for monitoring-based spray schedules. Conventional light traps may also be installed. Followed by adult catches, crop need to be surveyed for presence and stages of larvae. In case of higher catches (2-5 adults) per week or 1 larva per meter row (which is also ETL of this pest), plant protection measures should be adopted.
3. Heaps of grasses on bund encourage congregation of larvae which should be mechanically destroyed.

3.6.3. Botanicals

Neem based water-soluble powder formulations having strength of 3 per cent azadirachtin sprayed @ 400 gm/ hec were found as effective as recommended synthetic insecticides such as Profenophos and Chlorantranilliprole against pod borer without affecting the most common parasitoids (*Campoletis chlorideae* and Tachinid fly) present in the chickpea ecosystem.

3.6.4. Biological Control

1. *H. armigera* in the early instar feed on the leaves (act as defoliator) so application of HaNPV @ 250 LE/ ha will help in managing the pest in the early stage.
2. Spray of commercial *Bt* products (BioBit or Delfin and Dipel) will help in minimising pod damage.

3.6.5. Chemical Control

1. Application of insecticide like Chlorantraniliprole 18.5 SC @ 0.15 ml/ litre of water or Flubendiamide 48 39.25 @ 0.15 ml/ litre of water or Emamectin benzoate 5 SG @ 0.3 g/ litre of water or Spinosad 45 SC @ 0.1 ml/ litre of water or Indoxacarb 14.5 SC @ 0.3 ml/ litre of water or Profenophos 50 EC @ 2.0 ml/ litre of water or HaNPV @ 250 LE/ ha @ 1.0 ml/ litre of water or NSKE 5% is recommended whenever damage is visible above economic threshold level (1 larva per metre row).
2. The second spray of insecticide may be carried out after 15-20 days of first spray. Application of appropriate botanical pesticides such as neem-based pesticides, insect hormones, microbial agents and pathogenic fungi can also be applied whenever possible.
3. Seed treatment: Mixing of 1 kg activated kaolin in 100 kg seeds. Pack in polythene lined gunny bags for storage.
4. Spray Emamectin benzoate 5 SG @ 0.3 gm/ litre or Novaluron 10 EC @ 1.0 ml/ litre or Chlorantrainiliprole 18.5 SC @ 0.30 ml/ litre against pod borer.

3.7. Insect Pests of Minor Pulses

The minor pulse crops include black gram, green gram, horse gram, cowpea, lablab bean, moth bean, faba bean, grass pea, lentil and pea. The probable geographical origin of some common pulses has been mentioned below (Table 3).

Table 3. Minor pulses and their geographical origin

Pulse crops	Scientific name	Geographical origin
Black gram	*Vigna mungo* (L.) Hepper	Indian subcontinent
Green gram	*Vigna radiata* L. Wildzek.	Indian subcontinent
Horse gram	*Macrotyloma uniflorum* (Lam.) Verdc.	Indian subcontinent
Cowpea	*Vigna unguiculata* (L.) Walp.	West Africa
Lablab bean	*Lablab purpureus* (L.) Sweet	Indian subcontinent
Moth bean	*Vigna aconitifolia* (Jacq.) Marechal	Indian subcontinent

The minor pulse crops are attacked by a series of insect pests which are mentioned below (Table 4).

Table 4. Insect pests associated with minor pulse crops

Pulse Crops	Lepidopteran Pests		Sap Feeding Insect Pests	
Black gram, Green gram,	Common name	Scientific name	Common name	Scientific name
	Pod borer	*Helicoverpa armigera*	Bean aphid	*Aphis craccivora*
	Spotted pod borer	*Maruca vitrata*	Leaf hopper	*Empoasca kerri*
Horse gram, Cowpea, Lablab bean, Moth bean	Spiny pod borer	*Etiella zinckenella*	Pod fly	*Melanagromyza obtusa*
	Blue butterfly	*Lampides boeticus*	Lablab bug	*Coptosoma cribraria*
	Grass blue butterfly	*Euchrysops cnejus*	Whitefly	*Bemisia tabaci*
	Cutworm	*Agrotis ipsilon*	Thrips	*Megalurothrips usitatus* *Scirtothrips dorsalis*
	Tobacco caterpillar	*Spodoptera litura*	Blister beetle	*Mylabris* spp.
	Bihar hairy caterpillar	*Spilosoma obliqua*		
	Semiloopers	*Thisanoplusia orichalsia* *Trichoplusia ni*		
	Stem fly	*Ophiomyia phaseoli*		
	Pod weevil	*Apion amplum*		
	Sphingid caterpillar	*Acherontia styx*		

3.8. Lepidopteran Pests

3.8.1. Borer: Helicoverpa armigera *(Hubner) (Lepidoptera: Noctuidae)*

Identification, biology and nature of damage same as described under pigeon pea.

3.8.1.1. Natural Enemies of Helicoverpa armigera

- Parasitoids: Trichogramma spp., Tetrastichus spp., Chelonus spp., Telenomus spp., Bracon spp., Ichneumon spp., Carcelia spp. and Campoletis spp.
- Predators: Lacewing, ladybird beetle, spider, red ant, dragon fly, robber fly, reduviid bug, praying mantis, black drongo (King crow), wasp, common mynah, big-eyed bug (Geocoris sp), earwig, ground beetle and pentatomid bug (Figure 30).

Figure 30. Some natural enemies of *Helicoverpa armigera.*

3.8.1.2. Natural Enemies of Maruca vitrata

- Parasitoids: *Trichogramma* spp., *Tetrastichus* spp., *Chelonus* spp., *Telenomus* spp. and *Bracon* spp.
- Predators: Lacewing, ladybird beetle, spider, red ant, dragon fly, robber fly, reduviid bug and praying mantis.

3.8.2. Spiny Pod Borer: Etiella zinckenella *Treitschke (Lepidoptera: Pyralidae)*

It is found in Southern and Eastern Europe and the tropics and subtropics of Africa and Asia. It feeds on pigeon pea, cowpea, lal lab, chickpea, horse gram, green gram, black gram and peas.

3.8.2.1. Biology

- Egg: Female lays about 100-300 eggs.
- Larva: The young larvae are greenish initially, turns pink before pupation. It has 5 black spots on the prothorax and a lot of white frass and silk.
- Pupa: Brown in colour, cocoon is white thick and covered with soil particles. Pupation takes place in the soil.
- Adult: Small, brownish grey moth with orange prothorax and wingspan of about 20 mm. Fore wing has a white stripe along the anterior margin.

E. zinckenella completes one generation in 4 weeks under favourable conditions (Figure 31).

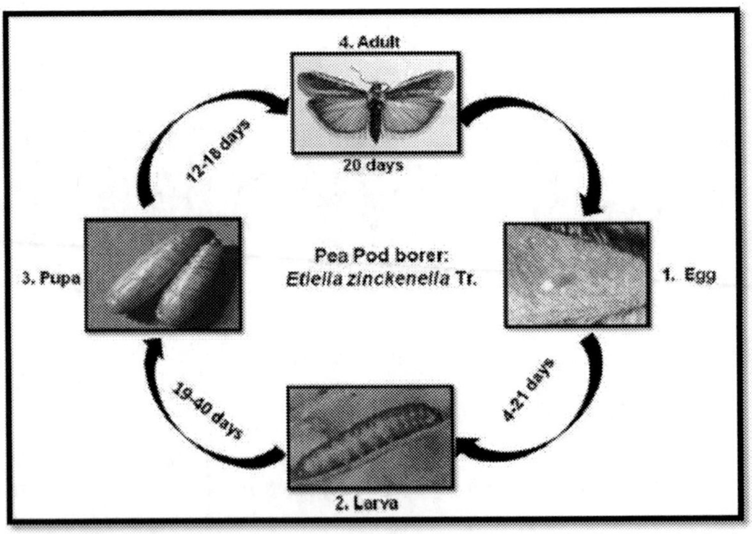

Figure 31. Life cycle of pea pod borer, *Etiella zinckenella.*

3.8.2.2. Damage Symptoms

The larvae bore into flowers, pods and feed on seeds results in dropping of flowers and young pods. Mature pods are marked with a brown spot where larva has entered and also due to build-up of faeces inside the pod (Figure 32).

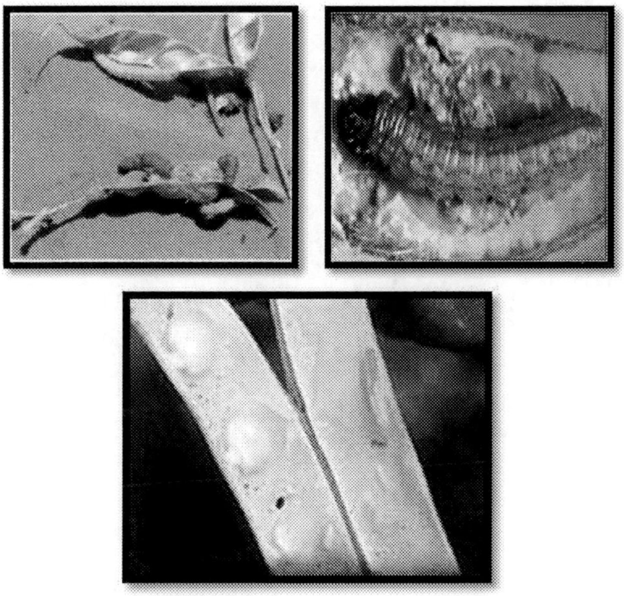

Figure 32. Pod damage caused by pea pod borer, *Etiella zinckenella.*

3.8.3. Blue Butterfly: **Lampides boeticus** *Linnaeus (Lepidoptera: Lycaenidae)*

L. boeticus is widely distributed in Asia, Africa and Mediterranean Europe. It is highly migratory and occasionally observed in south England.

3.8.3.1. Biology

- Egg: Female lays pale blue eggs singly on buds.
- Larva: Larvae are about 12 mm long, green, oval, flat and slug like. They are relatively sedentary.
- Pupa: Pupation occurs in the soil or in plants debris.
- Adult: Adults are slender, bluish-gray with a wingspan of 30 mm and are sexually dimorphic. Males have blue violet upper face of the wings with the brown edges, while the females have only a small amount of blue colour in the centre of the wings (sexual dimorphism). Both sexes have a thin, long tail in the hind wings and two black spots in the anal angle.

The pest completes one generation in five weeks under field conditions.

3.8.3.1. Damage Symptoms

Larvae chew leaves, buds, flowers and pods which can be easily observed.

3.8.4. Grass Blue Butterfly: **Euchrysops cnejus** *Fabricius (Lepidoptera: Lycaenidae)*

E. cnejus is spread throughout India except at very high elevations and is widely distributed in the Malayan sub region, extending to Australia and the South Sea Islands.

3.8.4.1. Biology

- Egg: Female lays single egg at a time under the leaves, flowers and pods. Eggs are greenish white in colour, round shaped with a slight depression on the top.
- Larva: Flat, slug like, green or yellowish green, red coloured mid stripe and few hairs.
- Pupa: This stage lasts for 7 – 8 days. The pupa is dark brown in colour with blunt end at the posterior end. The anterior end is broad with a small bulge on the top side. Towards anterior side, three prominent stripes are present. Pupation takes place in soil or on the plants
- Adult: Blue coloured butterfly with five black spots on the dorsal surface of the hind wing and two black spots on the ventral surface. The entire life cycle completes in 23-30 days.

3.8.4.2. Damage Symptoms

The larvae bore into the buds, flowers and green pods just within couple of hours after hatching and feed the developing grains inside the pods. There is presence of regular big circular borehole on the flowers and pods.

3.8.5. Cut Worm: **Agrotis ipsilon** *(Hufnagel) (Lepidoptera: Lycaenidae)*

The origin of black cutworm is uncertain; it is now found in many regions of the world, except in some cold areas. Cutworm is more widespread and damaging in northern hemisphere than in the southern hemisphere. It annually reinvades temperate areas, overwintering in warmer or subtropical regions. It feeds on green gram, black gram and chickpea.

3.8.5.1. Biology

- Egg: A female can lay up to 1500 cream-coloured eggs singly or in small clusters (2-3) on plants or on the soil surface.
- Larva: The full-grown larva is up to 40 mm long and brown-black in colour. Larvae pass through 6-7 instars and are nocturnal, staying in the soil during the day time.
- Pupa: Pupation takes place in the soil.
- Adult: Adults are large moths 40-50 mm long with brown forewings and white hindwings (Figure 33).

Depending on favourable environmental conditions, a single generation can be completed in 4 to 5 weeks.

Figure 33. Adult of *Agrotis ipsilon.*

3.8.5.2. Damage Symptoms

Caterpillars cut seedling stem at ground level resulting in death of the plant. Cutworm larvae are found in the soil or in leaf litter near damaged plant. Larvae feed on the foliage (Figure 34).

Figure 34. Damage to seedling caused by *Agrotis ipsilon.*

3.8.6. Tobacco Caterpillar: **Spodoptera litura** *(Fabricius) (Lepidoptera: Noctuidae)*

3.8.6.1. Biology

- Egg: Eggs are spherical and slightly flattened. Each individual egg is around 0.6 mm in diameter. They are orange-brown or pink in colour. Eggs are laid on the surface of leaves in big batches, with each cluster usually containing several hundred eggs. Females have a typical fecundity of 2000 – 2600 eggs. Eggs are covered with hair scales provided by the female, which gives off a golden brown colour.
- Larva: Body length ranges from 2.3 – 32 mm. The colour of larvae is variable based on age. Younger larvae are light green in colour, while older ones are dark green or brown. A bright yellow stripe along the dorsal surface is a characteristic feature of the larva. Newly hatched larvae can be found by looking for scratch marks on leaf surface. The pest is nocturnal and feed at night. During day time, they can usually be found in the soil around the plant.
- Pupa: Pupation lasts around 7 – 10 days and takes place on the soil near the base of the plant. The pupa is typically 15–20 mm long and its colour is red-brown. A characteristic feature is the presence of two small spines at the tip of the abdomen.
- Adult: Adult moths are gray-brown in colour. The forewings are patterned with dark gray, red and brown colours. The hindwings are grayish-white with a gray outline (Figure 35). The mean longevity for female and male is 8.3 and 10.4 days, respectively.

Figure 35. Adults of *Spodoptera litura.*

Although the length of a life cycle varies slightly throughout the different regions, a typical *S. litura* will complete 12 generations in one year. Each generation lasts about a month, but temperature causes slight variations: life cycle in the winter tends to be slightly more than one month but in the summer it is less than one month.

3.8.6.2. Damage Symptoms

Young larvae feed gregariously on the chlorophyll of the leaves and skeletonise foliage. The eaten leaves show the appearance of whitish yellow web. At maturity, larvae become solitary and make large irregular holes in foliage (Figure 36).

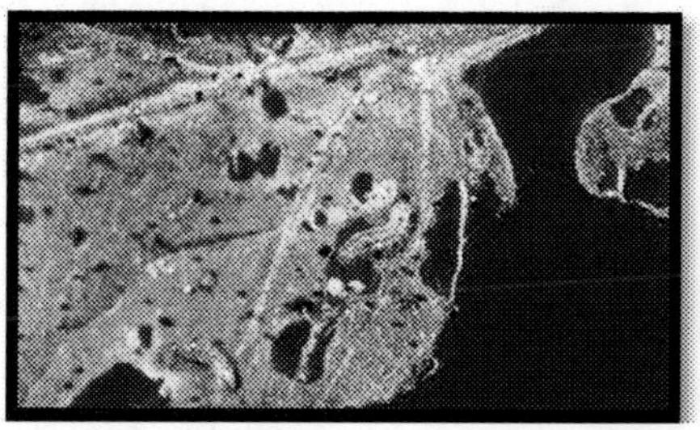

Figure 36. Leaf damaged by *Spodoptera litura.*

3.8.7. Bihar Hairy Caterpillar: Spilosoma obliqua *Walker (Lepidoptera: Arctiidae)*

3.8.7.1. Biology

- Egg: Female lays eggs in masses on the leaves.
- Larva: Larvae are pale yellow coloured with yellow hair over the body. They are polyphagous, feed on leaves and cause loss by way of defoliation. In severe cases only stem is left behind.
- Pupa: Pupates in leaf litter close to the plants.
- Several generations are completed in a year.
- Adult: Adult moth is reddish brown with black spots. Both the wings are pinkish and possess black spots (Figure 37).

Figure 37. Adult of Bihar hairy caterpillar, *Spilosoma oblique*.

3.8.7.2. Damage Symptoms

Young larvae feed gregariously mostly on the under surface of the leaves. Caterpillars feed on leaves and in severe infestation the whole crop is defoliated. The main symptom is drying up of infected foliage (Figure 38).

Figure 38. Leaves damaged by Bihar hairy caterpillar, *Spilosoma oblique*.

3.8.8. Stem Fly: **Ophiomyia phaseoli** *Tryon (Diptera: Agromyzidae)*

3.8.8.1 Biology

- Egg: The stem fly inserts eggs on the underside of young leaves. Eggs are elongate, oval and white in color with a smooth and transparent surface. Ovipositing sites appear as pale pinprick spots when infested leaves are held up to the light.
- Larva: Light yellow-coloured maggot moves in the leaf and then bore deeper into the stem. It travels downwards in the young plants.

In older plants, the maggots do not move much. It cuts an exit hole in the stem for the adult before pupation.

- Pupa: Pupation takes place in the underground portion in the young plants, whereas in the older plants, it is usually in the main stem or branches. Pupal stage lasts for 6–12 days.
- Adult: Freshly emerged adults are light brown in colour but when fully developed they attain metallic-bluish or greenish-black colour with light brown eyes. Wings are transparent. Female is slightly bigger than the male (Figure 39).

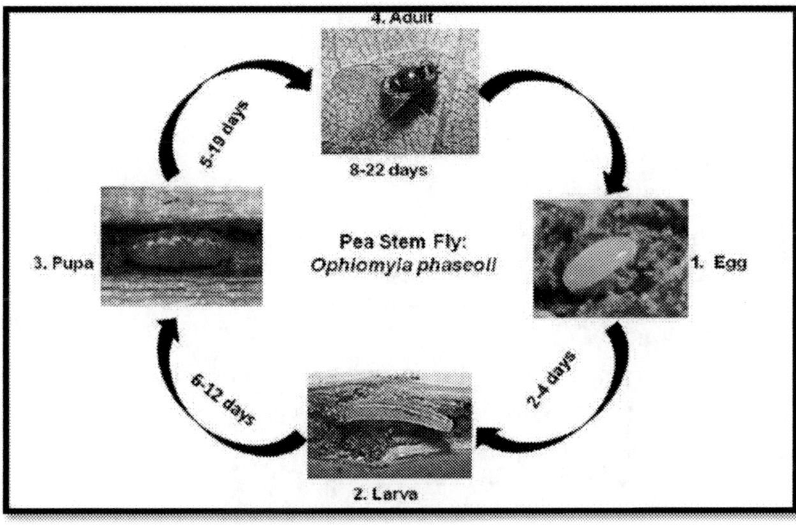

Figure 39. Life cycle of *Ophiomyia phaseoli.*

3.8.8.2. Damage Symptoms

Maggot is the damaging stage. It mines sub-epidermally through the leaves. Plants are most seriously affected at the seedling stage, where stem is tunneled. Infected stems are often red inside (sometimes pale) and a distinct zig-zag tunnel may be observed with maggots or pupae inside. Apart from the exit holes, the plant initially appear healthy. Large infestations (3 or more maggots/ plant) may cause wilting and may even cause plant death, especially in younger plants, if damage occurs in the plant's hypocotyl (basal stem) region. Older plants show stunting but are not usually killed (Figure 40).

A B

Figure 40. Damage caused by *Ophiomyia phaseoli* A. Symptoms on leaves; B. Damage on hypocotyl of plant.

3.9. Sap Feeding Insect Pests

3.9.1. Bean Aphid: Aphis craccivora *Koch (Hemiptera: Aphididae)*

Several species of aphids are associated with pulse crops, however *A. craccivora* is most prevalent in Asia and Africa.

Figure 41. Adults of *Aphis craccivora.*

3.9.1.1. Biology

- Egg: Very tiny, shiny-black, found in the crevices of bud, stems, and barks of the plant.
- Nymph: Similar to adult but smaller in size. Nymphs become adult within 7-10 days.
- Adult: Adults are small, back to dark brownish in colour, shiny, up to 2 mm long, soft -bodied with two long antennae that resemble

horns. Most aphids have two short cornicles (horns) towards the rear of the body. Some adults are winged. *A. craccivora* can reproduce without mating in Asia creating one generation in a week under optimum conditions. When the population density in the colony reaches a certain limit, winged individuals are found among the wingless forms. They fly away to form new colonies (Figure 41).

3.9.1.2. Damage Symptoms

Both nymphs and adults cause the damage by sucking the plant sap. Infestation on young seedlings results in twisted shoots under heavy infestation; seedlings may wilt, particularly under moisture stressed condition (Figure 42). Infested pods become misshaped, withered and malformed. Severe infestation may result in complete drying of affected pods. In addition, they secrete a sticky fluid (honeydew) on the plant, which turns black due to fungal infection (sooty mould growing on honeydew). Aphids act as vectors in transmitting bean common mosaic virus.

Aphids are considered important in both rainy and post rainy seasons for pulse crops. Heavy rainstorms reduce aphid population densities and favour the development of fungal diseases that suppress the population build up. Aphids are susceptible to predators such as coccinellids, syrphids and lacewings.

Figure 42. Leaves damaged by *Aphis craccivora.*

3.9.2. Jassids or Leaf hopper: **Empoasca kerri** *Latreille (Hemiptera: Cicadellidae)*

The common jassid that attacks minor pulses in Asia is *E. Kerri*.

3.9.2.1. Biology

- Egg: Yellow in colour, laid on the underside of the leaf, embedding into the leaf veins close to the midrib or into the petiole. Under normal conditions the eggs hatch in about one week.
- Nymph: Nymphs and adults have a similar shape (wedge-shaped) and green in colour, but the nymphs do not have wings and run sideways when disturbed. They are very active, suck cell sap from underside of the leaves. Nymphs develop into adults in 10 days. About 40 nymphs can be expected from a single female.
- Adult: Adults are 2.5-3.0 mm long and green in colour (Figure 43). They move briskly, forward and lateral, jump or fly away at slight disturbance, attracted to light at night.

High and well distributed rainfall and low temperature encourage jassid reproduction. One generation completes in 2 weeks under optimum conditions.

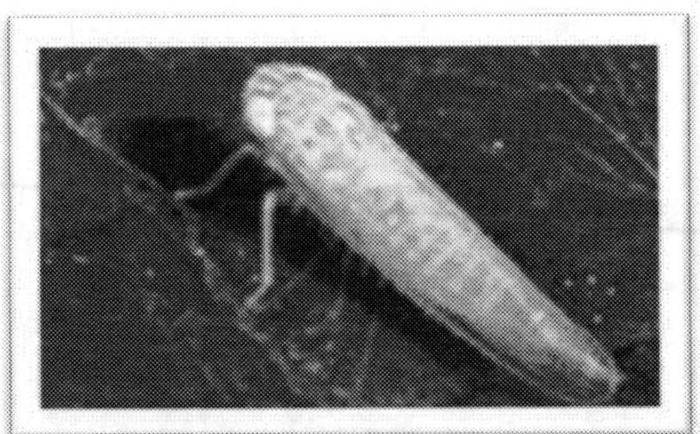

Figure 43. Adult of *Empoasca kerri.*

3.9.2.2. Damage Symptoms

Both nymphs and adults suck sap from new leaves, mostly from the lower surface. The first symptom of attack is whitening of the leaf veins. Chlorotic patches appear, especially at the tip of leaflets. Under severe infestation, leaflets are cup-shaped with yellow edge and tip with subsequent defoliation and stunting. In advance stage, the leaves turn brown, dry and brittle, a condition called "hopper burn" (Figure 44).

Figure 44. Leaves showing hopper burn symptom caused by *Empoasca kerri.*

3.9.3. Whitefly: **Bemisia tabaci** *(Gennadius) (Hemiptera: Aleyrodidae)*

White fly is a cosmopolitan pest. It is a vector of yellow mosaic disease. The host pulse crops of white flies are black gram, green gram, red gram and lobia.

3.9.3.1. Biology

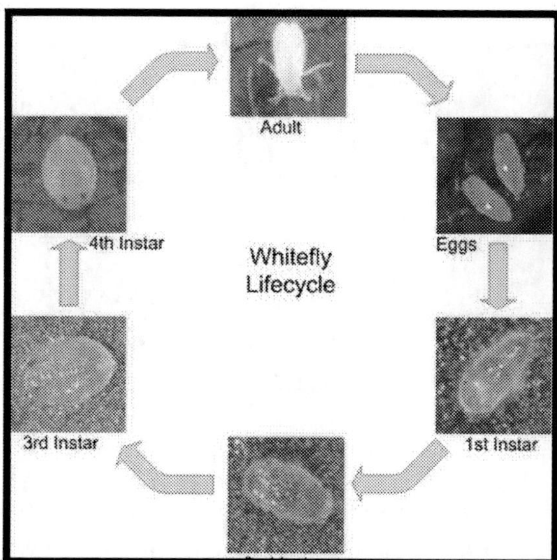

Figure 45. Life cycle of white fly, *Bemisia tabaci.*

- Egg: The female lays stalked eggs, singly on the under surface of the leaves. On an average, 119 eggs are laid by a single adult female. Initially, eggs are light yellow in colour but later turns brown. Eggs hatch in 3-5 days.

- Nymph: Oval or elliptical in shape, greenish yellow in colour, fix their mouth parts into plant tissues. Nymphs feed on cell sap and grow into three stages to form the pupa.
- Pupa: Oval in shape. They can be observed on the under surface of the leaves.
- Adult: Minute with yellow body and hyaline wings dusted with a waxy powder (Figure 45).

The lifecycle completes in 18-28 days and 11 generations are completed in a year.

3.9.3.2. Damage Symptoms

Nymphs and adults remain in clusters under the leaf surface and suck plant sap. Under severe infestation premature defoliation, development of sooty mould and shedding of flower and pods may be observed. *B. tabaci* also acts as vector of mungbean yellow mosaic virus (Figure 46).

Figure 46. Yellow mosaic virus disease on mungbean.

3.9.4. Thrips: **Megalurothrips usitatus** *(Bagnall) (Thysanoptera: Thripidae),* **Scirtothrips Dorsalis** *Hood (Thysanoptera: Thripidae)*

3.9.4.1. Biology

- Egg: Female lays 40-50 eggs that are white or yellow in colour and microscopic in size. Adult females prefer to lay their eggs in leaf, cotyledon, or flower tissues. Eggs are inserted one by one by the female in the plant tissue. Only one end of the egg is near the surface of the tissue to allow the immature to emerge.

- Nymph: Very small in size, pale yellow to brown in colour, possess short antennae, wing buds are visible but short and non functional (Figure 47).
- Adult: Approximately 2 mm in size and pale yellow to dark brown in colour. Adults have fully developed wings. The wings have a single longitudinal vein in which there are several hairs connected perpendicular to the vein. The wing appears as fringe with hairs. When at rest, the wings are folded along the back of the insect. Adults are attracted to yellow and white colours.

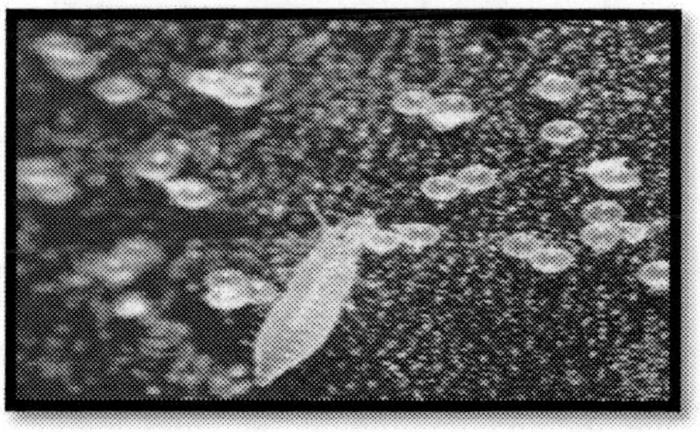

Figure 47. Nymph of Scirtothrips dorsalis.

3.9.4.2. Damage Symptoms

Thrips have a very peculiar feeding behaviour. They start feeding by piercing and rasping the leaf surface with their mouth parts to release the liquid into the plant cells. During the process, thrips release substances that help in predigesting the plant tissues. Later, they suck up the plant contents which leads to drying up of the small tender leaves, thus hindering the vegetative growth and development of the plant.

3.9.5. Lablab Bug: **Coptosoma cribraria** *(Fabricius) (Hemiptera: Plataspididae)*

3.9.5.1. Biology

- Eggs: Female lays ivory white sculptured eggs in double rows in batches of 35-50 on the tender pods. Incubation period lasts for 7 days.

- Nymph and Adult: Sub globular, oval and greenish shield bug (Figure 48). It has a characteristic buggy odour.

3.9.5.2. Damage Symptoms

Both nymphs and adults cluster on the tender shoots and suck the sap. Heavily infested plants dry and shed away. Moderately infested plants exhibit weak and stunted growth.

Figure 48. Nymph of *Coptosoma cribraria.*

3.9.6. Blister Beetle: Mylabris spp. *(Coleoptera: Meloidae)*

Several species of genus *Mylabris* have been found associated with pulses, however, *M. pustulata* is very common on pulse crops in Asia. They are polyphagus and flower feeders. *M. pustulata* have been found to occur throughout the year in red gram, cowpea, green gram and black gram.

3.9.6.1. Biology

- Eggs: Eggs are usually laid in the soil.
- Larva: Larvae feed on soil insects, including few major pests.
- Adult: Adults of *M. pustulata* are about 25 mm in length and possess red and black alternating bands on the elytra. Other species may vary in size but all are bright coloured.

3.9.6.2. Damage Symptoms

Adult beetles feed on flowers and tender pods and may have a significant impact on crop yields, especially of short duration pulse genotypes. During peak incidence, the adults may cause 95 per cent damage to flowers. However, larvae are beneficial as they feed on other soil insects.

3.9.7. Sphingid Caterpillar: **Acherontia Styx** *(Sphingidae: Lepidoptera)*

The adult is large, wings grey with waxy markings. Abdomen crimson coloured with black stripes (Figure 49). The larva is stout, green with yellowish oblique stripes and possesses curved anal horn (Figure 50). Larva feeds on leaves and causes severe defoliation.

Figure 49. Adult of *Acherontia styx.*

Figure 50. Larva of *Acherontia styx.*

3.9.8. Bruchids

- *Callosobruchus analis* (Fabricius) (Coleoptera: Bruchidae)
- *Callosobruchus chinensis* (L.) (Coleoptera: Bruchidae)
- *Callosobruchus maculates* (Fabricius) (Coleoptera: Bruchidae)

Bruchids are one of the major constraints, causing severe loss to pulses during storage. The three species of bruchids which cause damage to pulse grains during storage are *Callosobruchus analis*, *C. chinensis* and *C. maculates* (Figure 51).

Callosobruchus analis Callosobruchus chinensis Callosobruchus maculates

Figure 51. Three species of *Callosobruchus*, the store grain pests of pulses.

3.9.8.1. Biology

- Eggs: Adult females bore a small shallow hole in the pod and lays eggs singly on the seed surface. Eggs hatch between 5-7 days. Each adult can lay up to 100 eggs in its life span.
- Grub: The young grubs burrow directly into the seed. They feed and pupate entirely within the seed.
- Adult: Adults are small, 3 mm long brown beetles with black spots on the elytra. Adults mature within 24-36 hours after emergence, but they do not need feed. Under these circumstances, adults may live for 12-14 days. During this period mating and oviposition takes place. Sexes can be distinguished by means of morphological differences. Females have dark stripes on each side of elytra, which are absent in males. One generation completes in 4-5 weeks under optimum conditions.

Conclusion

Insect pests are one of the main constraints in the production of pulse crops. Use of chemicals has been used since a long time in the management of insect pests but in the recent past years, several pests have developed resistance against some pesticides. Under such circumstances, integrated approach of pest management should be adopted so as to utilize all the available techniques available at farming level and reduce the cost of pest management as well as the hazardous effect of chemicals. Use of biological control agents such as entomopathogenic fungi, entomopathogenic nematodes and HaNPV, botanicals such as neem cake and castor cake, pheromone traps, change in the date of sowing, intercropping with non-host crops and of course, the development of insect resistant cultivars will minimize the use of chemicals and protect the crop from insect pests and ultimately help in increasing the production of pulse crops.

Acknowledgments

The authors are indebted to all those who have helped in providing the information present in this chapter.

References

Anonymous (2021). *Annual Report. Indian Institute of Pulses Research.* Kanpur, India.

Gopali, J. B., Sharma, O. P., Yelshetty, S. and Rachappa, V. (2013). Effect of insecticides and biorationals against pod bug (*Clavigralla gibbosa*) in pigeon pea. *Indian Journal of Agricultural Sciences*, 83: 582-585.

Gopali, J. B., Raju, T., Mannur, D. M., and Suhas, Y. (2010). Web forming lepidopteran, *Maruca vitrata* (Geyer): an emerging and destructive pest in pigeon pea. *Karnataka Journal of Agricultural Sciences*, 23: 35-38.

Hossain, M. A. (2009). Field screening of chickpea genotypes against pod borer. *Bangladesh Journal of Agriltural Research*, 34: 517–521.

Kranthi, K. R., Jadhav, D. R., Kranthi, S., Wanjari, R. R., Ali, S. S. and Russell, D. A. (2002). Insecticide resistance in five major insect pests of cotton in India. *Crop Protection*, 21: 449–60.

Lateef, S. S. and Reed, W. (1990). Insect pests on Pigeon pea. In: *Insect Pests of Food Legumes* (Ed. Singh, S. R.). John Wiley and Sons Limited, New York, USA, pp. 193-242.

Muddukumar (2007). *Organic Nutrient Management and Plant Protection in Green Chickpea Production System. M.Sc.* (Ag.) Thesis, Department of Entomology, University of Agricultural Sciences, Dharwad, India.

Patel, I. S., Patel, P. S., Patel, J. K. and Acharya, S. (2010). Pod boring pattern of gram pod borer, *Helicoverpa armigera* Hub. in chickpea varieties. *Insect Environment*, 16: 41–42.

Rao, K. R. (2003). Influence of host plant nutrition on the incidence of *Spodoptera litura* and *Helicoverpa armigera* on groundnut. *Indian Journal of Entomology*, 65: 386–392.

Reddy, A. A. (2009). Pulses production technology: status and way forward. *Economic and Political Weekly*, 44: 73-80.

Saxena, H. P. (1978). Pests of grain legumes and their control in India. In: *Pest of Grain Legumes: Ecology and Control* (Eds. Singh, S. R., van Emden, H. F. and Taylor, T. A.). Academic Press, London, UK, pp. 15–23.

Shanower, T. G., Rameis, J. and Minja, E. M. (1999). Insect pests of pigeon pea and their management. *Annual Review of Entomology*, 44: 77-96.

Sharma, D. R., Thapa, R. B., Manandhar, H. K., Shrestha, S. M., and Pradhan, S. B. (2012). Use of pesticide in Nepal and impacts on human health and environment. *The Journal of Agriculture and Environment*, 13: 67-74.

Sharma, O. P., Bhosle, B. B., Kamble, K. R., Bhede, B. V. and Seeras, N. R. (2011). Management of pigeon pea pod borers with special reference to pod fly (*Melanagromyza obtusa*). *Indian Journal of Agricultural Sciences*, 81: 539–543.

Singh, B. K. and Singh, R. P. (2007). Effect of host plant nutrition and pesticidal management on larval population of *Helicoverpa armigera* (Hubner) in chickpea *Cicer arientium. Indian Journal of Entomology*, 69: 345–349.

Singh, R. N. and Singh, J. (1987). Pattern of boring of chickpea pods by *Heliothis armigera* Hubn. *Indian Journal of Agricultural Sciences*, 57: 376–377.

Chapter 4

Insect Pests of Major Spice Crops in Kashmir Valley of India

Akhtar Ali Khan*, Baber Parvaiz, Tahmina Mushtaq, Muneer Sofi, S. S. Pathania and Danish Mushtaq Wani

Division of Entomology, Faculty of Horticulture,
Sher-e-Kashmir University of Agricultural Sciences and Technology of Kashmir,
Srinagar, Jammu and Kashmir, India

Abstract

In India, spices grown in the region of Kashmir valley are of high quality and therefore famous all over the world. Saffron, kalazeera, chili and coriander are the main spice crops of this region. Annual production of spices in Jammu and Kashmir has been reported 1,180 thousand tons. However, there are several constraints in the production and productivity of spices such as lack of availability of good quality seed material, low soil fertility, lack of assured irrigation, infestation by insects, rodents and diseases and poor post-harvest management. Among these, insect pests are one of the major factors that infest the food and other stored products in godowns, bins, storage structures and packages causing a huge amount of loss to the stored food and also deteriorating the quality of products. This chapter deals with the insect pests associated with spice crops of Kashmir valley. Moreover, the techniques applied to manage the insect pests have also been described.

---

* Corresponding Author's Email: akhtaralikhan47@rediffmail.com.

In: Pest Management: Methods, Applications and Challenges
Editor: Tarique Hassan Askary
ISBN: 979-8-88697-268-9

Keywords: spice, saffron, kalazeera, coriander, chilli, insect pests, management

1. Introduction

Spice is more of a culinary term that refers to extracts from parts of plants, may be roots, leaves, bark, and blooms that are added to food to enhance flavour and smell of food (Farrell, 1990). The aromatic characteristics of essential oils from spices provide various functions for the plants including attracting or repelling insects, protecting themselves from heat or cold; and utilizing chemical constituents in the oil as defense materials. In general, they are complex mixtures of 20-60 organic compounds that give characteristic odour and flavour to leaves, flowers, fruits, seeds, barks and rhizomes (Farrell, 1990). Spices and their extracts are known to have various effects on insects of stored product (Owolabi, 2009). Spices have characteristic flavours and odours due to the essential (volatile) oils. Essential oils of spices and their constituents have also been reported as a potent source of botanical pesticides. Many essential oils extracted from various plant species belonging to different genera, contain relatively higher number of monoterpenes. Jointly or independently, they may contribute to the protection of plants against herbivores, although some herbivores have counter adapted to them (Devi and Maji, 2011).

India is the largest producer, consumer and exporter of spices and spice products in the world. Over 100 plant species are known to yield spices and spice products and among them 50 plant species are grown in India. The total spices production in India has been estimated to be around 10.4 million metric tons. They contribute 41 per cent of the total export earnings from all horticulture crops in India (Anonymous 2022). The state of Madhya Pradesh is the largest producer of spices followed by Rajasthan and Gujarat. Annual production of spices in Jammu and Kashmir is 1.180 thousand tons (Anonymous, 2020).

The present chapter is focused upon four important spice crops of Kashmir valley *viz*., saffron, kalazeera, chili and coriander, the major insect and nematode pests associated with them and their management.

2. Saffron (*Crocus Sativus* Kashmirianus; Family: Iridaceae)

Saffron is one of the most important spice crops grown in the Kashmir valley, which has the status of the most expensive spice crop by weight in the world. The world's best quality saffron is produced in the Kashmir valley of India. Commonly known as Zafran (Urdu), Kong (Kashmiri) or Kesar, it is the most expensive spice in the world used as a natural colouring and flavouring material in culinary preparations, bakery and confectioneries. It is expensive but its health benefits are priceless. Saffron is also used in folk medicine and Ayurvedic health system as a sedative, expectorant, anti-asthma, emmenagogue, and adaptogenic agent. It has the ability to treat depression, prevents loss of vision and improves memory. The stigma of the flower helps in relieving from digestive issues with the help of anti-inflammatory and antioxidant properties. Modern medicine has also discovered saffron as having anticarcinogenic, antimutagenic, immunomodulating and antioxidant like properties.

2.1. Insect Pests of Saffron

2.1.1. Thrips

The following species of thrips which cause infestation on saffron crop in Kashmir valley are:

- *Thrips flavus* Schrank,
- *Microcephalothrips abdominalis* Crawford, and
- *Anaphothrips obscures* (Muller).

2.1.1.2. Description

The female lays 80–100 eggs within plant tissues. Eggs hatch after 4 days. Till maturity, pupae remain underground and larvae feed on plant tissues. When mature, the second stage larvae fall on the ground, moult into pre-pupae and pupae in the soil. The adults are winged with silver stripes, glossy pale or black in colour provided with piercing and rasping type of mouth parts (Veer, 1985).

2.1.1.3. Damage Symptoms

Due to sucking of the plant sap, leaves develop a crinkled silver appearance as a result of damage to cells below the surface, especially alongside the mid rib and veins. Silvering and browning of leaves, stunting of young leaves and terminal growth, scarred and deformed fruits, may be observed in heavily infested plants. Developing leaves become distorted at the growing tips. The key symptom of thrip infestations is white and yellow spots on saffron leaves (Morse and Hoddle. 2006). Usually, thrips simply feed on the pollen and also help to pollinate the plant (Palmer, 1990; Marullo and De Grazia, 2012). However, they also attack flower petals. Damaged wet spots or discoloured blotches appear on petals, when creased or fold. Flower buds may stunt and grow deformed if too many thrips are present (Gao et al., 2021).

2.1.2. Mites

The two mite species that attack saffron are:

- *Rhizoglyphus robini* Claparede, and
- *Rhizoglyphus echinous* Fumouze and Robin.

2.1.2.1. Description

On an average, life cycle duration (egg to adult) on saffron corm mite ranges from 13–14 days with 15 generations per year. The colour of mites is almost transparent with prominent long hairs (0.6–0.8 mm). They are found in both the soil and in the crowns of susceptible plants. Immature and adult are similar to each other except in size. Immatures are smaller than adults (Diaz et al., 2000). Eggs are laid in the innermost parts of the plant which are transparent and spherical in shape.

2.1.2.2. Damage Symptoms

Mites suck the plant sap and produce yellow to white spots on the saffron leaves. Leaves become yellow, thin and reduced in size. Severely infected leaves turn yellow that ultimately results in reduced saffron production. Due to infestation, wound formation takes place in saffron corm leading to tunnels and cavities in corm (Bayram and Çobanoglu, 2006).

2.1.3. Nematodes

The plant parasitic nematodes associated with saffron crops in Kashmir valley are root lesion nematode *Pratylenchus* sp., stunt nematode, *Tylenchorhynchus* sp., dagger nematode, *Xiphinema* sp. and spiral nematode, *Helicotylenchus vulgaris*.

2.1.3.1. Description

Nematodes are unsegmented, triploblastic, bilaterally symmetrical, pseudocoelomate vermiform multicellular animal. Those which are parasitic on plants possess stylet at their anterior end. The stylet helps in piercing the plant tissues and absorbing the cell contents (Askary et al., 2012).

2.1.3.2. Damage Symptoms

Nematode causes injury with the help of stylet that punctures the cell wall. It is followed by the secretion of enzymes by nematode and extraction of cell sap due to which plant cells get damaged (Askary et al., 2018). This process leads to the symptoms like stunted plant growth, yellowing, curling and twisting of leaves, loss of vigour, excessive root branching, disorder in flower development/ sterility, dead or incapacitated root tips and patches of poor plant growth in field (Askary, 2017).

2.1.4. Rodents

Indian mole rat (*Bandicota bengalensis*), the house rat (*Rattus rattus*), the house mouse (*Mus musculus*), the soft furred field rat (*Rattus meltada*), the field mouse (*Mus booduga*) and the saffron mouse (*Pitymys leucurus*) and Porcupines belonging to the class Mammalia are some of the rodents associated with saffron.

2.1.4.1. Description

Rodents are economically important pest that cause damage to the saffron besides serving as reservoirs for some important human diseases such as plague, murine typhus, leptospirosis and food-poisoning. They are cosmopolitan in distribution, exhibiting a greater adaptability for diverse habitats from subterranean to terrestrial and arboreal modes of life. *Porcupines* possess soft hair but on their back sides and tail it is usually mixed with sharp *quills*. These *quills* typically lie flat until a *porcupine* is threatened, it leap to attention as a persuasive deterrent. Porcupines are fond of tuber crops like potatoes, saffron corms, cabbages and fruit barks.

2.1.4.2. Damage Symptoms

The damage to saffron corms by rodents varies from 10-50 per cent. It attacks the vegetative primordial at the base of attachment with the corms leaving the radical leaves detached, resulting in a complete loss of the saffron plants.

Rodent damage is evident from patches that have withered dry foliage resembling symptoms of corm rot.

2.1.5. Management of Saffron Insect Pests

2.1.5.1. Pre-Sowing

a. Deep ploughings 2–3 times at an interval of 10–15 days particularly in summer months.
b. Soil solarisation.
c. Manure and fertilizer application as per soil test recommendations.
d. Timely sowing.
e. Avoid dense plant populations.
f. Adopt raised bed technique at the time of field preparation.
g. Application of neem cake @ 80 kg/ acre for nematode control.
h. Destruction of alternate or collateral host plants.
i. Collection and destruction of plant debris.

2.1.5.2. Sowing

a. Apply fertilizers as per soil test recommendations.
b. Timely line sowing.
c. Adopt proper spacing.
d. Avoid high plant population.
e. Adopt crop rotation.
f. Use healthy corms for cultivation.
g. Set up light traps @ 1 trap/ acre.
h. Set up yellow sticky traps @ 4–5 traps/ acre coated with grease/ sticky oily materials.
i. Conserve natural enemies such as coccinellids, anthocorids, lygaeid, predatory thrips for controlling blossom thrips.
j. Biological control through release of pathogens, *viz.*, entomopathogenic nematodes (EPN), entomopathogenic fungi (EPF) and nuclear polyhedrosis virus (NPV).
k. Ecological engineering by flowering plants that attract natural enemies as well as plants that repel pests can be grown as border crop/ intercrop.

2.1.5.3. Vegetative Stage

a. Set up light traps @ 1trap/ acre.
b. Set up yellow sticky traps @ 4–5 traps/ acre coated with grease/ sticky oily materials.
c. Conserve natural enemies such as coccinellids, anthocorids, lygaeid and predatory thrips for controlling blossom thrips and mites.
d. If infestation of thrips are high i.e., above economic threshold level (ETL), apply Dimethoate @ 1.0 ml/ litre of water.
e. If infestation of mites are noted above ETL, apply Fenazaquin @ 0.4 ml/ litre of water.

2.1.5.4. Reproductive Stage
Same as described under vegetative stage.

2.1.5.5. Rodent Management

a. Plough the fields to demolish the rodent habitat and maintain weed free fields to reduce alternate source of food and habitat.
b. Growing of garlic on the peripheries of the beds/ field will act as repellents of the rodents and reduce their population.
c. Conserve the biological control agents, *viz.*, snakes, cats, dogs, owls, eagles in order to regulate rodent population.

3. Kalazeera (*Bunium Persicum* Bioss; Family:Apiaceae)

Black cumin, also known as Shahi or Kaala Jeera, is an integral part of Kashmiri cuisine and is used in a variety of dishes to enhance the taste. However, the cultivation of spice is only limited to a handful places in Kashmir valley. The insect pests infesting kalazeera are aphids, thrips, cut worm, drugstore beetle, tobacco caterpillar and root-knot nematodes.

3.1. Insect Pests of Kalazeera

3.1.1. Aphids
Aphid species infesting saffron in Kashmir valley have been identified as *Myzus percsicae* and *Aphis gossypii*.

3.1.1.1. Description

Aphids are soft-bodied insects which lay very tiny and shiny-black eggs in the crevices of host plant (Shah and Khan, 2013). The nymphs are similar to the wingless adults. During the summer months, aphid continues to produce wingless offspring by parthenogenesis i.e., the production of offspring without the involvement of sex. The winged offspring are produced intermittently and they fly away to colonise new plants (Khan and Shah, 2013). In the autumn season, females produce wingless off springs which are mated by the males and as a result hard-shelled eggs are laid on the tree branches. In spring season, the eggs hatch to wingless females producing winged offspring by parthenogenesis. The winged offspring fly to other plants where they feed and establish new colonies.

3.1.1.2. Damage Symptoms

Aphids damage plants by puncturing them and sucking their cell sap. They damage the young and soft plant parts, such as new leaves and shoots. The damaged leaves are smaller in size and do not open properly. In case of severe infestation, shoots wilt and dry out. The winged aphids move from one plant to another, spreading several viral diseases that are picked up from infected plants (Khan, 2015). Aphids secrete honeydew that stimulates black sooty mold growth. They may cover the surface of leaves and block photosynthesis. Natural enemies can restrict the development of aphids (Khan and Rather, 2012; Khan, 2017).

3.1.2. Thrips

Thrips tabaci is considered as one of the important pests of saffron in Kashmir valley.

3.1.2.1. Description

Eggs are white to yellow, kidney-bean shaped, microscopic in scale. Grow in leaf tissues with one end at the surface of the leaf. The feeding stages of thrips are first and second instar larvae. Nymphs are white to light yellow in colour, elongated with slender body. Nymphs appear like adults, but without wings. Antennae short, eyes dark in color. They crawl quickly when disturbed (Palmer, 1990; Marullo and De Grazia, 2012). The duration of larval stage is 10–14 days. The third and fourth instar larvae are inactive, non-feeding, and they are known as pre-pupa and pupa, respectively. They are pale yellow to brown in colour and possess more stout body than younger instars. Found in the soil these stages last for 5–10 days. Females are parthenogenic (asexually reproducing), males are extremely rare. Females insert eggs individually into leaves and continue to lay eggs for about three weeks. Adult life span is about one month.

3.1.2.2. Damage Symptoms

Thrips damage the undersides of leaves by sucking the sap. Damage is caused to the young and soft parts of plant such as new leaves and shoots. As a result, leaves curl downwards and change to a blackish silver in colour. Severe infestation causes young leaves to wilt and dry out. Thrips can act as vector of viral diseases too.

3.1.3. Cutworms

Agrotis ipsilon (Hufn.) (Noctuidae: Lepidoptera).

3.1.3.1. Description

Female moth comes out at dusk and lays creamy white, dome-shaped eggs (200–350) in clusters of around 30 each, either on the underside of the leaves of the host plants or in the soil. The newly formed young larva is yellow in colour, 1.5 mm long, with a glossy black head and a black shield on the prothorax. The fully grown larva is about 42–45 mm long and is black to dark brown in colour with a plump, greasy body. The larvae live in the soil and are yellow or blackish-green in colour. They possess striped markings running down the sides of their legs. The larval stage ranges from 30–34 days. Pupae are brown to medium brown, about 1.5–2.0 cm long and are typically found in or on piles of leaf mould. Pupation occurs in an earthen chamber underground (Joshi et al., 2020). The completion of pupal period takes place between 10–30 days. The adult measures about 25 mm from the head to the tip of abdomen and appears black on the back with a few greyish patches and dark lines on the forewings. The adults live for 7–10 days and overall life cycle completes in 36 days (egg to adult). Typically, the moths emerge at night. Generally, the pest completes three generations in one year.

3.1.3.2. Damage Symptoms

Both adult and caterpillars become active during night and in the daytime caterpillars hide in cracks and crevices of the soil. They attack young plants by severing their stems, pulling all parts of the plant into the ground and devouring them. Plants with severed stems have difficulty in growing again.

The pest can cause serious damage to crops particularly 25–35 days after planting. Persistent dry weather with lesser or no rainfall, reduced humidity and temperatures ranging between 16 °C–23 °C favour the development of cutworm.

3.1.4. Tobacco Caterpillar (Spodoptera Litura)

3.1.4.1. Description

It is found in the tropical and sub-tropical parts of the world, and also reported in Kashmir valley as well as other parts of India. It feeds on cotton, castor, groundnut, tomato, cabbage, cumin and various other cruciferous crops in addition to tobacco (Soni et al., 2001). The female lays about three hundred eggs in clusters. The eggs are covered over by brown hairs and they hatch in about 3–5 days. When fully grown, the caterpillar measures 35–40 mm in length. It is velvety black with yellowish green dorsal stripes. Larval stage lasts for 15–30 days and pupation takes place inside the soil and lasts for 7–15 days. Moth is medium sized and stout bodied, colour is wavy, white crisscross markings, forewings are light grey to dark brown. The hind wings are whitish, brown spots present around the edge of the wing. Pests breed throughout the year. Moths are active during night and live for 7–10 days. Adults live for 32–60 days. Eight generations may be completed in one year (Deepak et al., 2020).

3.1.4.2. Damage Symptoms

The caterpillars are gregarious in early stages and scrape the chlorophyll content of leaf lamina, giving it a papery white appearance. They are voracious feeders, build irregular leaf holes, and skeletonize the leaves, leaving only veins and petioles and cause heavy defoliation.

3.1.5. Nematodes

Pratylenchus sp., *Tylenchorhynchus* sp., *Xiphinema* sp. and *Helicotylenchus* sp. are the nematodes associated with kalazeera.

3.1.5.1. Description

Same as described above under section 1.3.1.

3.1.5.2. Damage Symptoms

Same as described above under section 1.3.2.

3.1.6. Management of Insect Pests of Kalazeera

The health of a plant is determined by its environment which includes physical factors such as soil, rain, sunshine hours and wind and biological factors such as pests, diseases and weeds. All these factors can play a role in the balance which exists between herbivore insects and their natural enemies. The methods adopted to manage insect pests of kalazeera are as follows:

a. Deep ploughings 2–3 times at an interval of 10–15 days particularly in summer month.
b. Soil solarisation.
c. Application of neem cake @ 80 Kg/ acre to manage nematode and cutworm.
d. Collection and destruction of plant debris.
e. Application of fertilizers as per soil test recommendations.
f. Setting up light traps @ 1 trap/ acre.
g. Setting up yellow sticky traps @ 4–5traps/ acre coated with grease/ sticky oily materials.
h. Conservation of natural enemies such as coccinellids, anthocorids, lygaeid, predatory thrips for the management of aphids and thrips.
i. Biological control through release of pathogens, *viz*., EPN, EPF and NPV.
j. Ecological engineering: Growing as border/ intercrop to those flowering plants that attract natural enemies or plants that repel pests.
k. Application of dimethoate @ 1.0 ml/ litre of water if infestation of aphids/ thrips are noted above ETL.
l. Application of chlorpyriphos @ 1.0 ml/ litre of water when infestation of any beetle/ caterpillars are noted above ETL.

4. Chilli (*Capsicum Annum* Linnaeus; Family: Solanaceae)

Chilli (*Capsicum annuum* L.) is an important spice and vegetable crop grown throughout India. It is an integral ingredient of Indian curry, characterized by tempting colour and tantalizing pungency. With over 293 insects and mites species weakening the crop in the field as well as in storage, the pest continuum of chilli is complex. *Scirtothrips dorsalis* Hood, *Myzus persicae* Sulzer, *Aphis gossypii* Glover, *Polyphagotarsonemus latus* Banks and *Helicoverpa armigera* Hubner are important pests that hamper chilli production in Kashmir valley.

4.1. Insect Pests of Chilli

4.1.1. Chilli Thrip

Scirtothrips dorsalis Hood (Thripidae: Thysanoptera).

4.1.1.1. Description

Eggs are minute and dirty white in colour. Nymphs and adults are short, slender, delicate, and yellowish-straw in colour; adults have heavily fringed wings that are uniformly grey in colour. Reproduction occurs by two ways i.e., sexual and parthenogenetic (Chintkuntlawar et al., 2015).

4.1.1.2. Damage Symptoms

The leaf tissues are lacerated by both nymphs and adults and the oozing sap is absorbed when the buds and flowers are targeted. The pest prefers tender leaves and rising shoots. Infested leaves begin to curl and crumble and are gradually shed, while buds become brittle and fall back. Loss of yield due to extreme thrip attack in chilli can range from 25–50 percent. Infested fruits exhibit light brown scars. The plants show stunted growth and flower development takes place at the early stage.

4.1.2. Cutworm

Agrotis ipsilon (Hufn.) (Noctuidae: Lepidoptera).

4.1.2.1. Description

The adult, with reddish tinge, is a dark brown stout moth, possesses wavy lines and spots on the fore wings. Larvae dark brown with red head, have greasy appearance. The full grown larvae pupate in an earthen cocoon.

4.1.2.2. Damage Symptoms

The cutworm is a serious pest of chilli which damages the crop mostly during the seedling stage. Larvae that hide in cracks in the soil during the day time become active at dusk, feed on leaves, and often cut off the tender stems of young and developing plants, resulting in a reduction in yield or total crop failure.

4.1.3. Spider Mites

Tetranhychus spp. (Tetranychidae: Acarina).

4.1.3.1. Description

Spider mites possess oval shaped eggs, white in colour, firmly attached to the leaf surface. Nymphs are white in colour. Adults are large, oval and yellowish in colour.

4.1.3.2. Damage Symptoms

Tetranhychus spp. cause 'Murda' disease of chillies. Infestation cause downward curling of leaves and affected leaves appear inverted boat shaped with elongation of petioles rolling down along the margin. In some cases, the affected leaves turn dark green and tips of branches get clustered (Kotresh et al., 2020).

4.1.4. Aphids

Myzus percsicae, Aphis gossypii (Homoptera: Aphididae).

4.1.4.1. Description

Eggs are very small, shiny-black, and are found in the buds, stems, and barks of the plant. The nymphs are young aphids, which look like wingless adults, but smaller in size. The adults are tiny, soft-bodied insects with two long antennas. Most aphids have two small cornices (horns) on the back of their body.

4.1.4.2. Damage Symptoms

Aphids generally appear on the tender shoots and on the lower surface of the leaves, suck the sap resulting in reduced plant vigour and crop yield. They also spread virus diseases indirectly by acting as vectors (Khan and Zaki, 2007). Aphids secrete honeydew that attracts ants, causing sooty mould due to which leaves turn black.

4.1.5. Tobacco Caterpillar

Spodoptera litura (Noctuidae: Lepidoptera).

4.1.5.1. Description

Larva is stout green-grey or dark brown in colour. They are spotted. Adult is stout dark moth with wavy white crisscross markings on the upper wings. Eggs are laid in masses on tender leaves. Eggs are covered with brown hairs.

4.1.5.2. Damage Symptoms

Caterpillars feed on leaves and pods. Moths are active at night. Many leaflets or the margin of the same broad leaf are webbed together giving the crop a shabby look. The young larvae live gregariously inside these webs, but later on they separate and spread out. Affected pods drop after getting drying (Kotresh et al., 2020).

4.1.6. Fruit Borer

Helicoverpa armigera Hubner (Noctuidae: Lepidoptera).

4.1.6.1. Description

Freshly hatched larvae are yellowish white, but eventually change and adopt a greenish tinge. Caterpillars are apple green in colour and possess whitish and dark-grey separated longitudinal stripes. The full-grown caterpillars slip down from the plants and burrow in the soil, where they pupate. Pupae are dark-brown in colour and the anal end has a sharp spine. Moths are medium sized, stout; males having olive grey forewings whereas females have reddish brown forewings with a circular dark brown mark.

4.1.6.2. Damage Symptoms

Newly hatched larvae feed initially on tender foliage. Fruits are invaded by larvae at a later stage. Larvae bore circular holes inside the fruit. They drive only a portion of their body inside the fruit and consume the inner contents. If the fruit is larger in size, the caterpillar partially damages it. The damaged fruits are invariably infected with fungi and bacteria and fully spoiled.

4.1.7. Drugstore Beetle

(*Stegobium paniceum*) (Coleoptera: Ptinidae).

4.1.7.1. Description

Females lay up to 75 eggs in the food or substratum. Larvae tunnel through the substrate and create a cocoon and pupate when fully grown. The pupal period lasts between 12–18 days. The beetles are cylindrical; length varies between 2.25–3.5 mm, colour brown to reddish brown, possess longitudinal rows of fine hairs on the elytra (wing cover). Elytra have rows of pits that give them a striated (lined) look. Adult females generally live for 13–65 days, however the entire life cycle is less than two months. Though, in some cases it may take up to 7 months.

4.1.7.2. Damage Symptom

Drugstore beetles cause damage during the storage of chilli. The products get contaminated when infested by drugstore beetles. The pests do the most damage in their larval form as they live, eat, and excrete waste within the products they have infested.

4.1.8. Nematodes

Root-knot nematode, *Meloidogyne hapla.*

4.1.8.1. Description

The life cycle consists of the egg, four juvenile stages and adult. First moulting occurs within the egg and second stage juvenile comes out which is the only infective stage. The second stage juveniles possess the feeding apparatus known as stylet which is present on the anterior part of their body. Nematode penetrates plant root tissues with the help of stylet, feed upon it, and deprives the plant of water uptake and nourishment.

4.1.8.2. Damage Symptoms

The symptoms appear as yellowing, curling and twisting of leaves, stunted plant growth, loss of vigor, excessive root branching, disturbance of flower growth/ sterility, dead or incapacitated root tips and patches of poor field growth.

4.1.9. Management of Insect Pests of Chilli

a. Deep ploughings 2–3 times at an interval of 10–15 days particularly in the days of summer month. It helps in exposing resting stages of pests and soil borne plant parasitic nematodes to direct sunlight due to which they get killed.
b. Soil solarization.
c. Application of neem cake @ 2.0 quintal/ ha to manage plant parasitic nematodes, cutworms and other soil insects.
d. Collection and destruction of plant debris.
e. Seed treatment with imidacloprid 70 WS (Water dispersible powder for slurry seed treatment) @ 7 g/ kg seed.
f. Seedling dip treatment with imidacloprid 17.8 SL (Soluble liquid) @ 0.3 ml/ litre of water at the time of transplanting.
g. Application of fertilizers as per the recommendations of soil test.

h. Setting up light traps @ 1trap/ acre.
i. Setting up yellow sticky traps @ 4–5 traps/ acre coated with grease/ sticky oily materials.
j. Conservation of natural enemies such as coccinellids, anthocorids, lygaeid, predatory thrips for the management of aphids and thrips.
k. Biological control through release of pathogens, *viz.*, EPN, EPFand NPV.
l. Ecological engineering: Flowering plants that attract natural enemies as well as plants that repel insect pests can be grown as border/ intercrop.
m. Spraying of NSKE (Neem seed kernel extract) @ 5% against sucking pest in early stages of the crop.
n. Single application of Azadirachtin 10000 ppm @ 2ml/ litre of water against sucking pests.

5. Coriander (*Coriandrum Sativum* L.; Family: Apiaceae)

An annual herb coriander also known as Chinese parsley or dhania is grown at various places in Kashmir valley. All parts of the plant are edible, but the parts most traditionally used in cooking are the fresh leaves and the dried seeds. The pests attacking coriander are whitefly, *Bemisia tabaci*; aphid, *Hyadaphis coriandri, Aphis* sp.; cutworm, *Agrotis* sp.; thrips, *Thrips* sp.; mite, *Tetranychus* sp., and nematodes.

5.1. Insect Pests of Coriander

5.1.1. Whitefly

Bemisia tabaci (Hemiptera: Aleyrodidae).

5.1.1.1. Description

Females of *B. Tabaci* lay eggs on the underside of leaves near the veins and in order to lay more eggs they prefer hairy leaf surfaces. Eggs are thin, pear-shaped, and are attached via a pedicel vertically to the leaf surface. The nymphs are flattened, oval shaped and greenish yellow in colour. Adults emerge through a T-shaped slit from puparia, leaving behind empty cases of pupae. The adult whitefly is a moth-like, soft-bodied fly, with powdery wax coat. The wings and the body is light yellow in colour. The wings are held over the body like a tent (Nissar, 2019).

5.1.2. Aphids

Hyadaphis coriandri (Homoptera:Aphididae)

5.1.2.1. Description

Eggs are very small, bright black, and are located in the crevices of the plant buds and stems. The nymphs (immature stages) are young aphids similar to adults but smaller in size. Adults are tiny, soft-bodied insects with two long antennas that mimic horns. They also have two small cornicles (horns) towards the rear end of the body.

5.1.2.2. Damage Symptoms

The aphid infests tender shoots and under surface of the leaves, resulting in leaf curling and crinkling and stunted plant growth. The secretion of honeydew by aphids leads to black sooty mould.

5.1.3. Cut Worm

Same as in Chili.

5.1.4. Thrips

Thrips tabaci (Thripidae: Thysanoptera).

5.1.4.1. Description

Eggs are microscopic and white or yellow in colour. Females lay their eggs in the tissues of leaves, cotyledon or flora. The nymphs are very small, light yellow to brown, and the pupae appear between the immature and the adult as an intermediate stage. There are short antennas in nymphs, wing buds are visible, but short and not functional. The adults have fully formed wings that are light yellow to dark brown in colour. The wing tends to be a fringe of hair. The wings are folded around the back of the insect when at rest.

5.1.4.2. Damage Symptoms

Thrips have a very characteristic feeding behaviour. They start feeding by rasping and sucking, and to expel the liquids from the plant cells, they lacerate the leaf surface with their mouth parts. Thrips release chemicals that assist in predigesting the cell sap. Silvery spots or stripes that shine in the sun are seen on the damaged leaves. When damage is severe, much of the surface of the leaf can be occupied by these small patches and the plant cannot photosynthesize adequately. The damaged tissues also act as entry points for plant pathogens.

5.1.5. Mites

Tetranhychus spp. (Tetranychidae: Acarina).

5.1.5.1. Description

Eggs are microscopic, hyaline and globular, laid under clods in masses. Nymphs are yellowish with pale yellow legs. Adult mites are very small, metallic brown to black in colour. Their forelegs are significantly longer than the three other pairs.

5.1.5.2. Damage Symptoms

Coriander is frequently attacked by mites and the whole plant becomes whitish yellow and looks sick. They feed mainly on young leaves, but in young inflorescence, infestation is more severe. Mites can be seen on the lower surface of leaves. In case of heavy infestation, they cause webbing and feed from within the web. Such infested plants show stunted growth.

5.1.6. Management of Insect Pests of Coriander

Same as described under section 3.8 for chili.

Conclusion

Spices are high value crops and to maintain the health of a crop is very important for quality and quantity production. The factors responsible for the good health of spice crops are physical factors (soil, rain, sunshine hours, wind) and biological factors (i.e., pests, diseases and weeds) which play a major role in the balance which exists between herbivore insects and their natural enemies. The knowledge of insect pests including their biology, nature of damage and understanding of their intricate interactions in an ecosystem play a significant role in pest management. Insecticides should be selected with care considering the impact on future pest populations and the environment. Broad-spectrum insecticides should be avoided when a narrow-spectrum or more specific insecticide is available. More preference should be given to green labeled insecticides for sustainable pest management of spices.

Acknowledgments

The authors are thankful to the spice workers of India, particularly of Kashmir valley.

References

Anonymous (2020). Production: Horticulture Crops: Spices: Jammu and Kashmir. Department of Agriculture and Cooperation, Jammu and Kashmir, India.

Anonymous (2022). *Estimated Production Volume of Spices across India in Financial Year 2021.* Statistical Research Department, India.

Askary, T. H. (2017). Diversity of plant parasitic nematodes in pulses. In: *Plant Diversity: Present Situation and Future Scenario* (Eds. Ansari, A. A., Gill, S. S., Abbas, Z. K. and Naeem, M.). CAB International, Wallingford, UK, pp. 239-274.

Askary, T. H., Banday, S. A., Iqbal, U., Khan, A. A., Mir, M. M. and Waliullah, M. I. S. (2012). Plant parasitic nematode diversity in pome, stone and nut fruits. In: *Agroecology and Strategies for Climate Change* (Ed. E. Lichtfouse). Springer, Heidelberg, Germany, pp. 237-268.

Askary, T. H., Khalil, A., Khan, A. A. and Nazir, N. (2018). Population fluctuation of plant parasitic nematodes associated with grapevine under hi-tech and low-tech polyhouse conditions. *International Journal of Current Microbiology and Applied Sciences*, 7: 2133-2140.

Bayram, S. and Çobanoglu, S. (2006). Astigmata and Prostigmata (Acari) of bulbaceous ornamental plants in Ankara, Turkey. *Acta Phytopathologica et Entomologica Hungarica,* 41: 367-381.

Chintkuntlawar, P. S, Pawar, U. A. and Saxena, A. K. (2015). Insect pest complex of chilli, *Capsicum annum* L. and their natural enemies in Jabalpur. *International Journal of Plant Protection*, 8: 270-278.

Deepak, Yadav, S. P., Yadav, S. and Puneet (2020). B i ology studies of tobacco caterpillar, *Spodoptera litura* Fabricius on castor (*Ricinus communis* L .). *Journal of Entomology and Zoology Studies*, 8: 1163-1168.

Devi, N. and Maji, T. K. (2011). Neem Seed Oil: Encapsulation and Controlled Release - Search for a Greener Alternative for Pest Control. In: *Pesticides in the Modern World: Pesticides Use and Management* (Ed. Stoytcheva, M.). *InTech Open*, pp. 191-322.

Diaz, A., Okabe, K., Eckenrode, C. J., Villani, M. G. and O'Connor, B. M. (2000). B i ology, ecology and management of the bulb mites of the genus *Rhizoglyphus* (Acari: Acaridae). *Experimental and Applied Acarology*, 24: 85-113.

Farrell, K. T. (1990). *Spices, Condiments, and Seasonings*. 2nd Edition. Springer, New York.

Gao, Y. F., Gong, Y. J., Cao, L. J., Chen, J. C., Gao, Y. L., Mirab-balou, M., Chen, M., Hoffmann, A. A. and Wei, S. J. (2021). Geographical and interspecific variation in susceptibility of three common thrips species to the insecticide, spinetoram. *Journal of Pest Science*, 94: 93–99.

Joshi, M. J., Verma, K. S. and Chandel, R. S. (2020). Feeding inhabitation with bioformulations in cutworms *Agrotis ipsilon* (Hufnagel). *Indian Journal of Entomology*, 82: 134-138.

Khan, A. A. (2013). Evaluation of the biological control efficiency of four spiders using functional response experiments. *The Bioscan*, 8: 1123-1128.

Khan A. A. (2015). Report of Department of Science and Technology on biodiversity and management of aphid in temperate horticulture ecosystem of Kashmir. Division of Entomology, Sher-e-Kashmir University of Agricultural Sciences & Technology, Shalimar Campus Srinagar-190025, India.

Khan, A. A. (2017). Effect of insecticides on biodiversity of aphidophagous syrphid flies in fruit ecosystem of Kashmir. *Journal of Entomology and Zoology Studies*, 5: 117-125.

Khan, A. A. (2017). Functional response of four syrphid predators associated with mealy cabbage aphid, *Brevicoryne brassicae* L. on cruciferous vegetables. *International Journal of Current Microbiology and Applied Sciences*, 6: 2806-2816.

Khan, A. A. and Shah, M. A. (2017). Records of aphids and their natural enemies in agro-ecosystem with special reference to horticultural ecosystem of Kashmir. *Journal of Entomology and Zoology Studies*, 5: 189-203.

Khan, A. A. and Rather, A. Q. (2012). Diversity and foraging behaviour of spider (Arachnida: Araneae) in the temperate maize ecosystem of Kashmir. *Journal of Biological Control*, 26: 179-189.

Khan, A. A. and Zaki, F. A. (2007). Predation rates of *Coccinella septempunctata* Linnaeus and *Chilocorus infernlais* Mulsant on aphids. *Asian Journal of Bio Science*, 2: 53-55.

Kotresh, S., Raghuraman, M. and Srushtideep, A. (2020). Seasonal incidence of mites and thrips in chilli. *Journal of Entomology and Zoology Studies*, 8: 569-572.

Marullo, R. and De Grazia, A. (2012). Thripidae: Pest thrips infesting citrus in the Mediterranean Region. In: *Integrated Control of Citrus Pests in the Mediterranean Region* (Eds. Vacante, V. and Gerson, U.). Bentham Science Publishers, Sharjah, UAE, pp. 109-118.

Morse, J. G., Hoddle, M. S. (2006). Invasion biology of thrips. *Annual Review of Entomology*, 51: 67-89.

Nissar, M., Khan, A. A., Khan, Z. H., Masoodi, K. Z., Khan, A. A., Wani, W. M., Hussain, K., Naik, H. R. and Nazir, N. (2019). Population dynamics of whitefly (*Bemisia tabaci*, Gennadius) on tomato (*Solanum esculentum*, Mill.) crop under protected conditions in Kashmir. *Journal of Entomology and Zoology Studies*, 7: 804-807.

Owolabi, M. S., Oladimeji, M. O., Lajide, L., Singh, G., Marimuthu, P. and Isidorov, V. A. (2009). Bioactivity of three plant derived essential oils against the maize weevils *Sitophilus zeamais* (Motschulsky) and cowpea weevils *Callosobruchus maculatus*

(Fabricius). *Electronic Journal of Environmental, Agricultural and Food Chemistry*, 8: 828-835.

Palmer, J. M. (1990). Identification of the common thrips of tropical Africa. *Tropical Pest Management*, 36: 27-49.

Shah, M. A. and Khan, A. A. (2013). Aphid in agroecosystem and their management. In: *Hill Agriculture: Prospects, Constraints and Mitigations* (Ed. Ahamad, S.). Daya Publishing House, New Delhi, India, pp. 314-330.

Soni, V. K., Dixit, S. A., Dubey, V. K. (2001). Host-biology relation of *Spodoptera litura* (Fab.) (Lepidoptera: Noctuidae). *Insect Environment*, 7: 111-112.

Veer, V. (1985). Observations on the host preferences and biology of *Thrips flavus* Schrank (Thysanoptera: Thripidae) from Dehradun, India. *Annals of Entomology*, 3: 39-47.

Chapter 5

Recent Advances and Current Status in the Rice Pest Management in Sri Lanka

Disna Ratnasekera[1,*], C. Kudagamage[2], Rohan Rajapakse[1] and L. K. W. Wijayaratne[3]

[1]Department of Agricultural Biology,
Faculty of Agriculture,
University of Ruhuna, Matara, Sri Lanka.
[2]Department of Agriculture, Peradeniya, Sri Lanka.
[3]Department of Plant Sciences,
Faculty of Agriculture,
Rajarata University of Sri Lanka,
Puliyankulama, Anuradhapura, Sri Lanka.

Abstract

Rice (*Oryza sativa* L.) is one of the key staple foods of almost half the world population, predominantly of Asians. It is also the staple diet of around 21.8 million people in Sri Lanka and provides almost all the energy requirements and some of the protein requirements of the people. Rice cultivation is the livelihood of more than 1.8 million farmers. More than 30 per cent of the labor force is either directly or indirectly employed in rice cultivation. Pest infestation is one of the key issues hampering rice production in Sri Lanka. Rice thrips (*Stenchaetothrips biformis*), gall midge (*Orsealia oryzae*), leaf folders (*Cnaphalocrocis medinalis* and *Marasmia* spp), yellow stem borer (*Scirpophaga incertulas*), brown planthopper (*Nilapavarvata lugens*) and paddy bug (*Leptocorisa oratorius*) are the major rice pests recorded in rice fields of Sri Lanka. Moreover, White backed plant hopper (*Sogatella furcifera*), cut-worm

* Corresponding Author's Email: disnaratnasekera@gmail.com.

In: Pest Management: Methods, Applications and Challenges
Editor: Tarique Hassan Askary
ISBN: 979-8-88697-268-9

(*Agrotis* spp.), ear cutting caterpillar (*Mythimma separata*), black bug (*Scotinophara lurida*), mites (*Stenotarsonemus spinsky*), nematodes (*Meloidogyne graminicola*) and rats (*Bandicota bengalensis*) are also reported in various locations of the country.

In addition, rice weevil (*Sitophilus oryzae*), grain moth (*Sitotroga cerealella*) and red flour beetle (*Tribolium castaneum* devastate stored grain. Now chemical pesticides management of pest infestation is most often relied on chemical insecticides are relied upon to manage pest infestation. Chemical pesticides are too costly and their indiscriminate use is a serious threat to human and environmental health. Consequently, emphasis came to be placed on the development of eco-friendly pest management techniques, known as Integrated Pest Management (IPM) which has gained popularity among the farming community. Host plant resistance, biological control, cultural control and traditional control methods along with chemical control are some available IPM methods recommended for rice. Though the rice ecosystems in Sri Lanka are very rich in diverse terrestrial and aquatic fauna and flora, only a few species of them have been found to be damaging insects. Most are predatory or parasitic while others are categorized as decomposers. Therefore, pest emergence is due to some external factors that affect this intricate balance, in particular the indiscriminate use of chemical insecticides. In 2021, Sri Lankan government has framed a policy to use organic inputs instead of chemical fertilizers and pesticides to minimize health risks, and lack of foreign exchange which could have proved a turning point for the production chain in whole agriculture sector. However, in 2022 this policy was reversed under pressure from farmers.

Keywords: paddy, pests, management, insecticides, environment, IPM

1. Introduction

Rice is the staple crop in Sri Lanka. It provides a livelihood to more than 1.8 million farm families. More than 30 per cent of the total labor force is directly or indirectly employed in the rice sector. According to a report published in the year 2018-19, the total area under rice cultivation was approximately 1.1 million hectares, producing 4.6 million metric tons of rice (AgStat, 2020).

Transition of rice cultivation from traditional cultivation methods to modern methods commenced around 1960s. Use of short stature, fertilizer responsive, lodging resistant varieties, annual double cropping, use of chemical fertilizers and agrochemicals for weed control and pest management are some of the changes that accompanied the modern cultivation methods.

Farmers who followed these techniques were able to get bumper harvests and high monetary returns. However, these favorable outcomes were short lived. By the early 1970s, following the introduction of modern cultivation practices, new pests emerged. These adverse effects including the emergence of new pests and diseases accompanying changes in cultural and ecological parameters were commonly observed in almost all Asian countries. The initial response for combating pests and diseases was to use insecticides, mostly Organochlorine insecticides like BHC, DDT, Aldrin and Dieldrin. However, DDT was less toxic to mammals but persisted in the environment for a long time. It was found accumulating in the fat tissues of mammals and fish. The use of DDT was banned in the early 1970s and subsequently other organochlorine insecticides were also banned for use in agriculture. However, chemicals continued to be used with insecticides like organophosphates and carbomates that did not persist in the environment as long. During the years 2013-2017, the pesticide imports to Sri Lanka were 28,640 metric tons worth US$ 225 million. The registrar of pesticides was appointed as the licensing authority to perform the following duties:

a. Control of imports
b. Designate authorized analyses
c. Setting the requirements for packaging, labeling, advertising, transportation, storage, pre-harvest intervals and residual limits.

It was felt that knowledge on bionomics and pest ecology was needed to formulate effective management methods of insects without relying solely on chemicals. Wickramasinghe (1978) studied the integrated pest management (IPM) methods in relation to existing control methods such as selective use of chemicals, preservation of natural enemies, cultivation of resistant varieties and cultural practices. A large amount of knowledge on non-chemical methods of pest management based on IPM was collected in the few decades that followed. In many instances, certain indigenous knowledge-based practices were blended with others.

Our attempt in this chapter is to discuss the existing system of pest management, its advantages and disadvantages, recent advances as well as future strategies for rice pest management that can be used without causing any adverse effects on the environment.

2. Rice Pests in Sri Lanka

The major rice pests in Sri Lanka include gall midge (*Orseolia oryzae*), rice thrips (*Stenchaetothrips biformis*), brown planthopper (*Nilaparvata lugens*), yellow stem borer (*Scirpophaga incertulas*), leaf folder (*Cnaphalocrocis medinalis* and *Marasmia* spp.), paddy bug (*Leptocorisa oratorius*), white mite (*Stenotarsonemus spinki*), white-backed plant hopper (*Sogatella furcifera*), cut-worm (*Agrotis* spp.), ear cutting caterpillar (*Mythimna separata*), black bug (*Scotinophara lurida*), nematode (*Meloidogyne graminicola*) and rat (*Bandicota bengalensis*) (Kudagamage and Nugaliyadde, 1995).

3. Management

3.1. Host Plant Resistance

Utilization of host plant resistance (HPR) is by far the most economical pest management method. The primary objective of a program on insect resistance in rice is to develop cultivars that are resistant to an insect pest while maintaining or improving basic agronomic characteristics. Breeding for resistance in rice commenced 30-40 years back and presently all the recommended varieties have resistance to a certain major insect pest (Table 1). However, one of the drawbacks of the HPR is the likely development of resistant biotypes. For example, in the 1980s new biotypes of rice gall-midge were observed. Continued breeding of new varieties is necessary in such instances. Therefore, future challenges are to recognize new biotypes in the field by following conventional techniques or molecular methods.

The introduction of genes through molecular means to transform *Bacillus thuringiensis* (Bt) so as to develop varieties against the leaf eating caterpillar were tried. However, the concern among the general public about genetically modified food hampered the progress. Furthermore, germplasm of about 2000 rice varieties had not been fully evaluated for pest resistance. In rice, the incorporation of insect resistance into modern varieties is an objective of Sri Lankan rice breeding program for three reasons. i. To increase income per hectare ii. To increase rice production iii. To provide host plant resistance varieties free to farmers. The economic advantage that insect resistant cultivars offer to producers is genetically incorporated insect resistance, which manages pests for the cost of seed alone. Secondly, insecticides are costly, they affect non target species, and cause health hazard and environmental pollution. Further, the use of insecticides requires farmers to gain knowledge about their effective use. Thirdly, HPR is compatible with other control measures such as chemical, biological and cultural. Breeding for pest resistance occupies a prominent place in the Sri Lankan rice breeding programs. Advances have been made in the development of varieties resistant to brown planthopper (BPH) (*Nilaparvata lugens*) and gall-midge (GM) (*Orseola oryzae*); breeding programs resistance to thrips (TH) (*Stenchetothrips biformis*) has not been found successful.

Table 1. Rice varieties bred for resistance to pests in Sri Lanka after green revolution

S. No.	Rice variety	Year/period of release	Level of Resistance		Reference
			GM*	BPH**	RRDI
1	H 9	Late 1957	Not available	Not available	RRDI
2	Bg 3-5	Early 1970s	Not available	R/MR	RRDI
3	Bg 38	1980-84	Not available	R/MR	RRDI
4	Bg 407	1980-84	Not available	MR	RRDI
5	Bg 745	1980-84	Not available	MR	RRDI
6	H 4	1957	Not available	Not available	RRDI
7	H 8	Late 1957	Not available	Not available	RRDI
8	Bg 11-11	Early 1970s	Not available	MS	RRDI
9	Bg 90-2	before 1975	S	MS	RRDI
10	Bg 379-2	1980-84	MR	MR	RRDI
11	Bg 400-1	1980-84	MR	MR	RRDI
12	Bg 380	1980-84	Not available	S	RRDI
13	Ld 66	before 1975	MR	Not available	RRDI
14	Bw 78	1975-79	MR	Not available	RRDI
15	Bw 100	1975-79	Not available	Not available	RRDI
116	Mi 273	1971	Not available	Not available	RRDI
17	Bw 400	Not Reported	MR	MR	RRDI
18	At 401	1990-94	Not available	Not available	RRDI
19	At 402	1990-94	Not available	Not available	RRDI
20	Bg 403	1990-94	MR	MR	RRDI
21	At 405	Mid-1990s	Not available	MR	RRDI
22	Bg 406	2005	Not available	MR	RRDI
23	Bg 407 H	2005	Not available	Not available	RRDI
24	Ld 408	2010	MS	MR	RRDI
25	Bg 409	Not Reported	MR/MS	MR	RRDI
26	Bg 450	1985-89	R/MR	MR	RRDI
27	Bw 451	1985-89	Not available	Not available	RRDI
28	Bw 452	1990-94	Not available	Not available	RRDI
29	Bw 453	1990-94	MR	Not available	RRDI
30	Bg 454	2005	Not available	R/MR	RRDI

Table 1. (Continued)

S. No.	Rice variety	Year/period of release	Level of Resistance		Reference
			GM*	BPH**	RRDI
31	Bg 455	2014	Not available	MR	RRDI
32	H 7	Late 1957	Not available	Not available	RRDI
33	At 16	1975-79	Not available	Not available	RRDI
34	Bw 266-7	1980-84	MR	S	RRDI
35	Bw 267-3	1980-84	Not available	S	RRDI
36	Bg 34-6	before 1975	Not available	Not available	RRDI
37	Bg 94-1	1975-79	MS	S	RRDI
38	Bg 94-2	1975-79	S	Not available	RRDI
39	Bg 350	1985-89	MR	Not available	RRDI
40	Bw 351	1985-89	Not available	Not available	RRDI
41	Bg 352	1990-94	Not available	MR/MS	RRDI
42	At 353	1990-94	Not available	Not available	RRDI
43	At 354	1990-94	Not available	Not available	RRDI
44	Ld 355	Not Reported	Not available	Not available	RRDI
45	Ld 356	Not Reported	MR	R/MR	RRDI
46	Bg 357	1995-99	R/MR	MR	RRDI
47	Bg 358	1995-99	MR/MS	MR	RRDI
48	Bg 359	1995-99	MR	MR/MS	RRDI
49	Bg 360	1995-99	MR	MR/MS	RRDI
50	Bw 361	2002	MR	Not available	RRDI
51	At 362	2002	MS	MR	RRDI
52	Bw 363	2002	MR	Not available	RRDI
53	Bw 364	2006	MR	MR	RRDI
54	Ld 365	2008	MR	MR/MS	RRDI
55	Bg 366	2009	MR/MS	MR/MS	RRDI
56	Bw 367	2011	MR	MS	RRDI
57	Ld 368	2011	MR/MS	MR/MS	RRDI
58	Bg 369	2012	MR	MR	RRDI
59	Bg 370	2013	S	MR	RRDI
60	Ld 371	2013	Not available	MR/MS	RRDI
61	Bw 372	2013	MR	MR/MS	RRDI
62	At 373	2014	R/MR	MR/MS	RRDI
63	Bg 374	2016	MS	MR	RRDI
64	H 10	Late 1957	Not available	S	RRDI
65	62-355	Not Reported	Not available	Not available	RRDI
66	Bg 34-8	before 1975	Not available	MR	RRDI
67	Bg 276-5	1975-79	MR	Not available	RRDI
68	Bw 272-6b	1980-84	Not available	MR/MS	RRDI
69	Bg 300	1985-89	MS	MR	RRDI
70	Bg 301	1985-89	MS	Not available	RRDI
71	Bw 302	1985-89	Not available	Not available	RRDI
72	At 303	1990-94	MS	MR	RRDI
73	Bg 304	1990-94	MS	Not available	RRDI
74	Bg 305	1995-99	MR	MR	RRDI

S. No.	Rice variety	Year/period of release	Level of Resistance		Reference
			GM*	BPH**	RRDI
75	At 306	2004	R/MR	MR/MS	RRDI
76	At 307	2005	MR	MR	RRDI
77	At 308	2008	MR/MS	MR	RRDI
78	At 309	2013	Not available	MR	RRDI
79	Bg 310	2014	Not available	MR	RRDI
80	At 311	2015	R/MR	MR	RRDI
81	Bw 312	Not Reported	Not available	MR	RRDI
82	Bg 750	1981	Not available	Not available	RRDI
83	Bg 250	2005	MR/MS	MR	RRDI
84	Bg 251	2014	MS	MR	RRDI
85	Bg 252	2016	S	MR	RRDI
86	Ld 253	2016	R/MR	MR/MS	RRDI

GM – Gall Midge, BPH – Brown Plant Hopper.

R – Resistant; MR – Moderately Resistant; MR/MS – Moderately Resistance/Moderately Susceptible; MS – Moderately Susceptible; S – Susceptible.

*Recommended Rice varieties and their Level of Resistance against Rice Gall midge (*Orseolia oryze)* (The level of resistance was scored according to the field screening in the Rice Research Development Institute (RRDI), Bathalgoda. Bg 94-1 was used as susceptible check.

**Recommended Rice varieties and their Level of Resistance against Brown Plant Hopper (N*ilaparvata lugens*). The level of resistance was scored by greenhouse screening. Bg 380 was used as a susceptible check and Ptb 33 as a resistance check.

Source: RRDI, Bathalagoda, Sri Lanka.

3.1.1. Brown Plant Hopper (BPH)

Of the many kinds of planthoppers infesting rice in Sri Lanka, brown plant hopper, *N. lugens* is the most damaging insect pest. BPH causes direct damage to the plant by sucking the phloem sap and transmitting the rice grassy stunt virus (RGSV) and the rice ragged stunt virus (RRSV). BPH infestation has become prominent after the introduction of semi- dwarf high yielding fertilizer responsive rice varieties. To develop a sustainable pest management system, it is important to find the right balance between breeding and management strategies to reduce the ecological fitness of BPH and to keep the pest under economic threshold.

Initially, we developed screening techniques in the field, in greenhouses and in the laboratory to evaluate germplasm for resistance to BPH (Kudagamage and Nugaliyadda, 1982). These techniques were standardized by the International Rice Research Institute. Since then, a large number of local varieties and germplasm accessions provided by International Rice Research Institute (IRRI) have been screened for resistance to BPH. Several resistant rice accessions have been found among the land races and wild varieties (Madurangi, et al., 2012). However, most often resistance derived from Ptb33 (Indigenous Indian variety) was used as a donor to develop resistance to BPH. The wider use of Ptb33 in the local breeding program was attributed to its compatibility in crosses with other improved varieties. Further, the high levels of resistance in Ptb33 make it a suitable parent in back crosses. It was found that Ptb33 possesses BPH2 (recessive gene) and BPH3 (dominant gene), both resistant genes and provides a major mechanism of resistance that curtails the infestation by BPH; the phytochemicals present in the plant can arrest the population growth, feeding, longevity survival and reproduction of the BPH. (Madurangi et al., 2013; Bhanu et al., 2014). Because of the high selection pressure exerted by some varieties possessing high antibiosis properties against populations of BPH, there is a common tendency to develop biotypes on these. Resistance to BPH in diverse germplasm has been evaluated using several methods including the survival of nymphs, nymphal duration, population growth, honey dew secretion, and plant reaction to artificial or natural infestation (Kudagamage and Nugaliyadda, 1982).

Apart from the techniques relating to plant phenotypic characters mentioned above, molecular markers can reveal differences among genotypes. Fernando et al. (2007) reported differences of micro satellite markers found in the virulent and non-virulent forms of BPH infesting the resistant variety Bg 379-2 cultivated in different locations in Sri Lanka. Further, Madurangi et. al. (2011) assessed the presence or absence of bph2, bph10 and bph13 gene in *Oryza nivara*, *O. eichingeri* derived accessions using molecular markers. Also, PCR (polymerase chain reaction) amplification with KAM primer demonstrated the presence of bph2 in the genome of Ptb33, although their studies did not indicate the presence of bph2 in Bg 379-2 which derived its resistance from Ptb33.

It has been demonstrated that virulent forms of BPH can be selected in the laboratory by infesting a resistant BPH variety with a diverse population of BPH. Under laboratory conditions, it takes place within 10-12 generations (Kudagamage and Nugaliyadda 1984). A similar selection process is expected to occur in the field but it may take a longer time due to the action of natural enemies, cultural control and climate related other factors, which limit the population build of BPH. It has also been observed that the field population of BPH in Sri Lanka showed adaptability to the varieties they infest (Claridge et al., 1982). Thus, long term planting of one variety may lead to the development of BPH populations, increasingly adapted to the variety, greatly increasing the possibility of serious outbreaks and hopper burns. These virulent forms are known as biotypes and are a limiting factor in the utilization of resistant varieties in the management of BPH. However, many possibilities exist for the development of more stable resistant varieties using more diverse sources of resistance found in closely related species such as wild rice (Madurangi, et. al., 2012). Screening of land race varieties and close relative species of rice are still being carried out in research programs. Up to now, 37 genes resistant to BPH have been identified (Sani Haliru et al., 2020). The resistant genes Bph1, Bph2, Bph3 and Bph4 have been used extensively in breeding programs in Southeast Asia. Improved cultivars carrying Bph1, Bph2, Bph3, and Bph4 genes however, lost their resistance against BPH due to the evolution of new biotypes. Four BPH biotypes are known to occur on rice. Biotypes 1 and 2 are widely distributed in Southeast and East Asia whereas biotype 3 was developed in the laboratory by rearing the insects on the resistant variety ASD7 which has the BPH2 gene for resistance (Panda and Heinrichs, 1983). The most destructive is biotype 4 which occurs in the Indian subcontinent and it is also called South Asia biotype. In order to keep ahead in biotype development, it is necessary to explore the potential of resistant varieties and breeding lines developed at IRRI as these germplasms are available to breeders in national research programs. Efficient introgression of various genes into elite rice varieties by marker aided selection and deployment of these resistant varieties in IPM program are important approaches to develop stable resistance against the development of biotypes. Jayathilaka et al. (2014) reported the use of Simple Sequence Repeats (SSR) and Single Nucleotide Polymorphism (SNP) makers in the development of rice varieties with the two traits BPH resistance and aroma.

3.1.2. Gall Midge (GM)

Rice gall midge (*Orseolia oryzae*) is a significant pest in rice in different parts of Sri Lanka. The major control methods of GM are the utilization of resistant varieties and insecticides. The resistant variety released by the department of agriculture has been recommended to farmers since about 1979. However, there were reports of the evolution of new biotypes of GM in the late 1980s. A study conducted by Kudagamage and Gunawardhana (1989) using GM populations from different parts of Sri Lanka revealed the existence of a virulent GM biotype in many parts of the country where varieties *viz.*, Bg 276-5 and Bg 400-1 (formally resistant varieties) were cultivated by farmers. Many new sources of resistance were also identified in this study that may serve as sources of resistance to this new GM biotype.

A defined set of varieties with known genes for resistance termed as differentials are used to determine the existence of biotypes (Fernando, 1972). New approaches such as Amplified Fragment Length Polymorphism (AFLP) are available to determine the degree of virulence in different populations of

GM in several countries (Katiyar et al., 2000). The existence of different biotypes in Rice Research Stations Batalagoda (Northwestern Sri Lanka) and Bombuwela (South Sri Lanka) was confirmed using this technique.

3.1.3. Other Pests

Two rice pests, where the employment of HPR was attempted were rice thrips, *Stenchaetothrips biformis* and rice leaf folder (RLF), *Cnaphalocrosis medinalis*. Although the presence of genetic resistance to thrips was found in several land race varieties, transferring resistance to improved varieties was not successful in the Rice Research and Development Institute, Sri Lanka. Gunasekera et al., (2017) reported genetic transformation of Bg94-1 (local improved rice variety) against RLF resistance by using *cry*2A gene of *Bacillus thuringiensis*, mediated through *Agrobacterium tumifaciens*. However, no further studies were conducted due to various reasons such as the non-availability of biosafety tests and the absence of safety protocols in field testing. Up to now transgenic rice with Bt genes are not available for general cultivation in Sri Lanka.

Future strategies in breeding for stable insect resistance involve i. identification of diverse sources of resistance comprising domesticated species and wild relatives and ii. The transgression of these genes into high yielding varieties. These will require further improvement of available techniques, comprising both conventional and novel methods used for the identification and breeding of resistance. Central research stations and regional research stations coming under the department of agriculture are equipped with greenhouses and other facilities for rearing insects and mass screening. However, many research stations do not have facilities for carrying out molecular breeding. Some of the universities with agriculture faculties have these facilities which need to be further strengthened with collaborative work among these institutions. This will allow sharing of resources among them in order to achieve objectives such as breeding high yielding stable pest resistant varieties within a short period of time. Collaboration with international research institutes needs to be further strengthened for sharing germplasm and breeding lines. Further, scientists should be trained in novel techniques and methods of breeding for resistance. With regard to transgenic rice and other genetically modified crops, a competent authority needs to be established who can monitor the risks and benefits of genetically modified crops and educate the policy makers and the general public on various issues such as environmental and health risks arising from the cultivation and the consumption of genetically modified crops.

3.2. Biopesticides

Biopesticides include microbes (fungi, bacteria, viruses, protozoa, and nematodes), semiochemicals (insect sex pheromones), botanicals and commercially produced natural enemies (predators and parasitoids). The rich natural enemy complex of pests has high potential in Sri Lanka as compared with other Asian countries. For instance, the egg parasitism of BPH is found to reach as high as 80 per cent in fields kept free of pesticides during the early vegetative stages. Furthermore, rice land spiders have also been found very effective in managing hopper pests.

3.2.1. Botanicals

Sri Lanka is rich with diverse plant species with pest control properties and these are popular among small scale and village level farmers. Neem products are one of the highly utilized plant-based products in pest management and a number of neem-based commercial products are available in the market under different trade names. The crude water extract of neem seed kernel was found to be effective in controlling BPH infestation (Sivasubramaniam et al., 1999). However, it is laborious to prepare crude extracts because large quantities of plant materials are required at the village level lack facilities to collect them. Volatile oils can be extracted from aromatic plants. The yield of volatile oil is usually low, but due to their repellence or toxicity, even small amounts of the concentrated essential extract can be highly effective in airtight or hermetic storage structures. Mixing plant materials in powder form is a simple technique that can be easily adopted by farmers. Many plants have been tested in the laboratory as powders to evaluate their possible effects.

3.2.2. Biocontrol Agents

Natural enemies are key components of IPM programs in paddy farming, providing a natural balance of conserving predators and parasitoids of the pests. Plenty of species with predatory and parasitoid action have been reported in diverse rice ecosystems throughout the country. The common predators recorded in Sri Lankan paddy fields are *Micraspis* spp., *Harmonia* spp., *Ophionea* spp., *Metioche* spp., *Microvelia* spp., *Mesovelia* spp., *Limnogonus* spp., *Cyrtorhinus lividipennis*, *Polytoxus* spp. and some spiders (Kudagamage et al., 1994). *Platigaster oryzae* has been reported to parasitize effectively on GM while the egg parasitoid of the paddy bug, *Gryon nixoni* was recorded on BPH and rice leaf folders (Kudagamage et al., 1994; Nugaliyadda et al., 1995).

3.3. Chemical Pesticides

Kudagamage and Nugaliyadda (1995) reported that the rice insect pest problem in Sri Lanka according their order of priority are leaf folder (22%), rice bug (17%), brown planthopper (16%), yellow stem borer (14%), rat (9%), thrips (5%), cut worm (5%), gall midge (4%) and army worm (4%). The leaf-feeding insects cause severe damage to rice and therefore it is suggested that insecticides should be sprayed early to avoid heavy crop losses. The government spends significant amounts on importing chemical pesticides. Survey results showed that the most commonly used insecticides in Sri Lanka are carbofuran (42%), chlorpyrifos (29%), phenthoate (28%), methamidophos (19%), monocrotophos (19%) and diazinon (13%) (Nugaliyadda et al., 1995). The insecticides recommended against rice pests by the Department of Agriculture, Sri Lanka are mentioned in below (Table 2).

Table 2. Insecticides against rice pests recommended in Sri Lanka

Pest	Common name, Formulation Type & Strength	Remarks
Rice Thrips,	Seed treatment	In endemic areas and in lately planted crops it is highly advisable to treat the soaked seeds with insecticide powder dissolved in water before incubation. Wear gloves while handling chemicals and broadcasting seeds. Apply foliar sprays when 50% of leaves/ hill show damage symptoms.
Stenchaetothrips biformis	Thiamethoxam 70% WS	
	Imidacloprid 70% WS	
	Foliar application	
	Buprofezin 25 SC	
	Pymetrozine 50 WG	
	Imidacloprid 17.8 SL	
	Imidacloprid 70 WG	
	Thiacloprid 240 SC	
	Ethiprole 100 g/l SC	
	Fipronil 50 SC	
	Carbosulfan 200 g/l SC	
Rice gall midge	Fipronil 0.3% GR	In endemic areas, as a regular practice, treat nurseries 5 days after seeding with 12g Fipronil/10 m2. Broadcast granules on wet mud or into 1 cm of standing water 1-2 weeks after transplanting or 1-3 weeks after broadcasting [Economic threshold level (ETL) - 5% Galls]
Orseolia oryzae		

Pest	Common name, Formulation Type & Strength	Remarks
Rice leaf folder	Flubendiamide 24% WG	Apply insecticides when 25% of the leaves show more than 50% leaf damage or 10 live larvae (in rolled leaves) in 10 randomly selected hills.
Cnaphalocrosis medinalis	Flubendiamide 20% WG	
	Chlorantraniliprole 20% +	
Marasmia patnalis	Thiamethoxam 20% WG	
Case worm	Tebufenozide 200 g/l SC	
Nymphula depunctalis	Methoxyfenozide 240 g/l SC	
	Azadirachtin 10 g/l EC	
	Chromafenozide 50 g/l SC	
	Chlorfluazuron 50 g/l EC	
	Novaluron 100 g/I EC	
	Bistrifluron 100 g/l EC	
	Spinetoram 25% WG	
	Fipronil 50 g/l SC	
Rice Stem borer	Chlorantraniliprole 20% +	Apply insecticides only when damage exceeds 10% dead hearts or 5% white heads.
Scirpophaga incertulas	Thiamethoxam 20% WG	
	Fipronil 0.3% GR	
	Carbosulfan 200 g/l SC	
	Thiocyclam 4% GR	
Brown Plant hopper	Buprofezin 10% WP	Apply insecticides only when number of BPH (nymphs+ Adults) exceeds 5/hill at tillering and 8/hill at reproductive stage. Spray directly to the base of the plant.
Nilaparvata lugens	Buprofezin 250 g/l SC	
	Thiocyclam (Hydrogen Oxalate) 50% SP	
	Pymetrozine 50% WG	
	Chlorantraniliprole 20% +	
	Thiamethoxam 20% WG	
	Imidacloprid 70% WG	
	Imidacloprid 200 g/l SL	
	Thiamethoxam 25% WG	
	Sulfoxaflor 50% WG	
	Etofenprox 100 g/l EC	
	Ethiprole 100 g/l SC	
	Fenobucarb 500 g/l EC	
	Carbosulfan 200 g/l SC	
	Thiocyclam (Hydrogen Oxalate) 4% GR	
	Acetamiprid 20% SP	

Table 2. (Continued)

Pest	Common name, Formulation Type & Strength	Remarks
Paddy bug	Thiocyclam (Hydrogen Oxalate) 50% SP	Apply pesticides when the bug density is 1/10 hill. Follow the same recommendation for pentatomid bugs (Paddy bug infestation is confined to canopy stage only).
Leptocorisa oratorius	Sulfoxaflor 50% WG	
	Ethiprole 100 g/l SC	
	Carbosulfan 200 g/l SC	
Field crab	Fenobucarb 500 g/l EC	Squirt insecticide solution into crab burrows after removing the swirl plate from the nozzle.
Paratelphusa ceylonensis		
Leaf mites	Hexythiazox 50 g/l EC	
Hemitarsonemus spp.	Sulphur 80% WP/WG	
Oligonychus spp.		
Tetranychus spp.		
Sheath mite	Fenpyroximate 50 g/l EC	Spray at the late booting stage of the crop. Do not apply insecticides after panicle emergence
Steneotarsonemus spinki	Hexythiazox 50 g/l EC	
Rats		Start baiting soon after transplanting and continue up to grain maturity. Place the bait inside a piece of bamboo about 1 foot long (40 baiting stations/ha).
Ratus spp.	Difenacoum 0.005% RB	
Bandicoota spp.	Brodifacoum 0.005% RB	

Source: Department of Agriculture, Sri Lanka, 2020

On the basis of surveys conducted in the early 1990s, it was concluded that most farmers believed chemicals were necessary to control pests in rice crops and that their use would increase crop yields (Sivayoganathan, 1992). For rice, farmers usually used 3 sprays. This may reflect differences in crop values as well as the degrees of pest infestations. Common rice pests targeted by farmers are worms, stem borers and plant hoppers. The damage caused by these insects is clearly visible and farmers insist on spraying insecticides. Thrips tend to be most damaging when they infest the crop during the early stages. High populations of thrips reduce the yield and the quality of crop and farmers tend to believe that insecticide treatment is the only way to protect the crop and obtain high yield (Nugaliyadda et al., 1995).

3.4. Indigenous Knowledge (IK) Based Insect Pest Management

Indigenous Knowledge (IK) refers to the local knowledge, wisdom and practices accumulated by those in society with agriculture, pest and disease management, cure ailments, preparing and preserving food, education and a wide range of other activities that sustain a society. IK is very unique and community specific, timely tested and proven and reliably well adapted to local conditions. In Sri Lanka, IK is still practiced in rural or traditional farming. In IK, people use natural resources and have done so over centuries in a sustainable manner.

Integrated crop protection measures such as the cultivation of a small plot of land area to attract birds, to set traps, to conduct kems, rituals, religious rites, the use of biological agents, the use of mechanical methods and use of plants and plant extracts are all employed in the management of pests in paddy fields. However, these indigenous methods prove inadequate in the case of improved high yielding varieties, which tend to be more prone to pest attacks. Therefore, IK needs to be integrated with modern knowledge and techniques so as to produce suitable technologies to develop sustainable development programmes.

3.4.1. Kem and Rituals

Kem and rituals are kind of techniques or customary practices use plants or recite "pirith", "suthra" or "gatha" to gain beneficial effects when confronted with certain problems. Sometimes kem methods are combined with astrology and with some plants or herbs. The Kem methods are applied in activities from land preparation to harvesting the crop. Illustrations of some popular kem and rituals in the cultivation of rice are shown in figures 1 and 2.

3.4.1.1. Pineapple Shoots Treatment

Slender pineapple shoots are planted randomly in paddy fields in the evening to protect the field from rodent damage. The person does not talk while establishing the pineapple to ensure the effectiveness of the treatment. The stinking odour of rotting tender pineapple shoots act as a repellent to rodents.

*3.4.1.2. Snake Plant (*Sansevieria zeylanica*) Treatment*

Snake plant shoots are planted in paddy fields randomly, the farmer keeping quiet. The strong and nasty odour that emanates from decaying shoots repels rodents from the paddy fields.

3.4.1.3. Crab Shell Treatment

Paddy bugs caught from the field are collected into crab shells. These shells are buried in paddy fields where paddy bugs tend to inflict most damage.

3.4.1.4. Lighting Torches (Pandam kaariya)

Locally prepared torches are lit on stalks of *Gmelina arborea* (English: white teak; Sinhala: Demata) and *Glycyrrhiza glabra* (English: Liqourice; Sinhala: welmee), established in the paddy field for three consecutive days to attract insect pests.

3.4.1.5. Sweeping with Bamboo Stems

Thin and long bamboo stems are tightly wrapped with ragged waste cotton rags. The bamboo stalks are then treated with oil extracted from *Dipterocarpus glandulosus* and an adhesive secretion collected from jack fruits. The stalk is then moved slowly over the paddy field as if being "swept." Paddy bugs and other insect pests adhere to the bamboo stem.

Figure 1. Traditional farming equipment, storage facility and water trap.
A. Traditional paddy seed storage; B. Water ghost (*Diya Holmana*);
C. Winnowing fan; D. Some traditional equipment.

3.4.1.6. Light Taps

After sunset, oil lamps are lit in the four corners of the field and sands charmed with "manthra" are spread over the field to control paddy bugs.

*3.4.1.7. Offering Cycas Flowers (*Madumal pideema*)*

At dawn, a naked farmer goes around the paddy field three times at dawn with a Cycas flower and stands the flower in the field to control paddy bugs.

3.4.1.8. Spraying a Mixture of Three Latexes (Thun-kirisaaththuwa)

Latex extracted from jack fruit and eggplant are mixed with coconut milk while chanting "Manthra." This mixture is sprayed in the paddy field to control hoppers.

3.4.1.9. Offering Milk Rice (Irima Pideema)

Balls of milk rice prepared with ghee are kept on poles planted in a paddy field. A part of the rice milk falls into the water to attract birds who also feed on insect pests in the water.

3.4.1.10. Offering Areca Nut Flowers with Oil Lamp (Puwak mal pahan pideema)

At the beginning of the season, all farmers collect areca nut flowers and oil lamps and randomly place them in the entire paddy land, preferably on Saturdays or Wednesdays. This set up is called "mal thatuwa," and is made by weaving tender coconut leaves and arranging them as a tray to keep all flowers and oil lamps together. A religious "manthra" is chanted in the evening while lighting oil lamps. This kem is believed to protect effectively the whole cultivated area from adverse biotic and abiotic events.

All these kem and rituals rely on the long-time practice by local farmers; they are from having been scientifically proven.

3.5. Traditional Biological Pest Management

Traditional farmers are well aware of their surrounding environment including natural resources, associated organisms, habitats, ecosystems and interactions among them. They have knowledge of various beneficial organisms such as vertebrates, reptiles, birds, mammals and their related habitats. The farmers protect, develop and maintain a microclimate that favours these organisms. Traditional farmers develop several strategies to attract birds that feed on insect pests of paddy.

*3.5.1. Planting Mee Trees (**Madhuca longifolia**)*

M. longifolia trees are planted on the boundary of the paddy fields to attract bats and owls, which feed on rodents and other pests that come to paddy fields.

3.5.2. Planting Coconut Fonds/Coconut Husks on Poles

Birds that feed on pests are attracted to paddy fields when they find perches (Figure 2-A). Flattened coconut husks are fixed to poles standing in fields to provide perches so that birds ca roost on them. Sometimes, food items like milk rice, crushed coconut kernels are served on the perches to attract such birds as the common babbler. These feed on bugs and worms in the field. Traditional farmers usually grow large trees around the paddy fields to provide safe habitats for nesting and breeding places to such beneficent creatures.

3.5.3. Setting Up Plantain Disks with Food

A mixture of roasted pulses, other food and flowers along with oil lamps are set up in the paddy field to attract birds to pulses and food, and insects to oil lamps. Insects become victims of burning oil lamps or birds.

The food placed on plantain disks falls down when birds land on them. Birds feed on fallen food when they see the caterpillars and other pests on ground, eat them up, as well.

3.5.4. Maintaining a Muddy and Watery Area

Traditional farmers left one corner at the edge of the drainage channel to form a muddy and watery pool so that rat snakes (*Ptyasmucosus*) can nest and breed there. Rat snakes are one of the best biocontrol agents in paddy fields as they are non-venomous and feed on rodents and other small destructive mammalian pests, especially during the period when grains ripen. One rat snake is capable of consuming over 20 rats per month (Senanayake, 1983). Similar habitats are favoured by the common water monitor, *Varanus salvator*, which consumes fresh water crabs that damage the paddy field when they dig holes in them.

3.5.5. Farming Indigenous Fish Species and Other Fauna

Farmers always keep the paddy fields flooded during the early stages of a crop so as to allow fish and other fauna to grow and at the same time suppress the emergence of weeds. The insectivorous fish species in the paddy field feed on insect pests that fall into the water. The spider species dwelling on plants also act as biological agents that feed on insect pests. In addition, farmers encourage the growth of frogs in the field as frogs feed on many harmful insect pests. The Swift birds around the paddy fields feed on BPH which cause serious damage to paddy crops.

Figure 2. Different traditional activities adopted for Pest control.
A. Coconut fonds to attract birds B. Farmer ready to plant poles; on top a cage woven with tender coconut leaves on which to light oil lamps after sunset; C. Planting poles with coconut husk randomly in the field; D. Broadcasting wood ash in early morning at the presence of dew; E. Preparation of robes treated with resins or bees wax; F. Dragging the bees wax or resin treated robe over paddy plants; G. Spraying "pirith" chanting, using an areca nut flower, a popular kem method; H. Ritual activity by a farmer.

3.6. Traditional Botanical Pest Management

From ancient times farmers were capable of identifying a wide range of plants that they used for pest management. Many such plants are resistant, toxic or repellent to insect pests. These plants naturally occur in nearby forests or scrublands. Farmers use different parts of these plants, including leaves, bark, roots and fruits to prepare formulations or use them to control insect pests. Parts of plants or formulations are worked into the soil or mixed with water in the paddy field. Sometimes leaves and twigs are hung in the field. The most common plant species that farmers use are *Diospyros affinin*, *Anamirta cocculus*, *Ananas comosus* and *Euphorbia antiquorum*, which effectively control *Orseolia oryzae* and *Atherigona oryzae*. The leaves of *A. comosus* and *E. antiquorum* are crushed and added to the water channels at the point of irrigation so that they may spread out in the field with water body and effectively control *Tryporisa insertulas.* The seeds of *Caryota urens* are crushed and added to the water in the field to control destructive worms. The dried leaves of *Pongamia pinnata* and *Crotolaria retusa* are mixed with paddy soil during land preparation to control destructive worms.

Traditional farmers use certain plant species with repelling odour such as *Cycas circinalis* or *Cymbopogon citrates,* which give out repelling odour. They are hung around paddy fields to repel rice pests. Sometimes farmers cut *C. circinalis* flowers into pieces and fix them to poles planted in the field randomly to spread repelling smells. Farmers plant yellow colored flowering plants like chrysanthemum in the bunds and around the field so that yellow color and the smell of flowers repel insect pests. Rodent attacks are controlled by spreading chopped stems and leaves of papaya in the field, particularly on the paths that rodents frequent. Papaya smell repels rodents and the chemical substances in the raw papaya causes wounds on their mouth when they feed upon it. Farmers believe that spreading of crushed leaves of *Coleus amboinicus* around the paddy field, on account of their strong smell, stops wild boar from entering fields.

3.7. Traditionally Used Mechanical Pest Management

The most popular traditional mechanical methods to control rice pests are food and light traps. Food traps are commonly used to control rodents while light traps control insect pests.

3.7.1. Oil Lamps

Oil lamps are hung at a point of rice plants with underlying water containers below them. Insects are attracted towards oil lamps and fall into the water.

3.7.2. Winnowing Fan Treatment

Neem or citronella oil is applied on the back side of a *winnowing fan*. Sometimes farmers apply a sticky substance extracted from jack fruit (*Artocarpus heterophyllus*) on the back of a winnowing fan and draged along a field so that insects get stuck to the fan. Normally, farmers do this twice a day to control paddy bug damage when grains ripen.

3.7.3. Traditional Resin/Bees Wax Treatment (Dummala or Mee-iti lanu adeema)

Resins extracted from *Vateria copallifera* or bees wax are applied on a rope and the paddy field is fenced with these treated ropes to prevent attacks from the paddy bug and other minor pests (Figure 2).

3.7.4. Broadcasting of Wood Ash

Farmers sprinkle wood ash before dawn when there is still dew on rice plants. This method is helpful in controlling a wide range of paddy pests. Because of dew, wood ash gets pasted on to the leaves. Sometimes farmers use torches made by wrapping sticks from *Calatropis gigantea*, *Pagiantha dichotoma*, or *Leea indica* trees with saffron robes as light traps. These selected sticks have insect controlling properties and can be considered as a combination of mechanical and botanical methods. In addition, traditional farmers use several sound generating devices to chase away harmful animals and birds. Some of these devices are operated with wind power or flowing water and are called 'wind ghost' or 'water ghost', respectively. Farmers also keep puppets resembling humans which help to keep away animals and birds.

3.8. Rice Pest Management at Post-Harvest Stage

Pest damage to paddy can occur at any stage, i.e., from harvesting to consumption. Damage occurs during bagging of harvested grains, transportation in the warehouse, storage as paddy, milling of paddy, storage of rice (milled paddy), processing as different products (eg. rice flour, rice noodles, rice biscuits) and the storage of processed products at retailer outlets and household level. Although farmers and other stakeholders do not seriously consider the losses during post-harvest stages, the pest damage to the stored paddy is indeed of great concern.

For insects to emerge and develop, the contribution from the ecosystem is indispensable. Temperature is the most critical environmental factor for insect development. In stored products, insects thrive at 25-33°C and so accounts for the higher abundance of insects and greater damage to the stored products that occurred in the tropical regions. In Sri Lanka, paddy cultivation (and thus the harvesting) is seasonal but the consumption is year around. To meet this regular demand, the surplus paddy has to be stored and the stored paddy has to be protected from insect infestation. Of old, farmers have used their experience-based knowledge to prevent temperature build up within the storage structures. Traditionally paddy/rice was stored in structures made of clay, wood or hay/straw. One feature of these structures was the lower temperature inside them, which was hostile to the development of insect populations. A few decades ago, paddy/rice were stored in jute sacks but now in modern storage practices, they have been replaced with polypropylene bags. At present, bagged storage dominates bulk storage of paddy in Sri Lanka. In a recent survey conducted in the Anuradhapura district in Sri Lanka, four different types of storage structures/places are being used by the farmers to store paddy; a room inside the house, a room outside the house, a temporary hut and the kitchen (Kumari et al., 2020a).

Palipane (2001) reported that in Sri Lanka the total post-harvest losses of durable crops were around 15%. Losses during storage were estimated at 4-6%. The same study found that 80% of the said losses occur due to insect infestation. Dissanayaka and Wijayaratne (2020a) reported that around 70% of the loss of stored paddy was due to insects in Trincomalee, a warm region in the North-East Sri Lanka. The most common insect species of stored paddy were *Rhyzopertha dominica*, *Sitophilus oryzae* and *Sitotroga cerealella*. In addition, rodents are common in stores. Kumari et al. (2020b) reported *Tribolium castaneum*, *S. oryzae*, *R. dominica* and *S. cerealella* as the dominant insect species of stored rice grains in the Anuradhapura district of Sri Lanka. They had been identified earlier too as major pests of paddy during storage in Sri Lanka (Dharmasena and Abeysiriwardena, 2003). Hettiarachchi et al. (2020a, b, c) found that the intensity of the infestation caused by *T. castaneum*, *S. oryzae* and *R. dominica* varied with the percentage of broken rice grains present in the medium.

These insect pests are a severe problem in rice mills as well. Dissanayaka and Wijayaratne (2020b) reported that insects found in a rice mill varied with the type of mill. In a large-scale mill, the intensity of infestation varied from the from the highest to the lowest from floor dust, polisher, paddy cleaner, packing area, elevator, conveyer and to the parboiling area. The recorded insect species, from the highest to lowest abundance, were *S. oryzae, T. castaneum, R. dominica* and *S. cerealella.* In large scale mills, the maximum number of insects are found in the floor dust; *S. oryzae* has been found more abundantly than *T. castaneum*. These large-scale mills are mechanized to a greater extent, making them difficult to clean which allows the buildup of insect populations. The *T. castaneum* populations are also found in elevators, packing areas, conveyers and paddy cleaners. In a medium-scale rice mill, insects are found with decreasing abundance in the paddy cleaner dust, elevator, separator, floor, silking machine, store house, grader and destoner. The recorded insect species in decreasing order are *R. dominica, T. castaneum, S. oryzae, S. cerealella, Cadra cautella* and grain mites. Mostly *T. castaneum* is found in separators, silking machines and elevators. *Rhyzopertha dominica* was abundant in the paddy cleaner dust and *S. oryzae* on the mill floor. In a small-scale rice mill, an abundance of insects from the highest to lowest are found in the de-husker, destoner, polisher, flour machine and store floor. The insect species were from the most abundant to the least; *R. dominica, T. castaneum, S. oryzae, S. cerealella* and *C. cautella*.

In the management approaches, chemical as well as non-chemical methods are adopted in Sri Lanka; the methods differ with the storage system. At the farmer level, the use of botanicals is popular. When the leaves of citrus (*Citrus aurantiifolia*), Chinese chaste tree (*Vitex negundo*), turmeric (*Curcuma longa*), neem (*Azadirachta indica*), eucalyptus (*Eucalyptus* spp.) is dried in the shade and mixed with stored paddy, it is protected paddy from insect infestation. Wijayaratne et al., (2009) reported successfully using smoke, generated from partially-burned paddy husk, to protect the freshly-harvested paddy stored in bags from insect pests.

The chemicals recommended in Sri Lanka for stored-product insects are Pirimiphosmethyl (contact insecticide) and phosphine (fumigant). As the alleged insect species become more active and come out of the storage bags in the evening hours, better results can be obtained by applying chemicals at that time. For a successful phosphine fumigation, the structure/area should be maintained sealed (airtight) during the entire duration (normally 72 hrs.). Usually, the grains in bags are covered by a gas-proof sheet pressed to the floor by using some weight. If applied at the correct dose (3-5 phostoxin tablets per ton of grains) and maintained in good, sealed condition, all the live insects present in that enclosed area are expected to be killed. However, there are practical limitations to achieving those results. If gas escapes or insects are exposed to inadequate doses, refugees from the premises raise the possibility of developing resistance to phosphine.

There have been many significant findings from the studies conducted in Sri Lanka on different management strategies for stored product insects. Airtight storage is a successful method where live insects are denied oxygen requirement. This method has been proved satisfactory at the laboratory level as well as small-scale and large-scale bulk storage systems. A series of research testing the chemical and non-chemical management of stored-product insects has been carried out at the Faculty of Agriculture, Rajarata University of Sri Lanka. Effects of high and low temperature exposures have showed promising management against *S. oryzae* and *R. dominica*. Sajeewani et al., (2020) reported new traps of different designs that capture *T. castaneum* more than twice as done by the commercially available dome trap. The aggregation pheromones of *T. castaneum* [4,8-dimethyldecanal or 4,8 DMD] and *R. dominica* [dominicalure-1 and dominicalure-2] have been tested to elucidate the effects in various avenues such as different concentrations, population densities, distances, heights and the influence of air flow. Furthermore, the sex pheromone of *C. cautella* [Z, E (9, 12) Tetradecadienyl Acetate] has been tested for mating disruption effect either alone or in combination with plant volatiles. Certain food oils such as *Madhuca longifolia* and *Cocos nucifera* have been found to augment the movement of *T. castaneum* to its aggregation pheromone. The reduced-risk insecticides spinosad and spinetoram have also been tested for their residual efficacy on traditional and new improved rice varieties for the protection from *R. dominica* and *C. cautella*. Furthermore, pre-exposure to spinosad or spinetoram, two bacterial formulations derived from the bacterium *Saccharopolyspora spinosa* was found to increase the trapping of *T. castaneum* by its aggregation pheromone 4,8 DMD.

3.9. Farmers Field School (FFS)

A concept that had its origin in Indonesia and few other Asian countries focus on participatory field-oriented approaches called Farmers Field School (FFS). This method brings together about 20-25 farmers groups in a Maya or tract. FFS gathers weekly with their facilitator to study and understand the dynamics of their crop ecosystem in an exercise called the ecosystem analysis. Attention is paid not only to crop pests, but the whole ecosystem comprising crops and their growth, existence of natural enemies and intensity of pest damage and effect on the crop growth.

Extension programs in rice commenced in 1984 and were implemented with training and visits. Messages were used to transfer technology to farmers using contact farmers. For this purpose, meetings among farmers were held at fortnightly intervals. A demonstration plot was established depicting the message to be transferred. Although this method was found suitable for transfer of simple messages such as the arrival of a new variety, it did not prove adequate in the case of complex messages such as pest management. Another system where groups of farmers from the same paddy tract participated in a periodic training class over several years and helped to adopt new technology in their tract. These farmers were given access to credit and other input and this approach was called block demonstration.

Assessment of quantitative outcomes of the broad-scale study undertaken by Van den Berg (2002) indicated a major impact in terms of increased production, reduced input costs and changed farmer practices. These impacts were durable over the covered period of 6 ½ years. However, the diffusion of the effects to other villages was insignificant. Extension programs through the participatory approach have many social benefits such as the development of critical thinking and increased creativity. The outcome of IPM increased with time; villages with longer post-field schooling have developed more outcomes such as empowering farmers to develop their own programs including animal husbandry and starting a rice processing unit.

Conclusion and Future Prospects

Rice has been cultivated in Sri Lanka since pre-historic times. The country is rich in diverse flora and fauna, among them most with beneficial properties in managing crop pests and diseases. Agronomic practices such as timely planting according to the lunar calendar and other traditions have great potential in managing rice pests. However, recent overloaded application of chemical fertilizers and chemical pesticides inflicted severe damage to the sustainability of the rice ecosystem, by destroying the ecological balance, soil properties and other natural resources. The government has implemented a policy to convert traditional agricultural practices into organic based approaches which can limit or totally ban the use of agrochemicals in 2021 and the policy was reversed in 2022. Within this platform, the use of integrated management components remain applicable and valid. However, these changers have not taken into account farmers perception on pest damage as well as avoiding damage due to sudden emergence of pests. In such instances, treatment with insecticides is vital to curtail the crop damage. However, more research is necessary to study various interactions, for example studies on the outcome of the interaction between biocontrol and plant resistance etc. The study of indigenous knowledge related to pest management should receive emphasis. Extra effort may be necessary to transfer technology based on non-chemical methods where farmers are more prone to applying chemicals on a routine basis. FFS methods which have been introduced to farmers may be still valid, although some fine tuning of FFS may be required when applying it into different farming systems.

Acknowledgments

The authors acknowledge the information sources of RRDI, Bathalagoda; Department of Agriculture, Sri Lanka; Biodiversity Secretariat, Ministry of Environment and Mahaveli Development, Sri Lanka and Dr. Geedreck Usvatte-Aratchi for editing English.

References

Bhanu, K. V., Lakshmi, V. J., Katti, G. and Reddy, A. V. (2014). Antibiosis and tolerance mechanisms of resistance in rice varieties carrying brown plant hopper resistance genes. *Asian Journal of Biological and Life Sciences*, 3 : 108–113.

Claridge, M. F., Hollander, J. D. and Furet, I. (1982). Adaptation of brown planthopper (*Nilaparvata lugens*) to the rice varieties of Sri Lanka. *Entomologia Experimentalis et Applicata*, 32 : 222-226.

den Berg, P., Thompson, K.J., Obremski-Brandon, K., Coovert, M., (2002). The Tripartite Influence model of body image and eating disturbance A covariance structure modeling investigation testing the mediational role of appearance comparison, Journal of Psychosomatic Research 53 :1007– 1020

Dharmasena, C. M. D. and Abeysiriwardena, D. (2003). Effect of variety, grain moisture and packing material on weight and storage pests of rough rice during storage. *Annals of the Sri Lanka Department of Agriculture*, 5 : 57-67.

Dissanayaka, D. M. S. K. and Wijayaratne, L. K. W. (2020a). Common agricultural practices and post-harvest losses in Trincomalee, Sri Lanka. *Journal of Tropical Forestry and Environment*, 10 : 51-58.

Dissanayaka, D. M. S. K. and Wijayaratne, L. K. W. (2020b). Common practices and abundance of insects in rice mills in Polonnaruwa district of Sri Lanka. *Journal of Science-FAS-SEUSL*, 1 : 1-9.

Fernando, H. E. (1972). Biology and laboratory culture of rice gall midge and studies on varietal resistance. In : Rice Breeding. International Rice Research Institute, Manila, Philippines. pp. 343-351.

Fernando, K. K. S., Ekanayake, E. M. D. S. B. and Ketepearchchi, Y. (2007). Identification of rice brown planthopper using microsatelite markers. *Annals of Sri Lanka Department of Agriculture*, 9 : 27-33.

Gunasekera, J., Jayasekera, G., Perera, K. and Wicramasooriya, A. (2017). Development of Sri Lankan rice variety Bg94-1 harbouring Cry2A gene of *Bacillus thuringiensis* to rice leaf folder (*Cnaphalocrosis medinalis*). *Journal of National Science Foundation of Sri Lanka*, 45 :143-157.

Hettiarachchi, J. M., Dissanayaka, D. M. S. K. and Wijayaratne, L. K. W. (2020a). Adult emergence of *Sitophilus oryzae* (L.) (Coleoptera: Curculionidae) in stored rice varieties and pulses with broken grains. *Journal of Food and Agriculture*, 13 : 37–46.

Hettiarachchi, J. M., Dissanayaka, D. M. S. K., Wijayaratne, L. K. W. (2020b). Influence of grain fractions on the progeny adult emergence of the lesser grain borer, *Rhyzopertha dominica* (F.) (Coleoptera: Bostrychidae) in stored rice varieties and pulses. *Journal of Science-FAS-SEUSL*, 1 : 18-26.

Hettiarachchi, J. M., Dissanayaka, D. M. S. K. and Wijayaratne, L. K. W. (2020c). Particle size of grains and pulses affects infestation by *Tribolium castaneum* (Herbst) (Coleoptera: Tenebrionidae). *Rajarata University Journal*, 5 : 91-95.

Jayathilaka, K. P. O. S., Kottearachchi, N. S., Kakulandara, D. S., Madanayake, M. A. R. A. (2014). Progress towards the marker assisted selection of rice lines for aroma and BPH resistance. *Tropical Agriculturist*, 162: 45–62.

Katiyar, S. K., Chandel, G., Tan, Y., Zhang, Y., Huang, B., Nugaliyadda, L., Fernando, K., Bentur, J. S., Inthavong, S., Constantino, S. and Bennett, J. (2000). Biodiversity of Asian rice gall-midge (*Orseolia oryzae* Wood and Mason) examined by AFLP analysis. *Genome*, 43 : 322-333.

Kudagamage, C. and Gunawardhana, D. L. A. (1989). Rice varietal reaction to new gall-midge biotype in Sri Lanka. *Tropical Agriculturist*, 145 : 129-131.

Kudagamage, C., Nugaliyadde, L. (1982). Laboratory and field investigations of brown planthopper *Nilaparvata lugens* resistance in rice varieties. *Tropical Agriculturist*, 138 : 149-159.

Kudagamage, C., Nugaliyadde, L. (1984). Development of a virulent form of brown planthopper on a resistant rice variety Bg 379-2. Annual Research Conference Cereal and Grain Legumes. Department of Agriculture, Peradeniya, Sri Lanka. pp. 49-50.

Kudagamage, C., Nugaliyadde, L. (1995). Present status and future direction of insect pest management in rice. Rice congress 1990, pp. 39-54.

Kudagamage, C., Ranasinghe, R. M. and Bandara, R. M. G. (1994). Effectiveness of Natural enemies of rice pests in IPM and non-IPM fields in Sri Lanka. *Krushi*, 14 : 21-28.

Kumari, J. M. P., Wijayaratne, L. K. W., Jayawardena, N. W. I. A., Egodawatta, W. C. P. (2020a). Quantitative and qualitative losses in paddy, maize and green gram stored under household conditions in Anuradhapura district of Sri Lanka. *Sri Lankan Journal of Agriculture and Ecosystems*, 2 : 99-106.

Kumari, J. M. P., Wijayaratne, L. K. W., Jayawardena, N. W. I. A. and Egodawatta, W. C. P. (2020b). Household level paddy storage and insect infestation of paddy, maize and mungbean in Anuradhapura district of Sri Lanka. *Sri Lankan Journal of Agriculture and Ecosystems*, 2 : 70-80.

Madurangi, S. A. P., Ratnasekera, D., Hemachandra, P. V. and Senanayake, S. G. J. N. (2012). Evaluation of brown planthopper *Nilaparvata lugens* (Stal) resistance in *Oryza nivara* wild rice accessions found in Sri Lanka. In : *Proceedings of the 15th International Forestry and Environment Symposium held on 26-27th November 2010. Published by Department of Forestry and Environmental Science, University of Sri Jayewardenepura, Sri Lanka.* pp. 172-175.

Madurangi, S. A. P., Samarasinghe, W. L. G., Senanayake, S. G. J. N., Hemachandra, P. V. and Ratnasekera, D. (2011). Resistance of *Oryza nivara* and *Oryza eichingeri* derived lines to brown planthopper, *Nilaparvata lugens* (Stal). *Journal of the National Science Foundation of Sri Lanka*, 39 : 175-181.

Madurangi, S. A. P., Samarasinghe, W. L. G., Senanayake, S. G. J. N., Hariachandra, P. V. and Ratnasekera, D. (2013). Antixenosis and antibiotic effects of *Oryza nivara* accession harbouring bph2 gene on brown planthopper (*Nilaparvata lugens*). *Journal of the National Science Foundation of Sri Lanka*, 41 : 147-154.

Nugaliyadda, L., Dissanayake, N. and Mithrasena, J. (1995). Pest management of rice farmers in Sri Lanka. In: *Proceedings of Publication by IRRI, Philippines* (Eds. Heong, K. L. and Escalada, M. M.). pp. 171-185.

Palipane, K. B. (2001). Current storage practices and quality improvement of stored grains. In : *Proceedings of the Seminar Organized by Sri Lanka Association for the Advancement of Science (SLAAS) in Association with the University of Kelaniya and Food and Agriculture Organization.* pp. 17-30.

Panda, N., Heinrichs, E. A. (1983). Levels of tolerance and antibiosis in rice varieties having moderate resistance to brown plant hopper, *Nilaparvata lugens* (Stal) (Hemiptera: Delphacidae). *Environmental Entomology*, 12 : 1204–1214.

Sajeewani, P. A. H., Dissanayaka, D. M. S. K., Wijayaratne, L. K. W., Burks, C. S. (2020). Changes in shape, texture and airflow improve efficiency of monitoring traps for *Tribolium castaneum* (Coleoptera: Tenebrionidae). *Insects*, 11 : 778. https://doi.org/10.3390/insects11110778.

Sani Haliru, B., Rafii, M. Y., Mazlan, N., Ramlee, S. I., Muhammad, I., Silas Akos, I., Halidu, J., Swaray, S. and Rini Bashir, Y. (2020). Recent strategies for detection and improvement of brown plant hopper resistance genes in rice: a review. *Plants (Basel)*. 9 : 1202. https://doi.org/10.3390/plants9091202.

Senanayake, R. (1983). The ecological, energetic and agronomic systems of Ancient and modern Sri Lanka. *The Ecologist*, 13 : 136-140.

Sivasubramaniam, W., Kudagamage, C. and Wijayagunasekara, H. N. P. (1999). Toxicity and mode of action of Neem (*Azadirachta indica* A. Juss) formulation on Rice brown plant hopper (*Nilapavata lugens* Stal.). *Tropical Agricultural Research*, 3 : 216-225.

Sivayoganathan, C. (1992). Usefulness of information sources on health hazards and protective measures in agro-pesticide application. *Sri Lankan Journal of Agricultural Sciences*, 29 : 109-117.

Socio Economics and Planning Centre, Department of Agriculture, Sri Lanka. (2020). *AgStat,* x Vol. 14.

Wicramasinghe, N. (1978). Approaches to Integrated control of pests. *Tropical Agriculturist*, 134 : 1-7.

Wijayaratne, L. K. W., Fernando, M. D. and Palipane, K. B. (2009). Control of insect pests under ware-house conditions using smoke generated from partial combustion of rice (paddy) husk. *Journal of the National Science Foundation of Sri Lanka*, 37 : 125-134.

Chapter 6

Chemicals in the Management of Plant Parasitic Nematodes: Challenges and Opportunities

Nishi Keshari*
Department of Nematology,
Rajendra Prasad Central Agricultural University,
Pusa, Samastipur, Bihar, India

Abstract

Plant parasitic nematodes are one of the major limiting factors in crop production. They cause losses directly by feeding on the plant roots or other parts of shoots or indirectly, by inviting the secondary plant pathogens at the site of injury, thus causing complex diseases. Among the several control measures, chemicals have been found very effective in reducing the population of plant parasitic nematodes to a significantly low level. Though, application of chemicals result in quick and effective control of nematodes but their harmful effects on fauna and flora are well known. Therefore, keeping in view the necessity of chemical application, advance planning should be done so that chemicals are applied at right time and with appropriate method, that help in minimizing its toxic effect and at the same time may prove cost-effective. Seed treatment and bare root dip treatment require very less amount of chemicals and it may prove helpful in protecting the plant from early infection of nematodes. Row, spot, strip and site application of chemicals are other effective and economical methods of protecting the plants from plant parasitic nematodes.

---

* Corresponding Author's Email: nishinema@gmail.com.

In: Pest Management: Methods, Applications and Challenges
Editor: Tarique Hassan Askary
ISBN: 979-8-88697-268-9

Keywords: plant parasitic nematode, injury, disease, chemical, management

1. Introduction

Plant parasitic nematodes (PPNs) are one the major limiting factors in agricultural crop production. Besides causing direct injury to plants, they are able to cause complex diseases in association with other plant pathogens such as bacteria, fungi and viruses. The losses to crop yield in monetary term has been estimated up to US $358 billion annually on a worldwide basis (Abd-Elgawad and Askary, 2015). PPNs are obligate parasites and therefore often require host for their multiplication. They cause injury to roots by penetrating the plant cell through their stylet and as a result the other plant pathogens like fungi and bacteria find a suitable place to multiply there easily. Thus, nematodes cause damage to roots directly by sucking the cell sap and also indirectly by predisposing the cells for secondary plant pathogens. Consequently, the disease severity increases and plants suffer complex disease. The plant physiology changes which sometimes break the resistance of those plant cultivars having resistant to certain fungal or bacterial pathogens (Powell and Nusbaum, 1960; Askary and Ali, 2012; Askary and Martinelli, 2015; Askary, 2017).

Although there are many nematodes causing diseases on the plants but root-knot nematode *Meloidogyne* sp., cyst nematode, *Heterodera* sp. and *Globodera* sp., wheat seed gall nematode, *Anguina tritici*, root lesion nematode, *Pratylenchus* sp., citrus nematode, *Tylenchulus semipenetrans*, burrowing nematode, *Radopholus* sp. and foliar nematode, Aphelenchoides are the most notorious which attack several plant species of economic importance. Some of these nematodes have a wide host range, so controlling their populations through other practices is often a challenging task (Crow et al., 2000, 2001). It is also true that once the nematode population is established in soil, it is nearly impossible to eradicate them because egg-laying capacity of some nematode species is very high. If the crops are planted or the seeds are sown in that field where the initial nematode population is very high then the crops get infection in early stages and the growers suffer a huge loss. Therefore, soil testing is advised prior to sowing or planting crop in a field. In case, the population is above the pathogenic level, the chemicals are the one of the best options left with the crop growers for effective control of these nematodes (Weingartener and shumaker, 1990a, b; Weigartener et al., 1993). One cannot avoid the fact that the chemicals have a toxic residual effect on soil and crop as well as the environment. But, still its importance cannot be ignored as it is an immediate solution to these problems.

Nematicides are chemical compounds which kills the nematodes directly and those who kill indirectly or through another means, are called, nematistat or nematistatic where the nematodes get paralysed and are not able to cause diseases in plants. It only becomes essential when all other methods like cultural, physical, resistant varieties and biocontrol agents are unable to protect the crops from pests and pathogens. The use of synthetic nematicides is still a very easy and effective method to manage the PPNs (Hajihassani et al., 2019; Oka and Saroya, 2019) and their application continues to be an important component in integrated nematode management (Ji et al., 2019; Hajihassani et al., 2019; Askary, 2015).

The focus of this chapter is to highlight the challenges faced in the application of chemicals for the management of PPNs and future strategies to overcome the problem.

2. Historical Background

The chemical used for the first time in the management of PPNs was CS_2. Kuhn (1881) applied it against sugarbeet cyst nematode in Germay. Bessey (1911) treated CS_2 for the control of root-knot nematodes in South Carolina, USA. Mathews (1919) in England, discovered the nematicidal properties of Chloropicrine, commonly known as tear gas. Carter (1943), an entomologist of Hawaii Pineapple Research Institute discovered the nematicidal value of DD mixture (1,3-dichloropropene 1,2-dichloropropane) by applying it against PPNs. In 1944, scientists from the USA reported the efficacy of Ethelene dibromide (EDB) against PPNs. This chemical was then started manufacturing by Dow chemical company and proved to be a successful soil fumigant against nematodes. The two nematicides, DD and EDB were the beginning of a new era of chemical management that helped in managing PPNs and ultimately increasing the yield and quality of crops.

The nematicides can be divided into two major groups i.e., fumigants and non-fumigants.

A. *Fumigant Nematicides*

Under this group, the nematicides after application form fumes due to volatization. Fumigants are injected in soil in liquid form where they are converted into gaseous state, penetrate through the soil pores and kill the nematodes as well as other soil microorganisms. Telone® (1,3-dichloropropene), Vapam® (metam sodium), K-Pam® (metam potassium), Paladin® (dimethyl disulfide), Basamid® (dazomet), Dominus® (allyl isothiocyanate) and chloropicrin (trichloronitromethane) are the fumigant nematicides used against PPNs (Gilreath et al., 2008; Noling, 2016). Fumigants have been placed under two groups, i. Halogeneted Hydrocarbons and ii. Methyl isothiocyanate.

- *Halogeneted Hydrocarbons*
  Under this group, five nematicides have been found very successful against PPNs. These are listed below (Table 1) and discussed briefly.

Table 1. List of halogenated hydrocarbons with their chemical and trade names

Common Name	Chemical name	Trade name
Methyle Bromide	Bromomethane	Embafume, Bed-Fume, Dowfume MC_2, Pest Master
Chloropicrin	Trichloronitromethane	Larvacide, Picfume
DD	1,2 dichloropropene and 1,3 - dichloropropane	DD, Vidden D, Telone
EDB	1,2-dibromoethane	EDB-85, Celmide
DBCP	1,2-dibromo-3-chloropropane	Nemagone, Fumazone

a. Methyl Bromide (CH_3Br): It is one of the major soil fumigants for managing nematodes as well as other plant pathogens that are soil borne. Methyl bromide has an excellent nematicidal activity due to its ability to diffuse in a vapour state within plant roots and kill nematodes and their eggs which survive within galls or cysts. Methyl bromide was one of the five most used chemical in the United States between late 1980s and early 1990s (Ristaino and Thomas, 1997). During the period, the cost of methyl bromide was much less than the higher potential returns (Duniway, 2002). The growers of high value crops relied completely on this chemical with an aim to get high returns as it was easy to apply and cheaper as compared to crop rotation and other methods (Mckenry et al., 1994).

b. Chloropicrin: It is less effective than CH_3Br (Duniway, 2002) and can be applied in combination with other fumigants like 1,3-dichloro-propene, iodomethane, metam sodium due to its fungicidal activity.

c. DD: It contain cis and trans isomers in equal quantities of 1,3 dichloropropene and a few other chlorinated compounds. Of these, dichloropropene is the most toxic component and among its two isomers, the trans isomer is twice as toxic as the cis-isomer (Jonathan, 2010). It is injected into the soil at a depth of 15-20 cm at 25 x 30 cm spacing.

d. EDB: It is applied in small quantities into the soil up to a depth of 15 cm or deeper with injectors. It is highly effective nematicide and applied as pre-plant treatment. Since EDB is phytotoxic, therefore there should be a time gap between the applications. This nematicide can not be used in bromine sensitive plants like, onion, garlic, lilly and amaryllis.

e. DBCP: It is highly efficient nematicide capable of protecting the plants from nematodes at a very low dose compared to other soil fumigants. It is slow in action but persist for long time in the soil, hence, beneficial for long term control of soil nematodes. It can be applied both as fumigant and through emulsifiable concentrate after mixing with water. It has high boiling point, and therefore can perform well at high soil temperature. DBCP is very less phytotoxic and hence safe for the plants which are to be treated. However, plants like, onion, potato, tomato, tobacco, are sensitive to this chemical.

- *Methyl Isothiocyanate*

  In soil, metam sodium and dazomet degrade to release methyl isothiocyanate. These nematicides when enter into the nematode, release cyanide which prevents the utilization of oxygen, presumably carried by oxygen-transporting globins, and thus prevents respiration. The nematicides like metam sodium and metam potassium as liquid formulations and dazomet (Basanid) as granular formulations, come under this group. These are broad spectrum nematicides and are very effective but have certain limitations like inadequate volatility that results in uneven distribution in soil and also poor penetration in woody roots or tubers.

3. Application Method of Fumigants

Fumigants have to be applied very carefully in the soil as careless application might result in loss of chemicals in environment which would be toxic to other live organisms. The following precautions should be taken during application of fumigants:

- The soil should be very fine and smooth, free from clods, stones and crop residues, so that fumigants move within the soil without any hinderance.
- The soil temperature should be optimum (21-27 °C) in the rhizosphere zone of 15 cm depth, from the above ground level.
- The soil moisture should neither be too wet nor very dry. It should be near to the range of field capacity, so that the fumigants could spread freely between soil particles.
- During application, perforated tubes are used for uniform spread of fumes. Therefore, after application, all the holes or crevices should be closed tightly to prevent the escape of fumes.
- After application, the field should not be disturbed for at least a week. After one week the field should be ploughed deeply and left for proper aeration. The planting should be done after soil temperature cools down and attain normal temperature, which generally occurs after 2-3 weeks.

B. *Non-Fumigant Nematicides*

The non-fumigant nematicides, organophosphates and carbamate compounds were developed in early 1960s. These are systemic nematicides and do not volatilize easily (Rich et al., 2004). They are easily soluble in soil water and kill or paralyse the nematodes in soil. Due to their less phytotoxic effect, they are recommended for foliar applications. Their formulations, whether granules or wettable powder have an adverse effect on PPNs but do not eradicate their population. In case of soil application, they are absorbed by the roots and taken to the foliage and branches by acropetal manner, i.e., from roots to foliage. But these nematicides have some limitations like high toxicity to non-target organisms including humans and therefore banned in developed countries.

Non-fumigant nematicides can be used to manage PPNs in an economically feasible manner (Ji et al., 2020; Oka and Saroya, 2019). Moreover, they can be applied during the growing season when nematodes become problematic for the crop. Below is the brief description of organophosphates and carbamates under separate headings:

- *Organophosphates:* The basic constituents of organophosphates are carbon, hydrogen and chlorine, however certain compounds may have oxygen and sulphur. The nematicides belong to the group orgnophosphate are listed below (Table 2).

Table 2. List of organophosphates with their chemical and trade names

Common Name	Chemical name	Trade name
Parathion	O,O-diethyl O-(p-nitrophenyl)-phosphorothioate	Ekatex, Folidol, Thiophos
Dichlofenthion	O, (2,4-dichlorophenyl) O,O-diethyle phosphorothioate	VC-13, Hexanema
Thionazin	O,O-diethyl-o-2-pyrazinyl- phosphorothioate	Zinophos, Nemaphos
Fensulfothion	O,O-diethyl-O-p(methyl sulfinyl)-phosphorothioate	Terracur – P, Dasanit
Fenamiphos	Ethyly-4-(methylthio)-m-tolyl isopropyle phosphoramidate	Nemacur
Ethoprophos	O-ethyl S, S-dipropyl- phosphorothioate	Mocap
Phorate	O, O-diethyl S-(ethylthiomethyl)-phosphorothioate	Phorate, Thimet

Table 3. List of carbamates with their chemical and trade names

Common Name	Chemical name	Trade name
Metam sodium	Sodium N-methyl dithiocarbamate	Vapam, Sistan, Vitafume, Unifume
Aldicarb	2-methyl-2-(methylthio) propionaldehyde-O-(methylcarbamoyl) oxime	Temik
Carbofuran	2,3-dihydro-2,2-dimethyl 7-benzofuranyl methyl carbamate	Furadan
Methomyl	S-methyl N-[(methylcarbamoyl) thioacetamidate	Lannate
Oxamyl	Methyl N, M'-dimethyl-N-[(methylcarbamoyl) oxyl-1-thioxamimidate	Vydate, D-1410

- *Carbamates:* They are derivatives of carbamic acid, reversible cholinesterase inhibitors and are highly cholinergic. The affected organisms show violent convulsions and other neuromuscular disturbances due to inhibition of cholinesterase (Jonathan, 2010). The nematicides belong to carbamate group are listed in Table 3 and discussed briefly.
  a. Metam sodium: It is water soluble and can be applied as drench in irrigation water or through injection in soil. It produces gas after application and disperses through the soil particles. Metham sodium is also used as pre-plant application but highly phytotoxic, so there should be more than two weeks interval between application and planting.
  b. Aldicarb: It is a systemic nematicide. It can be used as seedling bare root dip treatment, nursery and field treatments. It is highly effective in managing root-knot nematodes in vegetable and tobacco crops under field conditions.
  c. Carbofuran: It is systemic nematicide for soil and foliar treatments (Tomlin, 2001). Carbofuran is effective against many ecto and endoparasitic nematodes infesting potato, tomato, brinjal and pepper (Damadzadeh and Haque, 1981; Overman and Jones, 1975; Krishnaprasad and Krishnappa, 1981; Rhodes, 1981). The chemical is also reported for its persistency for longer time (Abu-Elamayem et al., 1982).
  d. Methomyl: It is systemic in nature and very much effective against PPNs.
  e. Oxamyl: It is also systemic nematicide and when mixed in soil, degrade rapidly and become extremely toxic to nematodes (Wagenet, 1985; Hayes and Lawn, 1990; Meister, 1991). Oxamyl has the capacity to translocate within root system and to the upper parts of the plant (Radewald et al., 1970; Atilano and Van Gundi, 1979; Potter and Marks, 1976).

3. Methods of Nematicide Application

The chemicals can be applied on plants mainly by three methods for obtaining maximum benefits.

1. *Gaseous diffusion*
   This is the most effective method where the gas/ fumes are formed after application. It reaches the nematodes easily and cause mortility. For its application, either hand applicator or tractor driven equipments are used for small and large areas, respectively.
2. *Water dispersion*
   In this method, the nematicides are applied by mixing in water and drenching near the furrows or near the roots. This method is useful for the non-phytotoxic nematicides like DBCP and Metham Sodium. It is highly effective when applied on ornamentals or turf grasses for controlling nematodes.
3. *Mechanical mixing*
   This method is useful for the granular nematicides like carbamates. The nematicide is mixed with soil during fertigation and thus easy to apply.
4. *Time of treatment*
   In order to cause maximum mortality to PPNs, appropriate time for nematicide application should be determined very carefully.
5. *Pre-plant treatment*
   Since many nematicides are highly toxic, they are advised to apply before planting. After application, the planting is done after a waiting period of 3-4 weeks, to escape the toxic effects of chemicals. Pre-plant application of nematicide is important to protect the high value crops as the initial nematode population is very crucial for infection. If the plants get infected then it becomes very difficult to manage the population below damaging level and protect the plants. The pre-plant soil fumigation is highly effective method for managing root-knot nematode population (Qiao et al., 2015). Traditionally, pre-plant soil fumigation is the primary method to manage M. incognita in vegetable production (Lembright, 1990).
6. *Planting time treatment*
   The less phytotoxic chemicals, like organophosphates and carbamates are applied at the time of planting.
7. *Post plant treatment*
   This method is also applicable for less toxic chemicals like DBCP, organophosphates and carbamates. This type of treatment is generally given in case of perennial crops.

8. *Bare root dip treatment*
   To protect the plants from PPNs in transplanting crops, the roots of the seedlings are treated with chemicals like Thionazin and aldicarb at 500 ppm for 15 minutes. This treatment is very much beneficial to protect the plant from root-knot nematode infestation.
9. *Seed treatment*
   This method is very profitable, cheap and also popular among farmers. Wheat seeds coated with furathiocarb (@ 10 g a.i./kg seed), carbofuran (@10 g a.i./ kg seed) or oxamyl (3.6 g a.i./ kg seed) was found effective in the management of cereal cyst and root lesion nematodes in Northern Negev region of Israel (Orion and Shleven, 1989). Seed dressing with oxamyl controlled the population of cereal cyst nematodes (Brown, 1984) whereas, seed dressing with carbofuran @ 2 kg a.i./ ha reduced the population of cereal cyst nematodes and increased plant growth parameters under field condition (Kaushal et al., 2001).
10. *Foliar treatment*
    Foliar treatment is useful when the nematodes are found infecting the aboveground plant parts or the nematicide is systemic as the chemical can pass through the stems to the roots. The systemic nematicide, oxamyl is used for foliar treatment.
11. *Soil application*
    Phorate, carbofuran and phenamiphos are generally recommended for soil application and they have been found very much effective in reducing nemtode population in soil.

3.1. Types of Nematicide Application

a. *Broadcast application:* This method is generally applied when plants are at 50-60 cm apart.
b. *Row application:* This method of application saves the quantity of nematicides.
c. *Spot application:* The chemicals are applied at a particular spot of crop, such as rhizosphere or around the tree trunk.
d. *Strip application:* The plants in strip are treated with nematicides only. This method is useful in saving the quantity and cost of nematicides.

e. *Site application:* Application of nematicides at the site where the infections are too high.

3.2. Nematicides – Mode of Action

The fumigants and non-fumigants have different modes of action. However, both the nematicides penetrate the nematode cuticle as they are highly impermeable. The fumigants kill the nematodes while the non-fumigants are nematostatic and they can only change the behaviour of nematodes.

a. *Halogenated hydrocarbons*
   This group of chemical act on nematodes by causing inhibition of certain enzymes by chemical reaction of relative halide with some required nucleophilic centre, such as sulfydryl (SH), amino (NH2) or dihydroxyl groups (OH) present in enzymes. These compounds induce hyperactivity that leads to paralysis and ultimately death of the nematode. The mortality also depends upon the product of concentration and time of exposure (Cremlyn, 1979).
b. *Alkyl isothiocynates*
   The isothiocynate group of chemicals act with nucleophilic centres i.e., thiol groups in vital enzymes of nematodes. The reaction causes hyperactivity followed by a gradual decrease in activity leading to eventual paralysis (Evans and Thomason, 1971).
c. *Organophosphates and carbamates*
   These nematicides inhibit cholinesterase with consequent disruption of nervous activity caused by accumulation of acetylecholine at the nerve endings. During cholinesterase inhibition, a complex between acetyle-choline and the phosphate or carbamate is formed; the complex reacts to deliver the phosphorylated or carbamylated enzyme and the original enzyme are formed by a hydrolysis reaction called dephosphorylation or decarbamylation. Organic phosphates in general cause irreversible enzyme inhibition whereas, carbamates are reversible inhibitors. Organophospahtes and carbamates impair neuromuscular activity interfering with the movement and invasion and probably also with hatching and feeding with consequent effects on development and reproduction (Wright et al., 1980). Due to inhibition of acetylcholinestrase, nematodes get paralyzed (Chitwood, 2003).

4. Factors Affecting Efficacy of Nematicides

The physiology and biology of nematodes are programmed for the environment extremities. The cuticle, cyst wall, seed galls, gelatinous matrix and egg cell are the structures which protect them from any adverse conditions. Moreover, soil qualities also affect the efficacy of nematicides. The other factors which have influence over the efficacy of nematicides are given below:

a. *Type of nematicide:* Fumigant and non fumigant nematicides have different mode of behaviour. The fumigants are in liquid state and after application, form gas. If gaseous phase is more, it is more effective because of high diffusion. On the other hand, if liquid phase is more, it is less effective due to low diffusion. The non fumigant nematicides require proper mixing, solubilize with soil water and then come in contact with the nematodes and other soil microorganisms.
b. *Soil type and organic matter:* Efficacy of nematicides is affected by soil type wherein the nematicides are applied and evaluated (Oka et al., 2013). The nematicides perform better in sandy soil because nematodes move easily there, whereas in clay soil, movement of gas or liquid is hindered due to very less space. High organic content of the soil also obstruct the movement of nematicides as it adsorbs the chemicals. Adsorption of nematicides by soil particles may result in reduced concentration or unavailability of the nematicides in the soil solution, thus reducing their efficacy against target nematodes (Morris et al., 2018). Variation in soil parameters and environmental factors may also be responsible for failure of the nematicides in managing nematode population (Oka et al., 2013).
c. *Soil moisture and temperature:* Soil moisture is an important factor which has an impact on the efficacy of nematicides. If moisture is very less, the movement is so quick that it spreads to distant places from the point of injection and if soil moisture is more than sufficient, it hinders the movement and thus become less effective. Hence, there should be optimum moisture in soil.
   Soil temperature is also very effective in increasing or decreasing the efficacy of nematicides. If temperature is high, the evaporation, hydrolysis and degradation of chemicals become high. This results in easy killing of nematodes. Generally fumigants are applied between the temperature range of 10-30 oC. Metham sodium is not recommended for cold soil because its degradation becomes very

low which reduce its efficacy. Excess soil moisture and low temperature increase the risk of phytotoxicity. At high soil temperature, the nematicides are in contact with nematodes for a longer time, which result in high mortality of nematodes (Haydock et al., 2012; Jones and Norris, 1998).

d. *Soil inhabiting microorganisms:* These organisms also influence the pesticide degradation (Smelt and Leistra, 1992). The efficacy of nematicides varies greatly in field conditions especially when microbial degrading bacteria are present. (Haydock et al., 2012; Norshie et al., 2017).
e. *Application time, method and planting time:* Longer time period between application of chemicals and planting time reduces the degree of phytotoxicity.

Conclusion

Most of the nematicides are not suitable for human and animal health and they are toxic to the other organisms of soil and environment. They may deplete the ozone layer and contaminate the ground water (Thomas, 1996; Chitwood, 2002). Because of long persistence in soil, the carbamates and organophosphates create toxic effects on the beneficial organisms, which in turn change the biosphere of the soil to the adverse side. Methyl bromide deplete ozone layer (Bell et al., 1996; Ristaino and Thomas, 1998) and therefore this chemical has been included in the list of substances having high ozone-depleting potential (UNEP, 1992). The non-fumigant nematicides such as carbamates and organophosphates have highly toxic chemicals (Gupta, 2011). They pose lesser side effects on the environment as compared to fumigants (Watson and Desaeger, 2019; Wram and Zasada, 2019). The carbamate group of pesticides are found in high quantity in ground water (Pionke and Glotfelty, 1989; Pionke et al., 1998; Garcia et al, 2001).

If the nematicides are used continuously in a field infested with root-knot nematodes, the nematodes develop resistance over the chemicals and also the nematicides are biodegraded by the soil micro-organisms (Huang et al., 2019; Meher et al., 2009). The development of resistance may vary with specific host, nematode species and strain, soil type and other related factors. Moreover, continuous use of one nematicide will affect the soil health adversely (Thoden et al., 2020; Waldo et al., 2019). The demerit with chemical application is that they reduce the population of beneficial organisms

(Sanchez-Moreno et al., 2010). Carbofuran residues were detectable in soil for 105 days when applied @ 2 kg a.i./ ha. The residues persisted for 90 days in plants and it increased gradually till 75 days in roots. In the shoots, maximum concentration was achieved in 7 days after germination (Meher et al., 1985). Fumigants have a broad spectrum of activity, often reduce populations of beneficial soil organisms (Sanchez-Moreno et al., 2010). They have frequently been associated with rapid and more prolific reinfestation of treated soil with plant-parasitic nematodes (Mazzola and Manici, 2012). Nematicide persistence in soil for 6-8 weeks is desired for effective plant protection (Karpouzas et al., 1999). The residues of carbosulfan and phorate persisted up to 90 days and cadusafos and triazophos beyond 120 days. Carbosulfan was rapidly converted into carbofuran (Meher et al., 2010). Residues of carbosulfan (de melo Plese et al., 2005) have been reported to persist less than cadusafos (Meher et al., 2005), triazophos (Li et al., 2008) and phorate (Das et al., 2003). Carbofuran is widely detected in groundwater and surface water (Pionke and Glotfelty, 1989; Pionke et al., 1998; Garcia et al., 2001; Parrilla, 1994; Chiron et al., 1993, 1995) because the degradation time of carbofuran has been found 7-10 times faster in alkaline soil (pH 7.9) than in acidic or neutral soil (pH 4.3-6.8) (Getzin, 1973).

To control this release in water, the best way is to apply it through controlled release system (CRS). It facilitates a gradual and controlled discharge of the pesticides that reduce the losses by evaporation and leaching, preventing both air and water pollution. For this system, controlled release formulations (CRFs) have been made. It has several advantages like safety to user, safety to non-target organisms in addition to least leaching and lesser amount used (Bahadir and Pfister, 1990; Wilkins, 1995; Hickman, 1999). Many CRS have been developed for use in agriculture (Ferna´ndez-Pe´rez et al., 1998; El-Nahhal, 2001).

The chemical methods do not only control the nematodes, but also manage the population of fungi, bacteria and weed seeds. Although the nematodes cannot be eradicated but its number can be lowered down to minimum level where it cannot cause damage to crops beyond the economic threshold. Nematicides are very few which are being used. It is the quickest means of managing the nematodes to reduce the population drastically in a very short span of time. The results are so spectacular that farmers mostly use the chemical methods without thinking the negative impacts of the chemical application. The nematicides are very useful for instant result and increase in yield but the residual toxicity in the air and water are numerous. Therefore, they are not always advisable until there is an emergency to protect the crops.

There is a need to focus on developing nematicides that should be less toxic, less persistent in soil, yet, harmful for the plant parasitic nematodes. Growers should use nematicides very carefully and judiciously. In the small level application, like in seed bed or seed treatment or seedling treatment, focus should be on economic and less toxicity of nematicides. If proper care is taken and nematicides are used judiciously, it will protect the crop and benefit the crop growers.

Acknowlegments

The author is indebted to all those who have supported and provided relevant information present in this chapter.

References

Abd-Elgawad, M. M. M. and Askary, T. H. (2015). Impact of phytonematodes on agriculture economy. In: *Biocontrol Agents of Phytonematodes.* (Eds. Askary, T. H. and Martinelli, P. R. P.). CAB International, Wallingford, UK, pp. 3-49.

Abu-Elamayem, M. M., El-Sebae, A. A., Rabie, R. S., Tantawy, G. A. and Ibrahim, I. K. A. (1982). Field evaluation of certain nematicides for the control of citrus nematode, *Tylenchulus semipenetrans* Cobb. *Beitrage Zur Tropischen Landwirtschaft und Veterinarmedizin*, 20: 85-87.

Askary, T. H. (2015). Limitations, research needs and future prospects in the biological control of phytonematodes. In: *Biocontrol Agents of Phytonematodes* (Eds. Askary, T. H. and Martinelli, P. R. P.). CAB International, Wallingford, UK, pp. 446-454.

Askary, T. H. (2017). Diversity of plant parasitic nematodes in pulses. In: *Plant Diversity: Present Situation and Future Scenario* (Eds. Ansari, A. A., Gill, S. S., Abbas, Z. K. and Naeem, M.). CAB International, Wallingford, UK, pp. 239-274.

Askary, T. H. and Ali, S. S. (2012). Effect of *Meloidogyne javanica* and *Fusarium udum* singly and concomitantly on wilt resistant accessions of pigeon pea. *Indian Journal of Plant Protection*, 40: 167-170.

Askary, T. H. and Martinelli, P. R. P. (2015). *Biocontrol Agents of Phytonematodes.* CAB International, Wallingford, UK.

Atilano, R. A. and Van Gundy, S. D. (1979). Systemic activity of oxamyl to *Meloidogyne javanica* at the root surface of tomato plants. In: *Soil Root Interface* (Eds. Harley, J. L. and Russell, R. S.) Academic Press, London. pp. 339-349.

Bahadir, M. and Pfister, G. (1990). Controlled release formulations of pesticides. In: *Controlled Release, Biochemical Effects of Pesticides, Inhibition of Plant Pathogenic Fungi. Chemistry of Plant Protection Vol. 6* (Eds. Haug, G. and Hoffmann, H.). Springer, Heidelberg, Germany, pp. 1-64.

Bell, C. H., Price, N. and Chakrabarti, R. (1996). *The Mthyl Bromide Issue*. John Wiley & Sons, New York, USA.

Bessey, E. A. (1911). Root-knot and its control. *USDA Bulletin*. 217: 89.

Brown, R. H. (1984). Cereal cyst nematode and its chemical control in Australia. *Plant Disease*, 68: 922-928.

Carter, W. (1943). A promising new soil amendment and disinfectant. *Science*, 97: 383-384.

Chiron, S., Ferna´ndez Alba, A. and Barcelo, D. (1993). Comparison of online solid-phase disk extraction to liquid-liquid extraction for monitoring selected pesticides in environmental waters. *Environmental Science and Technology*, 27: 2352-2354.

Chiron, S., Valverde, A., Fernandez-Alba, A. and Barcelo, D. (1995). Automated sample preparation for monitoring groundwater pollution by carbamate insecticides and their transformation products. *Journal of AOAC International*, 78: 1346-1352.

Chitwood, D. J. (2002). Phytochemicals based strategies for nematode control. *Annual Review of Phytopathology*, 40: 221-249.

Chitwood, D. J. (2003). Nematicides. In: *Encyclopedia of Agrochemicals Vol. 3* (Ed. Plimmer, J. R.). John Wiley and Sons, New York, USA, pp. 1104-1115.

Cremlyn, R. J. W. (1979). *Pesticides, Preparation and Mode of Action*. John Wiley and Sons, New York, USA.

Crow, W., Weingartner, D. and Dickson, D. (2000). Effects of potato-cotton cropping systems and nematicides on plant-parasitic nematodes and crop yields. *Journal of Nematology*, 32: 297–302.

Crow, W., Weingartner, D., Dickson, D. and McSorley, R. (2001). Effect of sorghum-sudangrass and velvetbean cover crops on plant-parasitic nematodes associated with potato production in Florida. *Journal of Nematology*, 33: 285–288.

Damadzadeh, M. and Haque, N. G. M. (1981). The efficacy of carbofuran against potato cyst nematode, *Globodera rostochiensis. Revue de Nematologie*, 4: 121-124.

Das, A. C., Chakravarty, A., Sukul, P. and Mukherjee, D. (2003). Influence and persistence of phorate and carbofuran insecticides on microorganisms in rice field. *Chemosphere*, 53: 1033–1037.

de Melo Plese, L. P., Paraiba, L. C., Foloni, L. L. and Pimentel Trevizan, L. R. (2005). Kinetics of carbosulfan hydrolysis to carbofuran and the subsequent degradation,of this last compound in irrigated rice fields. *Chemosphere*, 60: 149–156.

Duniway, J. M. (2002). Status of chemical alternatives to methyl bromide for pre-plant fumigation of soil. *Phytopathology*, 92: 1337–43.

El-Nahhal, Y., Undabeytia, T., Polubesova, T., Mishael, Y. G., Nir, S. and Rubin, B. (2001). Organo-clay formulations of pesticides: Reduced leaching and photodegradation. *Applied Clay Science*, 18: 309-326.

Fernandez-Perez, M., Gonzalez-Pradas, E., Urena-Amate, M. D., Wilkins, R. M. and Lindup, I. (1998). Controlled release of imidacloprid from a lignin matrix: Water release kinetics and soil mobility study. *Journal of Agricultural and Food Chemistry*, 46: 3828-3834.

Garcıa de Llasera, M. P. and Bernal-Gonzalez, M. (2001). Presence of carbamate pesticides in environmental waters from the northwest of Mexico: Determination by liquid chromatography. *Water Research*, 35: 1933-1940.

Getzin, L. W. (1973). Persistence and degradation of carbofuran in soil. *Environmental Entomology*, 2: 461-467.

Gilreath, J. P., Santos, B. M., Motis, T. N. (2008). Performance of methyl bromide alternatives in strawberry. *HortTechnology*, 18: 80–83.

Gupta, R. C. (2011). Toxicology of Organophosphate and Carbamate Compounds. Elsevier Academic Press, USA.

Hajihassani, A., Davis, R. F. and Timper, P. (2019). Evaluation of selected nonfumigant nematicides on increasing inoculation densities of *Meloidogyne incognita* on cucumber. *Plant Disease*, 103: 3161–3165.

Haydock. P. P. J., Ambrose, E. L., Wilcox, A. and Deliopoulos, T. (2012). Degradation of the nematicide oxamyl under field and laboratory conditions. *Nematology*, 14: 339-352.

Hayes, W. J. and Laws, E. R. (1990). *Handbook of Pesticide Toxicology*, Vol. 3, *Classes of Pesticides*. Academic Press, New York, USA.

Hickman, M. V. (1999). Controlled-release formulations from cornstarch. In: *Controlled-Release Delivery Systems for Pesticides* (Ed. Scher, H. B.). Marcel Dekker, New York, USA, pp. 153-172.

Huang, B., Wang, Q., Guo, M., Fang, W., Wang, X., Wang, Q., Yan, D., Ouyang, C., Li, Y. and Cao, A. (2019). The synergistic advantage of combining chloropicrin or dazomet with fosthiazate nematicide to control root-knot nematode in cucumber production. *Journal of Integrative Agriculture*, 18: 2093–2106.

Ji, X., Li, J., Dong, B., Zhang, H., Zhang, S. and Qiao, K. (2019). Evaluation of fluopyram for southern root-knot nematode management in tomato production in China. *Crop Protection*, 122: 84-89.

Ji, X., Li, J., Meng, Z., Li, N., Dong, B., Zhang, S. and Qiao, K. (2020) Fluopimomide effectively controls *Meloidogyne incognita* and shows a growth promotion effect in cucumber. *Journal of Pest Science*, 93: 1421-1430.

Jonathan, E. I. (2010). *Nematology: Fundamentals and Applications*. New India Publishing Agency, New Delhi, India.

Jones, R. L. and Norris, F. A. (1998). Factors affecting degradation of aldicarb and ethoprop. *Journal of Nematology*, 30: 45-55.

Karpouzas, D. G., Ginnakou, I. O., Walker, A. and Gowen, S. R. (1999). Reduction in biological efficacy of ethoprophos in a soil from Greece due to enhanced biodegradation: comparing bioassay with laboratory incubation data. *Pesticide Science*, 55: 1089–1094.

Kaushal, K. K., Sharma, G. L. and Paruthi, I. J. (2001). Nematode diseases of wheat and barley and their management. *National Congress on Centenary of Nematology in India – appraisal and future plans* held from December 5-7, 2001 at Division of Nematology, Indian Agricultural Research Institute, New Delhi, India, pp. 23-24.

Krishnaprasad, K. S. and Krishnappa, K. (1981). Effect of bare root dip treatments with pesticides on *Rotylenchulus reniformis* affecting brinjal. *Indian Journal of Nematology*, 11: 154-158.

Kuhn, J. (1881). Die Ergebnisse de Versuche zur Ermittelung der Ursache der Rubenmudigkeit und zur Erforschung der Natur der Nematoden [*The results of the experiments to determine the cause of the*

Rubenmudigkeit and to study the nature of nematodes]. Ber. Physiol. Lab. Landwirtsh. Inst. Univ. Halle 3, pp.153.

Lembright, H. W. (1990). Soil fumigation: principles and application technology. *Journal of Nematology*, 22: 632-644.

Li, W., Qiu, S. P. and Wu, Y. J. (2008). Triazophos residues and dissipation rates in wheat crops and soil. *Ecotoxicology and Environmental Safety*, 69: 312–316.

Matthews, D. J. (1919). Report of the work of the W. B. Randall Research Assistant from October 1st, 1918 to December 31st, 1919. Report. Experiment. Research. Station. *Cheshunt*, 5: 18-21.

Mazzola, M. and Manici, L. M. (2012). Apple replant disease: role of microbial ecology in cause and control. *Annual Review of Phytopathology*, 50: 45–65.

McKenry M. V. (1994). Nematicides. In: *Encyclopedia of Agriculture Science* (Ed. Arntzen, C. J.). Academic Press Incorporation, California, USA, 3: 87–95.

Meher, C. H., Gajbhaiye, V. T., Singh, G., Kamra, A. and Chawla, G. (2010). Persistance and nematicidal efficacy of carbosulfan, cadusafos, phorate and triazophos in soil and uptake by chickpea and tomato crops under tropical conditions. *Journal of Agricultural and Food Chemistry*, 58: 1815-1822.

Meher, H. C., Agnihotri, N. P., and Sethi, C. L. (1985). Persistence of carbofuran in soil and pea crop and its bioefficacy against soil nematodes. *Indian Journal of Nematology*, 15: 155-158.

Meher, H. C., Gajbhiye, V. T., Chawla, G. and Singh, G. (2009). Virulence development and genetic polymorphism in *Meloidogyne incognita* (Kofoid & White) Chitwood after prolonged exposure to sublethal concentrations of nematicides and continuous growing of resistant tomato cultivars. *Pest Management Science*, 65: 1201–1207.

Meher, H. C., Gaur, H. S. and Singh, G. (2005). Persistence of cadusafos in alluvial soil under tomato crop. *Pesticide Reserch Journal*, 17: 85–87.

Meister, R. T. (1991). *Farm Chemicals Handbook '91*. Meister Publishing Company, Ohio, USA.

Morris, K. A., Li, X., Langston, D. B., Davis, R. F., Timper, P. and Grey, T. L. (2018). Fluensulfone sorption and mobility as affected by soil type. *Pest Management Science*, 74: 430–437.

Noling, J. W. (2016). Nematode Management in Strawberries. University of Florida publication Series no. ENY-031, USA, pp. 12.

Norshie, P. M., Grove, I. G. and Back, M. A. (2017). Persistence of the nematicide fluensulfone in potato (*Solanum tuberosum* ssp. *tuberosum*) beds under field conditions. *Nematology*, 19: 739–747.

Oka, Y. and Saroya, Y. (2019). Effect of fluensulfone and fluopyram on the mobility and infection of second-stage juveniles of *Meloidogyne incognita* and *M. javanica*. . *Pest Management Science*, 75: 2095-2106.

Oka, Y., Shuker, S. and Tkachi, N. (2013). Influence of some soil environments on the nematicidal activity of fluensulfone against *Meloidogyne javanica*. *Pest Management Science*, 69: 1225–1234.

Orion, D. and Shlevin, E. (1989). Nematicide seed dressing for cyst and lesion nematode control in wheat. *Supplimentary Journal of Nematology*, 21: 629-631.

Overman, A. J. and Jones, J. P. (1975). Carbofuran in a system approach to pest management in tomato. *Nematropica*, 5: 26-27.

Parrilla, P., Martınez-Vidal, J. L., Martınez-Galera, M. and Frenich, A. G. (1994). Simple and rapid screening procedure for pesticides in water using SPE and HPLC/DAD detection. *Fresenius Journal of Analytical Chemistry*, 350: 633-637.

Pionke, H. B. and Glotfelty, D. E. (1989). Nature and extent of groundwater contamination by pesticides in agricultural watershed. *Water Research*, 23: 1031-1037.

Pionke, H. B., Glotfelty, D. E., Lucas, A. D. and Urban, J. B. (1998). Pesticide contamination of groundwaters in the Mahantango Creek Watershed. *Journal of Environmental Quality*, 17: 76-84.

Potter, J. W. and Marks, C. F. (1976). Persistence of activity of oxamyl against *Heterodera schachtii* on cabbage. *Journal of Nematology*, 8: 35-38.

Powell, N. T. and Nusbaum, S. J. (1960). The black shank root knot complex in flue cured tobacco. *Phytopathology*, 50: 899-906.

Qiao, K., Wang, Z., Wei, M., Wang, H., Wang, Y. and Wang, K. (2015). Evaluation of chemical alternatives to methyl bromide in tomato crops in China. *Crop Protection*, 67: 223–227.

Radewald, J. D., Shibuya, F., Nelson, J. and Bivens, J. (1970). Nematode control with 1410, an experimental nematicide-insecticide. *Plant Disease Reporter*, 54: 187-190.

Rhodes, H. L. (1981). Effect of nematicides on yield of pepper in *Belonolaimus longicaudatus* infested soil. *Nematropica*, 11: 121-127.

Rich, J. R., Dunn, R. A. and Noling, J. W. (2004). Nematicides: past and present uses, In: *Nematology: aAdvances and Perspectives. Vol 2: Nematode Management and Utilization* (Eds. Chen, Z. X., Chen, S. Y. and Dickson, D. W.). CAB International,Wallingford, UK, pp. 1179–1200.

Ristaino, J. B. and Thomas, W. (1997). Agriculture, methyl bromide, and the ozone hole: can we fill the gaps. *Plant Disease* 81: 964–77.

Sanchez-Moreno, S., Jimenez, L., Alonso-Prados, J. L., Garcia-Baudin, J. M. (2010). Nematodes as indicators of fumigant effects on soil food webs in strawberry crops in Southern Spain. *Ecological Indicators*, 10: 148–156.

Smelt J. H. and Leistra M. (1992). Availability, movement and (accelerated) transformation of soil-applied nematicides. In: *Nematology From Molecule To Ecosystem* (Eds. Gommers, F. J. and Maas P. W. T.). Dekker and Huisman, Netherlands, pp. 266–280.

Thoden, T. C., Alkader, M. A. and Wiles, J. A. (2020). Biological attributes of Salibro, a novel sulfonamide nematicide. Part 2: impact on the fitness of various non-target nematodes. *Nematology*, 22: 1–17.

Thomas, W. B. (1996). Methyl bromide: effective pest management tool and environmental threat. *Journal of Nematology (S),* 28: 2586-2590.

Tomlin, C. (2001). *The Pesticide Manual*. British Crop Protection Council, UK.

UNEP (1992). Synthesis report of the methyl bromide interim scientific assessment and methyl bromide interim technology and economic assessment. In: *Montreal Protocol Assessment Supplement* (Eds. Watson, R. T., Albritton, D. L., Anderson, S. O., Lee-Bapty, S. and UNEP). United Nations Environmental Program (UNEP), New York, USA, pp. 33.

Wagenet, L. P. (1985). *A Review of Physical-Chemical Parameters Related to the Soil and Groundwater Fate of Selected Pesticides in New York State*. Cornell University Agricultural Experiment Station, New York State College of Agriculture and Life Sciences. Ithaca, New York.

Waldo, B. D., Grabau, Z. J., Mengistu, T. M. and Crow, W. T. (2019). Nematicide effects on non-target nematodes in bermudagrass. *Journal of Nematology*, 51: 1–12.

Watson, T. T, Desaeger, J. A. (2019). Evaluation of non-fumigant chemical and biological nematicides for strawberry production in Florida. *Crop Protection*, 117: 100–107.

Weingartner, D., McSorley, R. and Goth, R. (1993). Management strategies in potato for nematodes and soil-borne diseases in subtropical Florida. *Nematropica*, 23: 233–245.

Weingartner, D. P. and Shumaker, J. R. (1990a). Control of nematodes and soil-borne diseases in Florida potatoes with aldicarb and 1,3-D. *Journal of Nematology*, 22: 775–778.

Weingartner, D. P. and Shumaker, J. (1990b). Effects of soil fumigants and aldicarb on bacterial wilt and root-knot nematodes in potato. *Journal of Nematology*, 22: 681–688.

Wilkins, R. M. (1995). Controlled release systems: Prospects for the future. In: *Proceeding of the 8th International Congress of Pesticide Chemistry* (Eds. Ragsdale, N. N., Kearney, P. C. and Plimmer, J. R.). American Chemical Society: Washington, DC, pp. 96-103.

Wram, C. and Zasada, I. A. (2019). Short-term effects of sub-lethal doses of nematicides on *Meloidogyne incognita*. *Phytopathology*, 109: 1605–1613.

Wright, D. J., Blyth, R. K. and Pearson, P. E. (1980). *Annual Applied Biology*, 96; 323-334.

Part II. Microbials in Pest Management

Chapter 7

Prospects of Biopesticides in Pest Management

Tarique Hassan Askary[1,*], Aasha Rana[2], Sumaya Mehraj[3], Tabassum Ara Khanum[4], Ulfat Jan[1], Summira Rafiq[1], Waseem Ahmad War[1] and Nune Akhil[1]

[1]Division of Entomology,
Faculty of Agriculture,
Sher-e-Kashmir University of Agricultural Sciences
and Technology of Kashmir, Jammu and Kashmir, India
[2]Department of Zoology,
Shrimati Kamla Devi Gauri Dutta Mittal PG College,
Churu, Rajasthan, India
[3]Department of Zoology,
University of Kashmir, Srinagar, Jammu & Kashmir, India
[4]National Nematological Research Centre,
University of Karachi, Karachi, Pakistan

Abstract

Biopesticides are products and by-products of naturally occurring substances that include plants and semiochemicals, microorganisms such as bacteria, fungi and nematodes. They are potential alternatives to synthetic pesticides, easily biodegradable, less expensive and generally less or non-toxic to humans and non-target organisms. The global market for biopesticides is valued at US$ 3.0 billion sharing 5% of the global pesticide market. Thus, their role is of much importance for crop protection in the agricultural-based economy of the world. Biopesticides are advantageous for safe environment and healthy food for human

* Corresponding Author's Email: tariq_askary@rediffmail.com.

In: Pest Management: Methods, Applications and Challenges
Editor: Tarique Hassan Askary
ISBN: 979-8-88697-268-9

consumption. However, limitations in complete adoption of biopesticides as pest management options are due to several factors such as lack of sufficient facilities and capital for their production especially in the developing countries, challenges faced in their formulation, registration and commercialization. Lack of awareness about biopesticides especially among the small-scale growers, stake holders and policy makers are other such hurdles. Therefore, there is a need of strong collaboration among research practitioners, industry and farmers in order to develop high quality products and to gain much acceptance in sustainable agriculture. Growers should also have prior information of the organisms they are utilizing as biopesticides and the target pest. Further research is required to guarantee stable and durable products by effective crop pest management under field conditions. Additionally, specific knowledge on optimizing application time, field rates, and application intervals will certainly help to include biopesticides tactfully as one of the components in integrated pest management programme.

Keywords: biopesticides, microorganisms, bacteria, fungi, nematode, limitations

1. Introduction

Excessive use of synthetic pesticides has led to several environmental and human health problems like ground water contamination, high and acute residual toxicity, detrimental effects on the natural enemies of the pests and hormonal imbalance. As reported by world Health Organization (WHO), 200,000 people are killed globally every year as a direct result of pesticide poisoning. Moreover, the repetitive use of synthetic pesticides upset the balance of nature through disruption of pollinators, other wild life and natural enemies that in turn lead to outbreaks of secondary pests (Dubey, et al., 2011; Pretty, 2009; Feng and Zheng, 2007; Khater, 2011). According to World Resources Institute, more than 500 insect and mite species have shown resistance to one or more insecticides (WRI, 1994) and the situation has worsened to such an extent that application rates of insecticides are increased double and triple (Stoll, 2000).

Biopesticides are pest management agents manufactured from living micro-organisms or natural products. They have been gaining attention particularly since the last two decades as they do not have any residue problem, generally pose no threat to the environment and can be effective in a very small quantity. They are also efficacious against those target pests which have developed resistance to synthetic chemical pesticides. The value of biopesticides in the world market is US$ 3.0 billion (Marrone, 2014) and they account 5% of the global pesticide market. It is estimated that the market share of biopesticides will equal that of synthetic pesticides between 2040 and 2050 (Olson, 2015; Dalmas and Koutroubas, 2018).

Biopesticides are obtained from organisms that include plants, bacteria, fungi, nematodes and other microbes. Based on active substance, biopesticides may fall into three different types that include (i) microorganisms, (ii) biochemicals and (iii) semiochemicals. However, The United States Environmental Protection Agency also classes some transgenes as biopesticides. Microbial biopesticides can reproduce on or in close vicinity to the target pest, giving an element of self-perpetuating control. Moreover, they do not require any specialized equipment for application and can be applied with farmers' existing spray equipment (Chandler et al., 2011). Additionally, biopesticides do not present the same regulatory problems as compared to chemical pesticides. They pose fewer risks than conventional pesticides, and that's why authorities generally require fewer data for their registration. Environmental Protection Agency (EPA) in the United States of America generally takes less than a year to register new biopesticides as compared to conventional pesticides where the average time is more than 3 years. By combining synthetic pesticide performance and environmental safety, biopesticides execute efficaciously with the tractability of minimum application limitations and with superior resistance management potential (Kumar, 2012; Senthil-Nathan, 2013).

The aim of this chapter is to highlight the challenges and opportunities that facilitate or hinder the utilization of biopesticides in Integrated Pest Management (IPM) programme. However, our focus is mainly on microbial pesticides and a little on bio-derived chemicals.

2. Microbial Pesticides

Microbial pesticides consists of fungi, bacteria, entomopathogenic nematodes, slug parasitic nematodes and viruses.

2.1. Fungi

2.1.1. Entomopathogenic Fungi

They comprise a wide range of fungal taxa that has the ability to infect, kill and digest insects. Infection begins with the adhesion of spores to the surface of the insect cuticle. During the process extracellular polysaccharides, lectins and some extracellular enzymes play an important role (Clarkson et al., 1998). Adhesion to the surface of the host allows further fungal development (St. Leger, 1993). The insect cuticle topography also plays a significant role in adhesion, germination and differentiation of appressoria (Butt, 2002).

Table 1. Some commercial products of entomopathogenic fungi, *Metarhizium anisopliae* and *Beauveria bassiana*

Fungus	Product	Formulation	Manufacturer	Country
Metarhizium anisopliae	Metatech®	Wettable Powder	Dudutech Division, Flamingo Horticulture (K) Ltd., Naivasha	Kenya
	Green Heal Meta	Liquid	Greenlife Biotech Laboratory	India
	BioBlast	Wettable Powder	Paragon Professional Pest Control Products	USA
	Bio Metaz	Liquid	Pioneer Agro Industry	India
	Sun Bio Meta	Liquid	*Sonkul Agro* Industries Private Limited	India
	Utkarsh Metarhoz	Liquid	Utkarsh Agrochem Pvt. Ltd.	India
	Dr. Bacto's Meta	Liquid	Anand Agro Care	India
Beauveria bassiana	Beauvitech®	Wettable Powder	Dudutech Division, Flamingo Horticulture (K) Ltd., Naivasha	Kenya
	Botanigard ES	Wettable Powder	LAM International Corp, USA 117 South Parkmont, Butte, MT59701	USA
	Dr. Bacto's Brave	Liquid	Anand Agro Care	India
	Green Beauveria	Liquid	Greenlife Biotech Laboratory	India

In case of *Verticillium lecanii* (Zare and Gams, 2001), the germ tubes directly penetrate the insect cuticle, whereas *Metarhizium anisopliae* and *Beauveria bassiana* form special penetration structures known as appressoria (Bidochka et al., 2000). The appressorium firmly links the fungus to the epicuticle of the insect and allows cuticle penetration through mechanical and enzymatic processes (Clarkson et al., 1998). The fungal secretion such as proteases and many other enzymes aids in the penetration of insect cuticle (Charnley, 1997; Butt, 2002). The fungus reaches the subcuticular epidermis and subsequently invades the insect's haemocoel (Clarkson et al., 1998). Finally, the host dies due to invasion of their vital organs, tissue destruction, lack of nutrients, or by the secretion of toxins by the invading fungus (Bidochka et al., 2000; Butt, 2002). The spores produced by the fungus within the insect cadaver have the ability to resist under unfavourable conditions

(Charnley, 1997; Butt, 2002). When the moisture is adequate, the fungus breaks the insect cuticle, produce aerial spores, which are dispersed by the wind, rain or by other arthropods. Thus, process of new infection begins upon dispersal of spores.

Several studies reveal that *Metarhizium* sp. increase plant growth when applied as a soil inoculation or seed treatment such as in case of onion (Maniania et al., 2003), tomato (Elena et al., 2011) and soyabean seedlings during salt stress (Khan et al., 2012). *M. anisopliae* has a wide host range whereas, *M. acridum* are specifically virulent against locusts and grasshoppers (Gottar et al., 2006), and are used as biological control agent against these two pests in Africa, Australia and Mexico.

In the past few years, various entomopathogenic fungi have been developed commercially and successfully exploited against insect pests in different countries of the world (Table 1). Though, slow mode of action, high costs and inconsistent results are still a challenge which needs to be solved in the coming days.

2.1.2. Nematophagous Fungi

They are found in soil ecosystem, produces trapping structures of various shapes and sizes such as constricting rings, two-dimensional or three-dimensional adhesive networks, adhesive knobs and adhesive hyphae (Askary, 1996). The nematodes moving in the soil are entrapped by these devices, killed and eventually digested by the fungus (Askary, 2015a). Royal 300®, the commercial formulation of nematode-trapping fungi, *Arthrobotrys robusta var. antipolis* has been successfully used for the management of the nematode, *Ditylenchus myceliophagus* in mushroom (Cayrol et al., 1978).

Some fungi are endoparasitic that produce spores (conidia or zoospores). The spores are ingested by the migrating nematodes together with the food or they adhere to the body surface of nematode (Askary. 1996). The encystment of spores take place around the natural orifices of nematode such as mouth, anus or vulva, the vegetative stage of fungus begins with the penetration of the host body by the encysted zoospores (Askary, 2015a; Askary and Martinelli, 2015). These fungi are reported to work naturally in soil ecosystem but unfortunately not commercialized successfully. However, alginate pellets of endoparasitic fungus, *Hirsutella rhossiliensis* was prepared by some research workers and successfully utilized against plant parasitic nematodes in laboratory and green house conditions (Lackey et al., 1993; Jaffee et al., 1996).

Egg and female-parasitic fungi use appressoria or zoospores to infect the eggs and sedentary females of plant-parasitic nematodes (Lopez-Llorca et al.,

2008). They are found in the rhizosphere, can survive saprophytically and are easy to mass culture (Moosavi and Zare, 2012). Several fungi belong to this group, however, *Paecilomyces lilacinus*, *Trichoderma harzianum*, *Pochonia chlamydosporia* and *Aspergillus niger* have drawn much attention to researchers and have been commercialized and successfully exploited against endoparasitic nematodes such as *Meloidogyne* and *Heterodera* species (Askary and Abd-Elgawad, 2019).

P. lilacinus enters its hyphae into the vulva and anus of the nematode (Jatala et al., 1979). After penetrating the egg, fungal hyphae grow rapidly and develop profusely inside and over the eggs and destroy the juvenile within it. In advanced stage, a large number of conidiophores are produced and the hypha moves towards adjacent eggs (Khan et al., 2006; Nasr Esfahani and Ansari Pour, 2006). *P. lilacinus* has been commercialized very successfully by different countries of the world. Some well-known formulations are BIOCON produced by Phillipines, Soybean Root Bio-Protectant, a product of China, Pl Plus®, a product of South Africa, Yorker, Nematofree, PAECILO®, Gmax bioguard and PAECILO™ are the products of Indian companies. *P. lilacinus* strain 251 (PL 251) has been commercialized as Melocon®WG in United States of America and BIOACT®WG in several countries of the world (Askary, 2015a; Moosavi and Askary, 2015).

P. chlamydosporia parasitizes the eggs and adult females of plant-parasitic nematodes (Stirling, 1991; Kerry, 2000). The fungus enters the cyst of the nematode, *Heterodera* or *Globodera* either through natural openings or by directly penetrating the wall of the cyst (Kerry, 1988). Cannayane and Jonathan (2008) reported that *P. chlamydosporia* effectively colonize the rhizosphere and interact with the nematode development process. A branched mycelia network is formed by the fungus when in close contact with the smooth egg shell (Morgan-Jones et al., 1983; Lopez-Llorca and Duncan, 1988). Appressorium of the fungus adheres to the egg shell by mucigens, forms an infection peg that penetrates the egg shell (Segers et al., 1996). As a result, disintegration of eggshell, vitelline layer and partial dissolution of chitin and the lipid layer takes place (Lopez-Llorca and Duncan, 1988; Stirling, 1991; Saifullah and Thomas, 1997). Toxins produced by the fungus are also reported to inhibit hatching of nematode eggs (Morgan-Jones et al., 1983). Some commercial products of *P. chlamydosporia* are Xianchongbike in China (Mo et al., 2005), PcMR-1 strain in Portugal (Moosavi and Askary, 2015), besides, several other products of *P. chlamydosporia* have been prepared by different countries of Europe, America and Africa (Askary, 2015a).

T. harzianum and *T. viride* are egg parasites of root-knot and several other plant parasitic nematodes (Dos Santos et al., 1992; Hafeez et al., 2000; Askary, 2008; Khattak et al., 2008). *T. harzianum* inhibits the invading nematodes by colonizing near the plant roots and growing on the roots (Inbar et al., 1994). The fungus causes direct parasitism to nematode eggs (Sahebani and Hadavi, 2008). The hyphae parasitize eggs and larvae of root-knot nematode, *Meloidogyne incognita* by dissolving the chitin layer through enzymatic activity. The fungus proliferates within the organism and produce toxic metabolites (Dos Santos et al., 1992). T. viride produces different antibiotics such as trichodermim, dermadin, trichoviridin and sesquiterpene heptalic acid that are involved in the suppression of nematodes (Jonathan, 2010). In Columbia, *T. harzianum* has been commercialized as Trichobiol and *T. viride* strain 2684 as Trifesol. Romulus is the commercial formulation of *T. harzianum* prepared by Dagutat Biolab, South Africa (Askary, 2015b). Commander Fungicide is a commercial product of *T. harzianum* in India. ECOSOM® is another product of *T. harzianum* based on strain IIHR-Th-2 and prepared by Agri Life, India. This product is useful in the management of root-knot and reniform nematodes (Askary, 2015a; Moosavi and Askary, 2015).

A. niger is egg parasite of nematodes. While coming in contact with a cyst or an egg mass, the fungus grows rapidly and preferably colonizes those eggs where larval formation has not been completed. Eggs containing larva are less vulnerable to fungus (Askary, 2015a). Kalisena SL, Pusa Mrida and Kalasipahi (capsules) are the formulations of *A. niger* applied in soil, whereas other formulations viz., Kalisena SD and Beej Bandhu are used for seed dressing. Kalisena SD helps in increasing the germination and seedling vigour in a variety of crops (Sen, 2000). Except Beej Bandhu, all the formulations have shelf life of more than 2 years at storage temperature of 15–35°C and relative humidity less than 80 per cent.

2.2. Bacteria

Bacteria that are antagonists or parasites kill their hosts through the production of toxins, enzymes, volatile compounds and antibiotics. *Bacillus thuringiensis* is considered a successful biocontrol agent that after sporulation produces the Cry (crystal) protein that has insecticidal and nematicidal properties (Bravo et al., 2011). Moreover, the Cry proteins of *B. thuringiensis* produce other virulence factors such as Cyt (cytolytic) inclusions, low molecular mass exotoxins, serine proteases, chitinases, collagenases and a metalloproteinase

(Guttmann and Ellar, 2000; Raymond et al., 2010; Jisha et al., 2013; Luo et al., 2013). These toxins may act in concert with the Cry proteins enhancing the toxicity against insects and free-living nematodes (Raymond et al., 2010; Luo et al., 2013). The formulation components of *B. thuringiensis* such as spores, crystals, protoxin, when mixed with insect diet renders the microbe available not only to defoliators but also to larvae that penetrate into the diet and feed on inner layers of the medium.

Bacillus firmus is another spore-forming bacterium which is being marketed with two types of products. One is drench product (Nortica®) and the other is seed treatment product (Poncho® VOTiVO®). Both the products are manufactured by the multinational company, Bayer Crop Science (Wilson and Jackson, 2013). The Poncho® VOTiVO® can be used for the control of insect pests and plant parasitic nematodes.

Avermectins are commercial product of *Streptomyces avermitilis* that have both nematicidal and insecticidal properties (Stretton et al., 1987). Avid 0.15 EC is commercial formulation of abamectin used for nematode control. (Martin, 2013).

The non-spore-forming bacteria are more difficult to formulate due to their low survival potential under environmental stresses (Islam et al., 2003). Such bacteria are formulated into various solid carriers such as wettable powder (WP). A good example is Sheathguard™, registered by Indian Pesticides Regulatory Authority – Central Insecticides Board, India.

2.3. Entomopathogenic Nematode (EPN)

EPNs comprising the families Steinernematidae and Heterorhabditidae are soil-dwelling microorganisms which cause high mortality to a broad range of insect pests belonging to different orders (Askary and Abd-Elgawad, 2017). The third stage juvenile (J_3) of EPN are free living in soil and the only infective stage. This stage is also known as infective juvenile (IJ). IJs of *Steinernema* and *Heterorhabditis* are symbiotically associated with bacteria *Xenorhabdus* and *Photorhabdus*, respectively. IJs move into the soil and upon finding the insect host, enter into it via natural openings such as mouth, anus or spiracles or by abrading the intersegment cuticle, reach haemocoel and release bacteria from their gut into the haemolymph of the insect. The bacteria upon finding a suitable medium in the form of haemolymph multiply very profusely, release toxins, which result in killing the host insect within 24-48 hours (Adams and Nguyen, 2002).

The insects killed by EPNs are loose bodied and do not putrify, those killed by *Steinernema* turns creamy, pale yellow or tan in colour, whereas the colour of insects killed by *Heterorhabditis* are brick red (Kaya and Gaugler, 1993; Sundarababu and Sankaranarayanan 1998; Cannayane et al., 2007; Fayyaz et al., 2017; Figure 1).

A variety of toxins produced by bacteria make the insect body mucilaginous and there is bioconversion of insect body into nutrient soup, vital for the growth and development of nematode. Moulting takes place, J_3 converts to J_4 and finally to adults. Adults multiply, lay eggs that eventually lay the formation of next generation. Generally, 2-3 generations take place depending on the size of the host. When the food starts depleting, IJs exit from the insect cadaver in search of new host, leaving the insect integument behind (Khan et al., 2007; Askary et al., 2017; Askary and Ahmad, 2017; Sankaranarayanan and Askary, 2017; Ramya, 2021).

EPNs have a short life cycle, easy to mass produce, easy to apply, safe to non-target organisms and have the ability to withstand under unfavourable environmental conditions (Askary, 2010; Abd-Elgawad et al., 2017; Askary and Abd-Elgawad, 2021). Besides, they are compatible with many chemical insecticides (Laznik and Trdan, 2017; Askary and Ahmad, 2020). The valid species of *Steinernema* and *Heterorhabditis* reported till now are 100 and 21, respectively (Bhat et al., 2020). The attributes of EPN in causing no harm to human health and the environment has led the US Environmental Protection Agency to exempt them from registration requirements. Similar exemptions have been provided to these beneficial organisms in many other countries of the world.

The development of EPNs in the past three decades can be attributed to several factors such as: i. discovery of numerous efficacious isolates/strains of EPNs; ii. awareness of harmful effects of pesticides and subsequent reduction in their usage, and iii. advancement in the mass production and formulation technology of EPNs. While preparing a commercial product, manufacturers should have prior knowledge about that particular EPN strain regarding their host finding ability, virulence potential against the host as well as reproductive potential within the host (Askary and Ahmad, 2017; Askary et al., 2018). The common species of EPNs that has been formulated and commercialized in different countries of the world are listed below (Table 2).

Figure 1. Larvae killed by entomopathogenic nematodes: (a) *Heterorhabditis* (*Helicoverpa armigera)*; (b) *Pieris brassicae*; (c) *Corcyra cephalonica* & Steinernema species; (d) *Helicoverpa armigera*; (e) *Pieris brassicae*; (f) *Corcyra cephalonica.*

The commercially produced EPNs are being used widely for the control of lepidopteran, coleopteran and dipteran pests of agricultural and horticultural crops that also include lawns and turf pests, fungus gnats in mushroom, black vine weevil in nursery plants and Diaprepes root weevil in citrus (Grewal et al., 2005; Abd-Elgawad et al., 2017; Lacey and Georgis, 2012). The native species of EPNs are considered a valuable resource, not only from a biodiversity perspective but also due their ability to adapt and work more efficiently under local conditions (Bhat et al., 2019; Askary and Ahmad, 2021a, b; Askary et al., 2021). In the past few years, the characteristics of some commercialized species of EPN–host matching have been reported (Abd-Elgawad, 2001; Shapiro-Ilan and Gaugler, 2015).

Table 2. Commercial products of entomopathogenic nematodes prepared by different countries

Nematode	Product name	Country
Steinernema carpocapsae	Boden-Niitzlinge Mioplant Nemabakt Biosafe Bouncer	Germany Austria Russia Columbia India
S. feltiae	Nemasys Entonem Owinema SC Exhibit Stealth Magnet	UK Netherlands Poland Switzerland UK USA
S. scapterisci	Proactant Ss	USA
S. riobrave	Biovector	Columbia
S. kushidai	SDS biotech	Japan
Heterorhabditis bacteriophora	Otinem E-Nema Gmbh	Switzerland Germany
H. megidis	Nemasys Larvanem NovoNem	UK Netherlands Germany
H. indica	Soldier	India

Source: Askary et al., 2017.

There are also several reports of EPN suppressing some species of plant parasitic nematodes in soil (Grewal et al., 1999; Somasekhar et al., 2002). On the basis of biochemical and molecular evidence it was suggested that EPNs induce components of systemic resistance in plants against plant parasitic nematodes (Jagdale et al., 2009).

3. Slug Parasitic Nematode

Slugs belonging to the families Agriolimacidae (mainly *Deroceras reticulatum*), Arionidae and Milacidae are important pests of agricultural and horticultural crops. Limited efficacy of molluscicidal bait pellets (Beer, 1989; Port and Ester, 2002; Glen and Moens, 2002) and their hazardous effect on grain-eating farm birds such as finches and sparrows (Clarke, 2014Fa) as well accidental poisoning of mammals such as dogs (Buhl et al., 2013), the most widely used molluscicide, have aroused the need to search for an alternative option that should be highly effective, non-hazardous and easy to apply against the target slug pests.

Phasmarhabditis hermaphrodita (Family: Rhabditidae) is a lethal nematode parasite of the slug species belonging to the families Arionidae, Agriolimacidae, Limacidae, Milacidae and Helicidae (Wilson et al., 1993; Koch et al., 2000; Askary et al., 2012). IJs of *P. hermaphrodita* have symbiotically associated bacteria, *Moraxella osloensis* in their gut. IJs are free living in soil which upon finding the host slug, enters into it via natural opening at the posterior end of the mantle. After entering into the host, the IJs release bacterial cells from their gut. Bacteria multiply very profusely inside the host and IJs start feeding on multiplying bacteria and develop into self-fertilizing hermaphrodites. The host slug dies within 7–21 days after infection.

The slugs are most active and destructive in wet weather conditions and this condition is also very suitable for the nematodes to work more actively. Contrary to this, poison pellets are less effective in wet conditions. Native isolates of *P. hermaphrodita* are more effective and adaptive than exotic ones but due to high price of commercial formulation of this nematode and also limited effect on some big (and often invasive) mollusc species, they are not in mass use in agriculture. In the Czech Republic, the cost of 1 hectare application can be about €4,000, which is an absolutely unacceptable price for most crop growers. Hence, application of *P. hermaphrodita* can be preferred in the most profitable crops or in the hobby segment of private gardens where the high price is not a problem. In 1994, MicroBio launched a commercial product Nemaslug® for the control of slugs in the home garden market in the UK which stimulated worldwide interest in the news media. It is suggested that use of *P. hermaphrodita* in integration with bait pellets will significantly reduce the chemical pesticides (Ester and Geelen, 1996; Wilson et al., 2000).

4. Virus

Viruses belonging to fifteen families are known to infect insects, but only those belonging to the family Baculoviridae have been used as biopesticides because they have narrow host range and are safe to vertebrates (Copping and

Menn, 2000; Lacey et al., 2002). Baculoviruses are double-stranded DNA viruses, highly pathogenic against a series of serious insect pests. (Moscardi, 1999). They are occluded, i.e., the virus particles are embedded in a protein matrix. The presence of occlusion aids in the survival of virus outside the host (Cory, 2000). There are two groups of baculoviruses (a) nucleopoly-hedroviruses (NPVs) and (b) granuloviruses (GVs). The occlusion bodies of NPVs comprise numerous virus particles, while in case of GVs there is just one virus particle. The infection begins with the ingestion of virus by the larva. Virus enters the body of insect via midgut and from there they spread throughout the body, although in some insects, infection can be limited to the insect midgut or the fat body. In the final stages of infection, the breakdown of host cuticle takes place and this aided by two viral enzymes, chitinase and cathepsin (Hawtin et al., 1997). The body of the infected larva liquefies. Polyhedra released from the cadaver of larva can infect other larvae by horizontal transmission which primarily occurs through larvae ingesting occlusion bodies present on leaves.

In vitro commercial production of Baculovirus generally faces technical problems, so not in wide use in agriculture. However, some products have been prepared by few companies. Certis has prepared Cyd-X™ which contains the codling moth GV and it is an efficient tool for codling moth management (Arthurs and Lacey 2004; Arthurs et al. 2005). Other products are Gemstar™ and Spod-X™ which contains *Heliothis zea* NPV and beet armyworm NPV, respectively. Gemstar™ is also registered for the control of lepidopteran pests, like the cotton bollworm and budworm, which are considered serious pests of corn, soybean, and other vegetables (Arthurs et al. 2005).

5. Bio-Derived Chemicals

Bio-derived chemicals include Pyrethrum, rotenone, neem oil and other essential oils controlling pests and microbial diseases. The chemical compounds synthesized by many plants serve in protecting the plants from different pests (Isman and Akhtar, 2007). Botanical extracts induce insecticidal activity, repellence or toxicity to insect pests, nematodes, mites, cause antifeedant effects and insect growth regulation, as well as antifungal, antiviral and antibacterial properties against pathogens (Prakash and Rao, 1986, 1997). These botanical pesticides are easily decomposed by a variety of microbes present in moist soils and their utilization help in maintaining the biological diversity of predators (Grainge and Ahmad, 1988) which in the long run results in reducing environmental contamination and human and animal health hazards. Moreover, they have little or no residual effects and have acceptability for use in organic farming.

6. Limitations

It is well known fact that microbial biopesticides have several advantages over chemical pesticides such as the former are selective, produce little or no toxic residue, and their development costs are significantly lower than those of conventional synthetic chemical pesticides, but still they have not yet achieved a broad level of commercial development and application particularly in many developing countries of the world. A number of difficulties exist, such as their culture and formulation, variable gap between laboratory and field performance as well as lack of awareness among the farmers, particularly those from the developing countries. Other reasons are i. product quality with low microbial count, resulting in poor field performance, ii. industries, lacking large scale production capacity, iii. short shelf-life of the product and iv. sale of non-registered products in the market (Alam, 2000; Arora et al., 2010; Gupta and Dikshit, 2010; Mishra et al., 2015).

Most of the countries have amended their policies with an aim to promote the use of biopesticides. This was done as a result of several factors such as tremendous potential of biopesticides to bring sustainability to agriculture and environmental safety. However, biopesticides are still under the regulatory system due to which there is market entry barrier, imposing burdensome costs on biopesticide industries. Technological and policy gaps are still causing hindrance for effective utilization of biopesticides. There is lack of a readily available market, which causes difficulties for the manufacturers to invest in biopesticides. Another important issue is imbalance between the costs and benefits of biopesticides as compared to synthetic pesticides. For example, biopesticides face high competition in comparison to synthetic pesticides as the former are generally produced for a small agricultural activity, which results in relatively higher costs and therefore not feasible. Some other serious concerns are relative immaturity of the policy network, limited resources and capabilities, lack of trust between regulators and producers which reflects weakness of the policy network.

7. Future Thrusts

Biopesticides cannot be a substitute for chemicals, which is meant for quick results, but it is environmentally safe and free from health hazards and so widely accepted among the growers (Askary, 2015b). In certain cases, biopesticides may replace chemical pesticides when used alone or in conjugation with them but are unlikely to dominate most pest control scenarios. In spite of this, they definitely have a place of their own in crop pest management programmes. In order to incorporate biopesticides with existing IPM modules for different crops, following points need immediate attention:

i. Efforts on isolating native biocontrol agents to expand the genetic base of native strains of biocontrol agent identifying those adapted to variable temperature.
ii. Utilization of recombinant DNA technology with an aim to enhance efficacy of biopesticides.
iii. Search for new biomolecules and enhancing efficiency of the known biopesticides.
iv. Small scale cottage industry for biopesticides production.
v. Proper training and knowledge about pests/ pathogens against which biopesticides can be applied successfully.
vi. The mode of action of biopesticides differs from conventional chemical pesticide and their actions are almost always specific. Therefore, users should have appropriate knowledge of the biopesticides they are going to apply, the target pest, suitable time for application, field rates, and application intervals.

Conclusion

Biopesticides have gained lot of interests in the last two decades in view of the growing demands for organic foods. Research practitioners are emphasizing mainly on quality and safety of the crop as compared to yield. The direct and indirect benefits of biopesticides may be seen in the reduction of external inputs such as inorganic fertilizers, pesticides and fuel by means of farm produced substitutes. However, still some areas need to be focused such as identifying native isolates, improving formulation and manufacturing technologies preferably with an aim to make the product cost-effective with increased shelf life. A strong collaboration among research practitioners, industry and farmers will certainly help in developing high quality products gaining much acceptance in sustainable agriculture. Undoubtedly, microbial biopesticides improve environmental and human health through the reduction of toxic chemical pesticides but they may be target specific in many cases. Therefore, users must have prior information of the organisms they are utilizing as biopesticides, so that they can be included tactfully as one of the components in IPM programmes. Organizational training to extension functionaries and crop growers with an aim to popularize biopesticides would be helpful for their better adoption.

Acknowledgments

The authors would like to thank all those who have generated the information present in this chapter. Trade names of biopesticides are based on the information received from various sources but they are neither intended for endorsement nor for criticism.

References

Abd-Elgawad, M. M. M. (2001). Entomopathogenic nematode–host matching. *Egyptian Journal of Agronematology*, 5: 91–104.

Abd-Elgawad, M. M. M., Askary, T. H. and Coupland, J. (2017). *Biocontrol Agents: Entomopathogenic and Slug Parasitic Nematodes*. CAB International, Wallingford, UK.

Adams, B. J. and Nguyen, K. B. (2002). Taxonomy and systematics. In: *Entomopathogenic Nematology* (Ed. Gaugler, R.). CAB International, Wallingford, UK, pp. 1–34.

Alam, G. (2000). *A Study of Biopesticides and Biofertilizers in Haryana, India*. International Institute for Environment and Development, London, UK.

Arora, N. K., Khare, E., and Maheshwari, D. K. (2010). Plant growth promoting rhizobacteria: constraints in bioformulation, commercialization, and future strategies. In: *Plant Growth and Health Promoting Bacteria* (Ed. Maheshwari, D. K.). Springer, Germany, pp. 97-116.

Arthurs, S. P. and Lacey, L. A. (2004). Field evaluation of commercial formulations of the codling moth granulovirus: persistence of activity and success of seasonal applications against natural infestations of codling moth in Pacific Northwest apple orchards. *Biological Control*, 31: 388–397.

Arthurs, S. P., Lacey, L. A. and Fritts, R. Jr. (2005). Optimizing use of codling moth granulovirus: effects of application rate and spraying frequency on control of codling moth larvae in Pacifi c Northwest apple orchards. *Journal of Economic Entomology*, 98: 1459–1468.

Askary, T. H. (1996). *Studies on Some Nematophagous Fungi in Agriculture Soil of Pusa Farm, Samastipur,* Bihar. M.Sc. (Ag.) Thesis, Department of Nematology, Rajendra Agricultural University, Pusa, Bihar, India.

Askary, T. H. (2008). *Studies on Root-Knot Nematode Infesting Pigeon Pea and Its Integrated Management.* Ph.D. Thesis, Department of Plant Protection, Faculty of Agricultural Sciences, Aligarh Muslim University, Aligarh, India.

Askary, T. H. (2010). Nematodes as biocontrol agents. In: *Sociology, Organic Farming, Climate Change and Soil Science* (Ed. Lichtfouse, E.). Springer, The Netherlands, pp. 347-378.

Askary, T. H. (2015a). Nematophagous fungi as biocontrol agents of phytonematodes. In: *Biocontrol Agents of Phytonematodes* (Eds. Askary, T. H. and Martinelli, P. R. P.). CAB International, Wallingford, UK, pp. 81-125.

Askary, T. H. (2015b). Limitations, research needs and future prospects in the biological control of phytonematodes. In: *Biocontrol Agents of Phytonematodes* (Eds. Askary, T. H. and Martinelli, P. R. P.). CAB International, Wallingford, UK, pp. 446-454.

Askary, T. H. and Abd-Elgawad, M. M. M. (2017). Beneficial nematodes in agroecosystem: A global perspective. In: *Biocontrol Agents: Entomopathogenic and Slug Parasitic Nematodes* (Eds. Abd-Elgawad, M. M. M., Askary, T. H. and Coupland, J.). CAB International, Wallingford, UK, pp. 3-25.

Askary, T. H. and Abd-Elgawad, M. M. M. (2019). *Key Nematode Pests of Pigeon pea: Systematics, Biology and Disease Management.* Lambert Academic Publishing, Germany.

Askary, T. H. and Abd-Elgawad, M. M. M. (2021). Opportunities and challenges of entomopathogenic nematodes as biocontrol agents in their tripartite interactions. *Egyptian Journal of Biological Pest Control*, 31: 42. https://doi.org/10.1186/s41938-021-00391-9.

Askary, T. H. and Ahmad, M. J. (2017). Entomopathogenic nematodes: mass production, formulation and application. In: *Biocontrol Agents: Entomopathogenic and Slug Parasitic Nematodes* (Eds. Abd-Elgawad, M. M. M., Askary, T. H. and Coupland, J.). CAB International, Wallingford, UK, pp. 261-286.

Askary, T. H., Ahmad, M. J., Wani, A. R., Mohiddin, S. and Sofi, M. A. (2018). Behavioural ecology of entomopathogenic nematodes, *Steinernema* and *Heterorhabditis* for insect biocontrol. In: *Sustainable Agriculture Reviews 31* (Ed. Lichtfouse, E.). Springer, The Netherlands, pp. 425-441.

Askary, T. H. and Ahmad, M. J. (2020). Efficacy of entomopathogenic nematodes against the cabbage butterfly (*Pieris brassicae* (L.) (Lepidoptera: Pieridae) infesting cabbage under field conditions. *Egyptian Journal of Biological Pest Control*, 30: 39. https://doi.org/10.1186/s41938-020-00243-y.

Askary, T. H. and Ahmad, M. J. (2021a). Survival and virulence capacity of native strain of entomopathogenic nematode, *Steinernema cholashanense* in different formulations. *Agricultural Research*, 10: 390-397.

Askary, T. H. and Ahmad, M. J. (2021b). Biocidal efficacy of some native isolates of entomopathogenic nematodes against oriental armyworm, *Mythimna separata* Walker (Lepidoptera:Noctuidae). *Indian Journal of Nematology*, 51: 67-73.

Askary, T. H., Bhat, A. H., Ahmad, M. J., Chaubey, A. K. and Spiridonov, S. E. (2020). *Steinernema feltiae* (Rhabditida: Steinernematidae) from hilly areas of Kashmir valley, India with a note on its geographical distribution. *Russian Journal of Nematology*, 28: 99 – 106.

Askary, T. H., Khan, A. A., Waliullah, M. I. S., Banday, S. A., Iqbal, U. and Mir, M. M. (2012). Slug pest management through nematodes in agricultural and horticultural crops. In: *Nematodes: Morphology, Functions and Management Strategies* (Eds. Boeri, F. and Chung, J. A.). Nova Publishers, USA, pp. 197–211.

Askary, T. H. and Martinelli, P. R. P. (2015). *Biocontrol Agents of Phytonematodes*. CAB International, Wallingford, UK, 470 pp.

Askary, T. H., Nermu, J., Ahmad, M. J. and Ganai, M. A. (2017). Future thrusts in expanding the use of entomopathogenic and slug parasitic nematodes in agriculture. In: *Biocontrol Agents: Entomopathogenic and Slug Parasitic Nematodes* (Eds. Abd-Elgawad, M. M. M., Askary, T. H. and Coupland, J.). CAB International, Wallingford, UK, pp. 620-627.

Beer, G. J. (1989). *Levels and Economics of Slug Damage in Potato Crops 1987 and 1988*. Monograph – British Crop Protection Council, 1989, 41: 101–105.

Bhat, A. H., Askary, T. H., Ahmad, M. J., Suman, Aasha and Chaubey, A. K. (2019). Description of *Heterorhabditis bacteriophora* (Nematoda: Heterorhabditidae) isolated from hilly areas of Kashmir valley. *Egyptian Journal of Biological Pest Control*, 29: 96. https://doi.org/10.1186/s41938-019-0197-6.

Bhat, A. H., Chaubey, A. K. and Askary, T. H. (2020). Global distribution of entomopathogenic nematodes, *Steinernema* and *Heterorhabditis*. *Egyptian Journal of Biological Pest Control*, 30: 31. https://doi.org/10.1186/s41938-020-0212-y.

Bidochka, M. J., Kamp, A. M. and de Croos, A. (2000). Insect pathogenic fungi: from genes to populations. In: *Fungal Pathology* (Ed. Kronstad, J. W.), Kluwer Academic Publishers, The Netherlands, pp. 171–93.

Bravo, A., Likitvivatanavong, S., Gill, S. S. and Soberon, M. (2011). *Bacillus thuringiensis*: a story of a successful bioinsecticide. *Insect Biochemistry and Molecular Biology*, 41: 423–431.

Buhl, K. J., Berman, F. W. and Stone, D. L. (2013). Reports of metaldehyde and iron phosphate exposures in animals and characterization of suspected iron toxicosis in dogs. *Journal of the American Veterinary Medical Association*, 242: 1244–1248.

Butt, T. M. (2002). Use of entomogenous fungi for the control of insect pests. In: *The Mycota XI: Agricultural Applications* (Ed. Kempken, F.). Springer, Germany, pp. 111–34.

Cannayane, I., Banu, J. G., Subramaniam, S. and Rajavel, D. S. (2007). Preliminary evaluation of the entomopathogenic nematodes on the root grub, *Basilepta fulvicome* in cardamom. *International Journal of Nematology*, 37: 213–214.

Cannayane, I. and Jonathan, E. I. (2008). Nematode egg parasitic fungus-II: Pochonia chlamydosporia for commercial exploitation – a review. *Agricultural Reviews*, 29: 48–54.

Cayrol, J. C., Frankowski, J. P., Laniece, A., Hardemare, G. D. and Talon, J. P. (1978). Control of nematodes in mushroom cultures. Statement of a method of biological control using a predatory hyphomycetes: *Arthrobotrys robusta* strain Antipolis (Royal 300). Pepinieristes Horticulteurs Maraichers. *Revue Horticole*, 184: 23–30.

Chandler, D., Bailey, A. S., Mark Tatchell, G., Davidson, G., Greaves, J. and Grant, W. P. (2011). The development, regulation and use of biopesticides for integrated pest management. *Philosophical Transactions of the Royal Society B: Biological Sciences, 12*, 366 (1573): 1987-1998. https://doi.org/10.1098/rstb.2010.0390.

Charnley, A. K. (1997). Entomopathogenic fungi and their role in pest control. In: *The Mycota IV: Environmental and Microbial Relationships* (Eds. Wicklow, T. D. and Soderstrom, B.). Springer, Germany, pp. 185–201.

Clarkson, J. M., Scree, S., Bailey, A., Cob, B. and Charnley, K. (1998). Fungal pathogenesis in insects. In: *Molecular Varability of Fungal Pathogens* (Eds. Bridge, P. D., Couteaudier, Y. and Clarkson, J. M.). CAB International, Wallingford, UK, pp. 83–94.

Copping, L. G. and Menn, J. J. (2000). Biopesticides: a review of their action, applications and efficacy. *Pest Management Science*, 56: 651- 76.

Clarke, A. (2014). EU votes to ban methiocarb slug pellets. *Farmers Weekly*, 22 January 2014.

Cory, J. S (2000). Assessing the risks of releasing genetically modified virus insecticides: progress to date. *Crop Protection*, 19: 779–785.

Charnley, A. K. (1997). Entomopathogenic fungi and their role in pest control. In: *The Mycota IV: Environmental and Microbial Relationships* (Eds. Wicklow, T. D. and Soderstrom, B.). Springer, Germany, pp. 185–201.

Dalmas, C. A., and Koutroubas, S. D. (2018). Current status and recent developments in biopesticide use. *Agriculture*, 8:13.

Dos Santos, M. A., Ferraz, S. and Muchovej, J. J. (1992). Evaluation of 20 species of fungi from Brazil for biocontrol of *Meloidogyne incognita* race-3. *Nematropica*, 22: 183–192.

Dubey, N. K., Shukla, R., Kumar, A., Singh, P. and Prakash, B. (2011). Global scenario on the application of natural products in integrated pest management. In: *Natural Products in Plant Pest Management* (Ed. Dubey, N. K.). CAB International, Wallingford, UK. pp. 1-20.

Elena, G. J., Beatriz, P. J., Alejandro, P. and Lecuona, R. E. (2011). *Metarhizium anisopliae* (Metschnikoff) Sorokin promotes growth and has endophytic activity in tomato plants. *Advances in Biological Research*, 5: 22–27.

Ester, A. and Geelen, P. M. T. M. (1996). Integrated control of slugs in a sugar beet crop growing in a rye cover crop. In: *Proceedings from the Symposium on Slug and Snail Pests in Agriculture* held from 24-26 September 1996 at University of Kent, Canterbury, UK, British Crop Protection Council No.66, pp. 445-450.

Fayyaz, S., Kazi, F. and Khanum, T. A. (2017). Status of entomopathogenic nematodes in integrated pest management strategies in Pakistan. In: *Biocontrol Agents: Entomopathogenic and Slug Parasitic Nematodes* (Eds. Abd-Elgawad, M. M. M., Askary, T. H. and Coupland, J.). CAB International, Wallingford, UK, pp. 383-408.

Feng, W. and Zheng, X. (2007). Essential oil to control *Alternaria alternata* in vitro and in vivo. *Food Control*, 18: 1126-1130.

Glen, D. M. and Moens, R. (2002). Agriolimacidae, Arionidae and Milacidae as pests in West European cereals. In: *Molluscs as Crop Pests* (Ed. Barker, G. M.). CAB International, Wallingford, UK, pp. 271–300.

Gottar, M., Gobert, V., Matskevich, A. A., Reichhart, J. M., Wang, C., Butt, T. M., Belvin, M., Hoffmann, J. A. and Ferrandon, D. (2006). Dual detection of fungal infections in Drosophila via recognition of glucans and sensing of virulence factors. *Cell*, 127: 1425–1437.

Grainge, M. and Ahmed, S. (1988). *Handbook of Plants with Pest Control Properties*. 1st Edn., John Wiley and Sons, New York, USA.

Grewal, P. S., Ehlers, R. U. and Shapiro-Ilan, D. I. (2005). *Nematodes as Biocontrol Agents*. CAB International, New York, USA.

Grewal, P. S., Lewis, E. E., Venkatachari, S. (1999). Allelopathy: a possible mechanism of suppression of plant-parasitic nematodes by entomopathogenic nematodes. *Nematology*, 1: 735–743.

Gupta, S., Dikshit, A.K. (2010). Biopesticides: an eco-friendly approach for pest control. *Journal of Biopesticides*, 3: 186-188.

Guttmann, D. M. and Ellar, D. J. (2000). Phenotypic and genotypic comparisons of 23 strains from the Bacillus cereus complex for a selection of known and putative *B. thuringiensis* virulence factors. *FEMS Microbiology Letters*, 188: 7–13.

Hafeez, U. K., Riaz, A., Waqar, A., Khan, S. M. and Akhtar, S. (2000). Evaluation of chemical vs biological control treatments against root-knot nematode (*M. incognita*) on tomato. *Pakistan Journal of Phytopathology*, 12: 118–120.

Hawtin, R. E., Zarkowska, T., Arnold, K., Thomas, C. J. and Gooday, G. W. (1997). Liquefaction of *Autographa californica* nucleopolyhedrovirus-infected insects is dependent on the integrity of virus-encoded chitinase and cathepsin genes. *Virology*, 238: 243-53.

Inbar, J., Abramsky, M., Cohen, D. and Chet, I. (1994). Plant growth enhancement and disease control by *Trichoderma harzianum* in vegetable seedlings grown under commercial conditions. *European Journal of Plant Pathology*, 100: 337–346.

Islam, Z., Pamplona, R., Atkinson, A. D. and Ezucena, E. J. (2003). *Biological Control of Rice Diseases*. http://www.vegetableipmasia.org/docs/Rice/Control_of_Rice_Diseases.pdf.

Isman, M. B. and Akhtar, Y. (2007). Plant natural products as a source for developing environmentally acceptable insecticides. In: *Insecticides Design Using Advanced Technologies* (Eds. Shaaya, I., Nauen, R. and Horowitz, A.R.). Springer, Heidelberg, pp. 235-248.

Jaffee, B. A., Muldoon, A. E. and Westerdahl, B. B. (1996). Failure of a mycelia formulation of the nematophagous fungus *Hirsutella rhossiliensis* to suppress the nematode *Heterodera schachtii*. *Biological Control*, 6: 340–346.

Jagdale, G. B., Kamoun, S., Grewal, P. S. (2009). Entomopathogenic nematodes induce components of systemic resistance in plants: Biochemical and molecular evidence. *Biological Control*, 51: 102-109.

Jatala, P., Kaltenbach, R. and Bocangel, M. (1979). Biological control of *Meloidogyne incognita acrita* and *Globodera pallida* on potatoes. *Journal of Nematology*, 11: 303.

Jisha, V. N., Smith, R. B. and Benjamin, S. (2013). An overview on the crystal toxins from *Bacillus thuringiensis*. *Advances in Microbiology*, 3: 462–472.

Jonathan, E. I. (2010). *Nematology: Fundamentals and Applications*. New India Publishing Agency, New Delhi, India.

Kaya, H. K. and Gaugler, R. (1993). Entomopathogenic nematodes. *Annual Review of Entomology*, 38: 181–206.

Kerry, B. R. (1988). Fungal parasites of cyst nematodes. *Agriculture, Ecosystems and Environment*, 24: 293–305.

Kerry, B. R. (2000). Rhizosphere interactions and the exploitation of microbial agents for the biological control of plant-parasitic nematodes. *Annual Review of Phytopathology*, 38: 423–441.

Khan, A., Williams, K. L. and Nevalainen, H. K. M. (2006). Interaction of plant parasitic nematodes by *Paecilomyces lilacinus* and *Monacrosporium lysipagum*. *Biological Control*, 51: 659–678.

Khan, A. L., Hamayun, M., Khan, S. A., Kang, S. M., Shinwari, Z. K., Kamran, Rehman, M. S., Kim, J. G. and Lee, I. J. (2012). Pure culture of *Metarhizium anisopliae* LHL07 reprograms soybean to higher growth and mitigates salt stress. *World Journal of Microbiology and Biotechnology*, 28: 1483–1494.

Khan, M. R., Khan, U., Askary, T. H., Mohiddin, F. A. and Khan, M. M. (2007). Pathogenicity and host suitability for *in vivo* mass production of *Steinernema masoodi* AMU EPN-1. *International Journal of Nematology*, 17: 151-157.

Khater, H. F. (2011). Ecosmart biorational insecticides: alternative insect control strategies. In: *Insecticides-Advances in Integrated Pest Management* (Ed. Perveen, F.). IntechOpen, Rijeka, Croatia, pp. 17-60.

Khattak, B., Saifullah and Stephen, M. (2008). Effect of some indigenous isolates of Trichoderma harzianum on root-knot nematode, *Meloidogyne javanica* (Treub) Chitwood. *Sharad Journal of Agriculture*, 24: 285–288.

Koch, R., Jackel, B. and Plate, H. P. (2000). Controlling pest slugs: new methods and the verification of their effects. *Gesunde Pflanzen*, 52: 1-10.

Kumar, S. (2012) Biopesticides: a need for food and environmental safety. *Journal of Biofertilizers and Biopesticides* 3:e107. https://doi.org/10.4172/2155-6202.1000e107.

Lacey, L. A. and Georgis, R. (2012). Entomopathogenic nematodes for control of insect pests above and below ground with comments on commercial production. *Journal of Nematology*, 44: 218-225.

Lacey, L. A., Vail, P. V. and Hoffmann, D. F. (2002). Comparative activity of baculoviruses against codling moth *Cydia pomonella* and three other tortricid pests of tree fruits, *Journal of Invertebrate Pathology,* 80: 64 - 68.

Lackey, B. A., Muldoon, A. E. and Jaffee, B. A. (1993). Alginate pellet formulation of *Hirsutella rhossiliensis* for biological control of plant parasitic nematodes. *Biological Control*, 3: 155–160.

Laznik, Z. and Trdan, S. (2017). Compatibility between Entomopathogenic Nematodes and Phytopharmaceuticals. In: *Biocontrol Agents: Entomopathogenic and Slug Parasitic*

Nematodes (Eds. Abd-Elgawad, M. M. M., Askary, T. H. and Coupland, J.). CAB International, Wallingford, UK, pp. 581-595.

Lopez-Llorca, L. V. and Duncan, G. H. (1988). A study of fungal endoparasitsm of the cereal cyst nematode (Heterodera avenae) by scanning electron microscopy. *Canadian Journal of Microbiology*, 34: 613–619.

Lopez-Llorca, L. V., Macia-Vicente, J. G. and Jansson, H. B. (2008). Mode of action and interacactions of nematophagous fungi. In: *Integrated Management and Biocontrol of Vegetable and Grain Crops Nematodes* (Eds. Ciancio, A. and Mukerji, K.G.) Springer, The Netherlands, pp. 54–76.

Luo, X., Chen, L., Huang, Q., Zheng, J., Zhou, W., Peng, D., Ruan, L. and Sun, M. (2013). *Bacillus thuringiensis* metalloproteinase Bmp1 functions as a nematicidal virulence factor. *Applied and Environmental Microbiology*, 79: 460–468.

Maniania, N. K., Sithanantham, S., Ekesi, S., Ampong-Nyarko, K., Baumgärtner, J. L., Löhr, B., and Matoka, C. M. (2003). A field trial of the entomogenous fungus Metarhizium anisopliae for control of onion thrips, *Thrips tabaci*. *Crop Protection*, 22: 553–559.

Martin, S. B. (2013). *Nematode Control.* Available at: http://media.clemson.edu/public/turfgrass/2013%20Pest%20Management/2013_nematode_cont.pdf.

Marrone, P. G. (2014). The market and potential for biopesticides. In: *Biopesticides: State of the Art and Future Opportunities* (Eds. Gross, A. D., Coats, J. R., Duke, S. O. and Seiber, J. N.). American Chemical Society, Washington, USA, pp. 245–258.

Mishra, J., Tewari, S., Singh, S., Arora, N. K. (2015). Biopesticides: where we stand? In: *Plant Microbes Symbiosis: Applied Facets* (Ed. Arora, N. K.). Springer, India, pp. 37-75.

Mo, M., Xu, C. and Zhang, K. (2005). Effects of carbon and nitrogen sources, carbon-to-nitrogen ratio, and initial pH on the growth of nematophagous fungus *Pochonia chlamydosporia* in liquid culture. *Mycopathologia*, 159: 381–387.

Moosavi, M. R. and Askary, T. H. (2015). Nematophagous fungi: Commercialization. In: *Biocontrol Agents of Phytonematodes* (Askary, T. H. and Martinelli, P. R. P.). CAB International, Wallingford, UK, pp. 187-202.

Moosavi, M. R. and Zare, R. (2012). Fungi as biological control agents of plant-parasitic nematodes. In: *Plant Defence: Biological Control* (Eds. Merillon, J. M. and Ramawat, K. G.). Springer, The Netherlands, pp. 67–107.

Morgan-Jones, G., White, J. F. and Rodriguez-Kabana, R. (1983). Phytonematode Pathology: Ultrastructural studies. I. Parasitism of *Meloidogyne arenaria* eggs by *Verticillium chlamydosporium*. *Nematropica*, 13: 245–260.

Moscardi, F. (1999). Assessment of the application of baculoviruses for control of Lepidoptera. *Annual Review of Entomology*, 44: 257 - 89.

Nasr Esfahani, M. and Ansari Pour, B. (2006). The effects of *Paecilomyces lilacinus* on the pathogenesis of *Meloidogyne javanica* and tomato plant growth parameters. *Iran Agricultural Research*, 24/25: 67–75.

Olson, S. (2015). An analysis of the biopesticide market now and where is going. *Outlook on Pest Management*, 26: 203–206.

Port, G. and Ester, A. (2002): Gastropods as pests in vegetable and ornamental crops in Western Europe. In: *Molluscs as Crop Pests* (Ed. Barker, G.M.). CAB International, Wallingford, UK, pp. 337–352.

Prakash, A. and Rao, J. (1986). Evaluation of plant products as antifeedants against the rice storage insects. *Proceedings from the Symposium on Residues and Environmental Pollution* held on 2nd October, 1986 at Muzaffarnagar, Uttar Pradesh, India, pp. 201-205.

Prakash, A. and Rao, J. (1997). *Botanical Pesticides in Agriculture*. 1st Edition, CRC Press, Florida, USA, pp. 461.

Pretty, J. (2009). *The Pesticide Detox, Towards A More Sustainable Agriculture*. Earth scan, London.

Ramya, A. R. (2021). Pathogenic and reproductive potential of some local strains of entomopathogenic nematode, *Heterorhabditis bacteriophora* against cabbage butterfly, *Pieris brassicae* (L.). M.Sc. (Ag.) Thesis, Division of Entomology, Faculty of Agriculture, Wadura, Sopore, Sher-e-Kashmir University of Agricultural Sciences and Technology of Kashmir (SKUAST-K), Srinagar, India.

Raymond, B., Johnston, P. R., Nielsen-LeRoux, C., Lereclus, D. and Crickmore, N. (2010). Bacillus thuringiensis: an impotent pathogen? *Trends in Microbiology*, 18: 189–194.

Sahebani, N. and Hadavi, N. (2008). Biological control of the root knot nematode *Meloidogyne javanica* by *Trichoderma harzianum. Journal of Soil Biology and Biochemistry*, 40: 2016–2020.

Saifullah and Thomas, B. J. (1997). Parasitism of *Globodera rostochiensis* by *Verticillium chlamydosporium* low temperature scanning electron microscopy and freeze fracture study. *International Journal of Nematology*, 7: 30–34.

Sankaranarayanan, C. and Askary, T. H. (2017). Status of Entomopathogenic Nematodes in Integrated Pest Management Strategies in India. In: *Biocontrol Agents: Entomopathogenic and Slug Parasitic Nematodes* (Eds. Abd-Elgawad, M. M. M., Askary, T. H. and Coupland, J.). CAB International, Wallingford, UK, pp. 362-382.

Segers, R., Butt, T. M., Kerry, B. R., Beckett, A. and Peberdy, J. F. (1996). The role of the proteinase VCP 1 produced by the nematophagous *Verticillium chlamydosporium* in the infection process of nematode eggs. *Mycological Research*, 100: 421–428.

Sen, B. (2000). Biological control: A Success Story. Jeensanidhi Award Lecture. *Indian Phytopathology*, 53: 243–249.

Senthil-Nathan, S. (2013). Physiological and biochemical effect of neem and other Meliaceae plants secondary metabolites against Lepidopteran insects. *Front Physiology*, 4: 359.

Shapiro-Ilan D. I. and Gaugler, R. (2015). Nematodes (Rhabditida: Steinernematidae & Heterorhabditidae). In: *Biological Control: A Guide to Natural Enemies in North America* (Ed. Shelton, A.). Available at: https://biocontrol.entomology.cornell.edu/pathogens/nematodes.php.

Somasekhar, N., Grewal, P. S., de Nardo, E. A. B. and Stinner, B. R. (2002). Non-target effects of entomopathogenic nematodes on the soil nematode community. *Journal of Applied Ecology*, 39: 735-744.

St. Leger, R. J. (1993). Biology and mechanisms of insect-cuticle invasion by deuteromycete fungal pathogens. In: *Parasites and Pathogens of Insects.* Volume 2:

Pathogens (Eds. Beckage, M. E., Federici, B. A. and Thompson, S. N.). Academic Press, New York, USA, pp. 211–29.

Stirling, G. R. (1991). *Biological Control of Plant Parasitic Nematodes*: *Progress, Problems and Prospects.* CAB International, Wallingford, UK.

Stoll, G. (2000). *Natural Crop Protection in the Tropics: Letting Information Come to Life*. 2nd Edition, Margraf-Verlag, Germany.

Stretton, A. O. W., Campbell, W. C. and Babu, J. R. (1987). Biological activity and mode of action of avermectins. In: *Vistas on Nematology* (Eds. Veech, J. A. and Dickson, D. W.). Society of Nematologists, Hyattsville, Maryland, pp. 136–146.

Sundarababu, R. and Sankaranarayanan, C. (1998). Biological control of insects using nematodes. In: *Recent Advances in Plant Nematology* (Ed. Trivedi, P. C.). CBS Publishers and Distributors, New Delhi, India, pp. 153–170.

Wilson, M. J. and Jackson, T. A. (2013). Progress in the commercialization of bionematicides. *Bio Control,* 58: 715–722.

Wilson, M. J., Glen, D. M. and George, S. K. (1993). The rhabditid nematode *Phasmarhabditis hermaphrodita* as a potential biological control agent for slugs. *Biocontrol Science and Technology*, 3: 503-511.

Wilson, M. J., Hughes, L. A., Hamacher, G. M. and Glen, D. M. (2000). Effects of *Phasmarhabditis hermaphrodita* on non-target molluscs. *Pest Management Science* 56: 711–716.

WRI (1994). *World Resources, 1994/1995*. Oxford University Press, USA, pp. 416.

Zare, R. and Gams, W. (2001). A revision of *Verticillium* section Prostrata. IV. The genera *Lecanicillium* and *Simplicillium* gen. nov. *Nova Hedwigia*, 73: 1–50.

Chapter 8

Entomopathogenic Nematodes in Crop Pest Management: Application Tactics and Challenges

Gitanjali Devi[*]
Department of Nematology,
Assam Agricultural University,
Jorhat, Assam, India

Abstract

Insect pests cause heavy losses to production of agricultural commodities. In the face of increasing demand of food, the yield loss due to insect pest should be minimized. Revamped interest in the biocontrol of insect pests has recently emerged for Integrated Pest Management (IPM) system as well as organic agriculture. In order to develop a safe and sustainable method of pest management, use of entomopathogenic nematodes as a biocontrol agent has increased over the last few years. Entomopathogenic nematodes (EPNs) in the genera *Steinernema* and *Heterorhabditis* are used against many insect pest species. However, field efficacy of EPNs towards targeted pests depends on environmental condition and their delivery in aboveground or belowground level. Various spray equipments and irrigation systems are available for application of entomopathogenic nematodes to soil as well as above ground parts. Besides application equipments, EPN application is influenced by biotic factors such as pathogens and predators and abiotic factors such as temperature and moisture. Other factors are rate of application and method of application. Successful application will facilitate more efficient and improved pest control measure through

---

[*] Corresponding Author's Email: gitanjali.devi@aau.ac.in

In: Pest Management: Methods, Applications and Challenges
Editor: Tarique Hassan Askary
ISBN: 979-8-88697-268-9

increased understanding of entomopathogenic nematode biology, ecology, target pest's biology and ecology, and crop morphology and phenology. This chapter envisages the latest findings on this topic with the expectation that such information will refine our understanding of application methods of entomopathogenic nematodes for better field performance.

Keywords: biological control, entomopathogenic nematode, insect pests, management, application, environmental condition.

1. Introduction

Insect pests cause direct damage to crop plants. They also act as vectors for several viral diseases. Several polyphagous pests are becoming severe and expanding their host-horizon. Many pests have developed resistance to insecticides and there is secondary outbreak thereby upsetting the ecological and environmental balance in nature. Arthropod pests destroy an estimated 18–20 per cent of annual crop production worldwide, which is valued at more than US $ 470 billion. Indian agriculture suffers an annual loss of about 15.7 per cent due to ravages by these insect pests which accounts for US $ 36 billion. (Sharma et al., 2017). Crop losses due to various biotic and abiotic stresses are rising in the face of increasing intensity of cultivation, reduced agro-ecosystem diversity, narrow genetic base of modern crop cultivars, intensive use of agrochemicals and the rapid changes in climate. This envisages that there is a need to relook into the cropping systems for reorientation for adequate insect pest management. Biopesticides are placed as an important tool in pest management strategies in Integrated Pest Management (IPM) system as well as in organic agriculture. Insect-parasitic nematodes are considered good candidates for IPM as well as sustainable agriculture (Grewal et al., 2005). Nematodes which are capable of killing, sterilizing or hampering the development of insect and completing at least one stage of their life cycle in the host are called entomopathogenic nematodes (EPNs) (Kaya and Gaugler, 1993). These nematodes are tiny, simple, soft bodied, unsegmented, hyaline, usually microscopic in size. They are ubiquitously distributed, isolated from every inhabited continent from a wide range of ecologically diverse soil habitats including cultivated fields, forests, grasslands, deserts, and even ocean beaches. Entomopathogenic nematodes, *Steinernema* and *Heterorhabditis* in the family Steinernematidae and

Heterorhabditidae of the order Rhabditida are obligate parasites of insect pests (Burnell and Stock, 2000, Askary, 2010). They have huge potential as a biocontrol agent against a wide range of insect pests (Askary and Abd-Elgawad, 2017). Due to their high reproductive potential, ease of mass production and application and their harmlessness to microbes, animals, humans and plants, EPNs are considered as one of the most significant non-chemical alternatives to insect pest control (Akhurst and Smith, 2002; Askary and Ahmad, 2017; Sankaranarayanan and Askary, 2017).

Applied biological control involves human contribution to increase natural enemy activities so that it maintains a host population at levels lower than would occur in its absence. Applied biological control can be further divided into classical biological control and augmentative biological control. In augmentative biological control, natural enemy activities increase through manipulation of the environment (i.e., conservation) or the natural enemies themselves. In augmentative biological control, two broad approaches can be employed. These are inoculative release, in which a small number of natural enemies of the same species are released and the progeny of the natural enemies is expected to increase, and in inundative release, large numbers of natural enemies are released with the expectation that these enemies will give immediate control. EPNs are the most promising biological control agents of insect pests applied in augmentative method, while in inundative approach their infective juveniles (IJs) are released in a uniform manner for rapid control of insect pest population.

In managed systems due to lack of physical disturbance and favourable soil conditions such as in greenhouse ornamentals and vegetables, citrus, cranberry, turf grass, inundative application of EPNs are successful (Shapiro-Ilan et al., 2002; Duncan et al., 2003). They have been most efficacious for insects in soil or cryptic habitats where there is protection from rapid desiccation and UV radiation. However, nematodes are not efficacious for an inundative approach in wide acreage crops such as cotton and soybeans due to their survival ability in field, susceptibility to environmental extremes, and of course high application cost. Inoculative releases may be more promising and feasible in situations where the moderately susceptible insect hosts are present throughout the year with a high economic threshold level and environmental conditions are favorable for nematode survival (Parkman et al., 1994). Though under laboratory condition EPN shows high efficacy against almost all insect topests, but under field condition their efficacy is not the same and this depends upon a variety of factors including application technology of EPNs (Hiltpold, 2015). Nematodes being a live material have to be handled very carefully so that there is minimum loss during transfer of nematodes to the target pest and deleterious effect of surrounding is reduced to the minimum and they work efficiently (Askary et al., 2012).

This chapter is an attempt to highlight the tactics involved and challenges faced during application of EPN against the target insect pests under field condition.

2. Factors Affecting Application Success

In the recent years many species of *Steinernema* and *Heterorhabditis* have been isolated from different countries of the world (Bhat et al., 2019; Askary et al., 2020). However, exploitation of these EPN species depends upon several factors which I have tried to summarize briefly under the following subheadings.

2.1. Matching of EPNs with Target Pest

Application success depends on the appropriate matching of nematode with the particular target pests. Better understanding of the host habitat finding, host finding, and host suitability permit better matches and better field performances (Gaugler, 1999; Khan et al., 2007; Askary et al., 2018; Askary and Ahmad, 2017, 2021b).

2.2. Time of Application

Timing of application is significant when applying nematodes against target pest. Nematode application should coincide with the target larval stage. Field application in apple orchard under tree canopy in the spring with *S. riobrave*, *S. carpocapsae*, and *S. feltiae* has potential to attack *Rhagoletis pomonella* pupae (Usman et al., 2020). The duration of contact between the host and the nematode and host defense mechanisms and behavior are important factors affecting nematode efficacy. It has been reported that application through in-furrow irrigation is more effective in causing target pest mortality (95%) than application before (56%) or after (84%) irrigation (Cabanillas and Raulston, 1996).

2.3. Rate of Application

It has been observed that specific rate responses for each species in various cropping system to control specific pests are different. Depending on the target pests, a higher rate or in some rare cases lower rates of application may be required (Shapiro-Ilan et al., 2006a). However, EPNs must be applied to soil at minimum rates of 2.5×10^9 IJs/ha (25 IJs/cm^2) or higher with a spray volume of 750-1890 litre/hectare (Georgis and Hague, 1991). Nematodes can also be delivered with low-volume sprayers, usually in the range of 200-400 litre/hectare. The equipment should be calibrated to ensure appropriate application rates. Pre and post application irrigation should be adjusted to compensate for reduced volume. Studies indicated that 0.25-0.65 cm of post-application irrigation is sufficient to move the nematodes into the soil (Shetlar et al., 1988).

Generally, as long as environmental conditions are favourable, nematode populations will remain high enough to provide effective pest control for 2-8 weeks after application (Duncan and McCoy, 1996). Twice-yearly application of *S. riobrave* through micro-sprinkler irrigation systems at 20-40 IJs/cm^2 under the tree canopies have reduced emerging adult weevil populations (Bender et al., 2014). Seasonal re-application or in multiple seasons or years was found to be effective (Klein and Georgis, 1992; Shields et al., 1999).

2.4. Relationship with Other Biocontrol Agents

Application success depends on the relationship such as antagonism, additivity, and synergism with other biocontrol agents and time or rate of application (Barbercheck and Kaya, 1990). EPNs can act synergistically with other entomopathogens such as *Paenibacillus popilliae* (Thurston et al., 1994), *Bacillus thuringiensis* (Koppenhofer and Kaya,1997; Gassmann et al., 2008), *Metarhizium anisopliae* (Ansari et al., 2004, 2006; Anbesse et al., 2008); other studies indicate antagonism with *Beauveria bassiana* (Brinkman and Gardner, 2000) or *Isaria fumosorosea* (Shapiro-Ilan et al., 2004).

2.5. Abiotic Factor

Successful application of EPNs depends on abiotic environment *viz.*, adequate moisture/relative humidity, temperature, and ultraviolet radiation. Besides abiotic factor, several biotic agents such as nematodes pathogens or predators can influence the outcome of their application (Kaya, 1990; El-Borai et al., 2011).

2.6. Formulation of EPNs

Nematode formulations that provide protection to the nematodes from environmental extremes during and after application can enhance nematode efficacy. Low post-application movement and survival in soil and on foliage reduces nematode efficacy (Smits, 1996). The formulations specifically intended to enhance post-application survival of nematodes in the soil and foliage is described below.

2.6.1. Applying Nematodes with Phoretic Hosts

Earthworms are considered as phoretic hosts for *S. carpocapsae* (Shapiro-Ilan and Brown, 2013).

2.6.2. Infected Host Cadaver

Cadaver application increase nematode dispersal, infectivity, survival/ persistence, and ease of handling (Welch and Briand, 1960; Janson and Lecrone, 1994; Shapiro-Ilan and Glazer, 1996; Shapiro-Ilan and Lewis, 1999; Perez et al., 2003; Gulzar et al., 2020). Nematodes applied in host cadavers were effective and persistent when added to bags of potting media for subsequent distribution to target pest sites (Deol et al., 2011). It has been reported that the time period of six to ten days between application and infection of *Galleria mellonella* on soil resulted in higher emergence of IJs (Shapiro-Ilan et al., 2001; Del Valle et al., 2008, 2009, 2011; Batista et al., 2010; Zhu et al., 2011; Dolinski et al., 2015).

2.6.3. Pre-Infected Hosts that Reach Cryptic Habitats

The pre-infected hosts have the ability to carry the next generation of emerging IJs that is hard-to-reach target sites, and thereby facilitate enhanced control in cryptic habitats. Last instar *G. mellonella* and *Spodoptera cilium* were used to acquire nematode, *S. carpocapsae* that resulted in 86 per cent mortality of *Coccus cossus* (Gumus et al., 2015).

2.6.4. Infected Host Macerate or Ascaroside-Based Nematode Pheromones

Ascaroside-based nematode pheromones can boost EPN infectivity and dispersal and have been shown to improve efficacy against soil pests (Wu et al., 2018; Oliveira-Hofman et al., 2019; Shapiro-Ilan et al., 2019).

2.6.5. Nematode-Impregnated Collars

These can be placed around hibernation sites on tree surfaces which provide a slow-release system against various orchard pests (Kaya et al., 1984; Nachtigall and Dickler, 1992).

2.6.6. Wettable Granule (WG)

An increase in the efficacy of EPN, *S. carpocapsae* (WG formulation) against cutworm and diamond backmoth was observed when the nematode was rehydrated for 48 hours prior to use (Baur et al., 1997a).

2.6.7. Desiccated Cadavers

A formulation based on desiccated cadavers coated with clay may allow application without cadavers rupturing or adhering together (Georgis, 1990).

2.6.8. Additives

Additives such as chitosan and wood flour foam are used to enhance aboveground EPN applications. A sprayable gel, Barricade®, can be used aboveground to kill target pests such as the wood boring pest, *Synathedon. pictipes* by providing protection against desiccation (van Niekerk and Malan, 2015; Shapiro-Ilan et al., 2017).

2.6.9. Capsules

Macrogels containing encapsulated nematodes have been suggested as delivery systems for the control of soil and foliar pests. Encapsulation of nematodes in calcium alginate gel beads was first reported by Kaya and Nelsen (1985). When the alginate capsules were placed in soil with adequate moisture, most nematodes migrated out of the granules within a week. Navon et al. (2002) evaluated an edible-to-insects, calcium alginate gel for *S. riobrave*. They reported high nematode survival for 48 hours in the gel at 61 per cent relative humidity. A gel-forming polyacrylamide used to enhance water-holding capacity of sandy soils, has been shown to enhance survival of *S. carpocapsae* when applied against the root weevil in citrus (Shapiro-Ilan et al., 2010; Hiltpold et al., 2012). Chang and Gehert (1992) developed macrogels containing encapsulated *S. carpocapsae* along with a water-holding acrylamide in gellan gum.

2.6.10. Baits

Baits containing IJs, an inert carrier and a feeding stimulant or a sex pheromone have been developed. Bait formulation can enhance persistence of EPNs (*S. carpocapsae* and *S.scapterisci*) and reduce the quantity of microbial agents required per unit area (Grewal, 2002).

2.6.11. Interaction with Other Agro-Chemicals

Nematodes are applied with fertilizers and chemical pesticides. Often it is desirable to tank mix one or more inputs to save time and money. The relationship between chemical pesticides/fertilizers and EPNs varies based on the specific chemical and nematode species or strain, dosages, and timing of application (Benz, 1971; Bednarek and Gaugler, 1997; Askary and Ahmad, 2017, 2020; Sankaranarayanan and Askary, 2017). Fertilizers that are applied at recommended rates have little impact on EPN efficacy. However, fresh manure or crop residue or high rates of chemical fertilizers can be detrimental to EPN persistence and efficacy (Bednarek and Gaugler, 1997; Shapiro-Ilan et al., 1996).

Some chemical pesticides are toxic to EPNs, whereas others tend to be compatible and, in some cases, may be synergistic when applied with EPNs (Koppenhofer and Kaya, 1998; Nishimatsu and Jackson, 1998; Alumai and Grewal, 2004; Koppenhofer and Grewal, 2005; Koppenhofer and Fuzy, 2008; Salame and Glazer, 2000). IJs are tolerant to short exposures (2-6 hours) to most agrochemicals including herbicides, fungicides, acaricides, and insecticides and therefore, can often be tank-mixed. Heterorhabditids tend to be more sensitive to physical challenges, including pesticides, than steinernematids. Thus, combinations should be tested before application.

2.6.12. Plant Volatiles

Certain plant volatiles may be used to enhance EPN efficacy by directing nematode behavior and enhancing infectivity (Willett et al., 2018).

3. Application Equipment

EPNs are generally applied to soil and foliage in water suspension using sprayers, mist blowers, or irrigation systems (Shapiro-Ilan et al., 2017). EPN require adequate soil moisture for survival and movement. Optimum moisture level required varies by nematode species and soil type. In a sandy loam soil *S. carpocapsae* infected insects at moisture levels as low as -5 Mpa (megapascal) and had the highest host establishment rates between - 0.1 and - 0.01 Mpa, whereas *S. glaseri* required a minimum water potential of - 0.3 Mpa to infect (Koppenhofer et al., 1995; El-Borai et al., 2016). IJ of entomopathogenic nematodes are usually applied using standard irrigation systems such as overhead, microjet, furrow, center pivot, and trickle irrigation (Georgis, 1990; Lara et al., 2008).

Nematodes are compatible with standard agrochemical equipment, including various devices available in the agricultural and urban environment (Wright et al., 2005; Shapiro-Ilan et al., 2006a). The application equipment and parameters associated with EPN distribution can have a direct impact on their efficacy. Simple watering cans or hose end or back-pack sprayers are used for small plot or home garden use. It is important to ensure adequate agitation during application. When nematodes are applied to larger plots, a suitable apparatus such as boom sprayer should be considered. Aerial applications cover large fields or orchards. The application equipment used depends on the cropping system as well as spray volume, agitation, pump, nozzle type, flow rate, pressure and recycling time, and environmental conditions (Hayes et al., 1999; Grewal, 2002; Fife et al., 2005).

Nematode viability during spraying depends on droplet sizes. Differences in nozzle characteristics along with pressure greatly influenced the droplet sizes. Conventional hydraulic nozzles i.e., fan nozzles, and cone type nozzles are known to produce a wide range of droplet sizes, many being too small to carry an IJ. Based on the flow field characteristics, the recommendation is that an appropriately sized (i.e., larger than the organism) cone nozzle is more suitable for spray application than a fan nozzle to avoid hydrodynamic damage to the EPNs (Brusselman et al., 2012). It is possible to spray EPNs without significant loss in their concentration and viability with equipment with those using hydraulic and rotary nozzle tips. Larger capacity nozzles are more suitable for applying entomopathogenic nematodes, particularly for soil applied treatments where a high volume of water is necessary to get the nematodes beyond the soil surface. It has been observed that a hydraulic spray gun used to concentrate nematode suspensions close to trunks produced higher infection rates of codling moth larvae (Unruh and Lacey, 2001). In the foliar environment, droplet sizes and spray distribution are important parameters whereas an increase in flow rate result in greater number of nematodes deposited per cm^2. The spinning disc uses centrifugal energy to generate droplets. The size of the droplets produced and spray coverage can be manipulated over a wide range of nematode concentrations and sizes.

Nematode delivery with the spinning disc can be improved by altering the flow rate with different or multiple discs by maintaining a low volume application (Mason et al., 1999). It has been reported that spinning disc was able to cause nearly 50 per cent mortality of *Plutella xylostella* when less than 9 per cent of the IJs of *S.carpocapsae* was applied on Chinese cabbage (Lello et al., 1996). Based on average size of *S. carpocapsae* IJ, it was calculated that the minimum droplet diameter needed to accommodate an IJ is about 178 μm, but larger openings are required for larger species such as *S. glaseri* and *H. megidis*. Therefore, removal of the filters and screens is usually recommended, although this requires recalibration of the equipment. IJs ranging in length from 500 to 1000 μm can be passed through this disc system. A significant decrease in viability of *S. feltiae* was observed as the length of pumping period with high pressure sprayer increased (Nilsson and Gripwall, 1999). High-pressure hydraulic pumps may damage the nematodes (Abd-Elgawad et al., 2017).

In order to maintain viability, operating pressures for some nematode species may reach up to 2000 Kilopascal (kPa) (eg. *H. bacteriophora* and *S. carpocapsae*) and 1380 kPa (*H. megidis*). Therefore, the diaphragm or roller pumps are better matched for use with nematodes than the centrifugal type. A large volume of liquid provides a high temperature in the spray tank. Temperatures exceeding 30°C either in the tank, delivery hose or nozzles should be avoided. Lack of oxygen in the spray tank and delivery hose can also inactivate nematodes are more prone to sunlight and desiccation damage. However, no adverse effects were observed for *Steinernema* spp. and *H.bacteriophora* after flow through several different pumps, nozzle types, and strainers (Klein and Georgis, 1994).

4. Application Tactics

Entomopathogenic nematodes have been applied against soil inhabiting insects as soil application as well as above-ground insects in cryptic habitats as foliar spray (Georgis, 1990; Arthers et al., 2004).

4.1. Seed Treatment

As prophylactic applications, the IJs of *H. bacteriophora* were applied at recommended rate for the management of western corn rootworm larvae. Suspension of the IJs was applied at sowing, directly into the seed furrows, at a depth of 5-6 cm using a special EPN application system. The EPN application system consisted of a 30 litre tank, a 12 V pump, and distribution pipes. The tube for the application of EPN was aligned to each row unit on the planter, which allowed the direct application into the vicinity of the sown maize kernel of the EPN suspension during sowing. The treatment with *H. bacteriophora* significantly decreased (46.2%) the numbers of emerging beetles (Modic et al., 2020). The efficacy of combination of a polymer gel (Barricade 1%) and Scanmask (*S. feltiae*) as seed treatment of canola resulted in highest yield and lowest leaf area damage against crucifer flea beetle (Antwi and Reddy, 2016).

4.2. Seed Bio-Priming

Seed bio-priming is the immersion of seeds in a microbial suspension and give protection against diseases and pests.

4.3. Seedling Dip

Dipping seedling roots or soaking plant cuttings into EPN suspension are good mean of prophylactic application to economize nematode application rates.

4.4. Applying to the Infection Site

Spraying nematodes just around the base of plants (Askary et al., 2017), Subsurface application of nematodes with an adapted seed-driller (Shetlar et al., 1993) or subsurface injection (Raja et al., 2015).

4.5. Soil Application

The most widespread practice for *Steinernema* and *Heterorhabditis* is the application of IJs in belowground agroecosystems, where they are naturally adapted in soil (Klein, 1990; Bruck et al., 2005; Garcia et al., 2008; Lacey et al., 2015). Application methods aim to maximize EPN survival in soils and movement towards their hosts. Efficacy of EPNs in soil is highly dependent on several biotic and abiotic factors (Stuart et al., 2015). Soil texture affects nematode movement and survival (Barbercheck, 1992; Shapiro-Ilan et al., 2006b; Toepfer et al., 2010). Soils with higher clay content have potential for reduced aeration, which can restrict nematode movement and thereby reduced nematode efficacy (Georgis and Poinar, 1983; Molyneux and Bedding, 1984; Kung et al., 1990). However, Kaya (1990) reported that activity and survival of EPNs is favored by clay rather than sandy loam soil. A soil pH of 10 or higher is likely to be harmful to EPN applications, can affect natural distribution, whereas a range of 4-8 is not likely to have any negative effect on EPNs (Kung et al., 1990). Nematodes require a thin film of water for movement but are not capable of movement under flooded conditions.

Soil temperature can also affect nematode efficacy (Askary and Ahmad, 2021a). Optimum temperature for infection and reproduction varies among nematode species and strain (Grewal et al., 1994). Some nematode species such as *H. indica, S. glaseri, S. riobrave* are relatively heat tolerant, whereas *H. megidis*, *S. feltiae* are more tolerant to cooler temperature. Warm temperatures reduce nematode survival (Askary and Ahmad, 2021a), while cooler temperatures reduce activity and infectivity (Grewal et al., 1994; Fitters et al., 2001). Soil temperatures between 12 to 28°C are considered favourable for application of most nematode species. If soil temperature is above 28°C, a pre-application irrigation is recommended to reduce soil temperature before nematode application (Grewal, 2002).

4.6. Foliar Application

IJs may be applied to foliage using common agrochemical equipment (Georgis, 1990; Arthurs et al., 2004). Entomopathogenic nematodes have now been applied in a range of habitats that include boreholes and galleries in stems or wood, leaf mines, curled leaves, reproductive structures (e.g., flowers, buds) as well as in the leaf surface. IJs are sensitive to inactivation by extremes of physical environment, i.e., desiccation (Begley, 1990; Womersley, 1990;

Lello et al., 1996), high temperatures (Grewal et al., 1994) and UV radiation (Gaugler and Boush, 1978; Gaugler et al., 1992; Shapiro-Ilan et al., 2015).

Most nematode species do not infect hosts at temperatures exceeding 32°C and differ in desiccation and UV tolerance. The effects of sunlight can be minimized by applying the nematodes at dusk but maintaining high relative humidity (>80%) and free water on the leaf surfaces is more difficult to achieve. The detrimental effects can be alleviated with antidesiccants such as certain commercial oils, plant-based products, and glycerin. Nickle and Shapiro (1994) demonstrated effective protection of *S. carpocapsae* from sunlight using stilbene brighteners, fluorescent brightener (Tinopal) and Blankophor BBH. Efforts to extend nematode survival with the addition of evapo-retardants and UV protectants resulted in partial success against leaf miners (*Liriomyza* spp.) and the corn earworm (*Heliothis* spp.) (Glazer and Novon, 1990). Direct observations showed that IJs of *S. carpocapsae* exposed to the adjuvants unsheathed more readily as compared to IJs stored in distilled water alone (Campbell and Gaugler,1993).

S. carpocapsae can tolerate desiccation stress better than other EPNs (Glazer and Navon, 1990) and a few foliar applications of this nematode against leaf-eating caterpillars such as *Tuta absoluta*, *Spodoptera littoralis*, *Helicoverpa armigera*, and *Pieris brassicae* on several crops have proved successful. IJs of *S. carpocapsae* nictate and have the ability to leap and disperse on the leaf (Ishibashi and Kondo, 1990; Lewis et al., 1992; Baur et al., 1995). Plant-boring insects have been controlled by injecting nematode suspensions directly into the borer holes or blocking the holes with sponges soaked with nematode suspensions (de altube et al., 2008; Dembilio et al., 2010). For the banana weevil, a nematode suspension can be placed into insect-attracting cuts in residual rhizomes of bananas. Baits containing IJs may offer a cost-effective way of controlling adult house flies.

Although adjuvants, instance gels and injections generally enhance nematode efficacy (MacVean et al., 1982; Glazer et al., 1992; Broadbent and Olthof, 1995; Baur et al., 1997b; de altube et al., 2008; Morton and Garcıa Del Pino, 2008; Dembilio et al., 2010; Shapiro-Ilan et al., 2010b), the level of increase is generally considered insufficient for recommending foliar applications (Baur et al., 1995; Mason et al., 1998). Therefore, application method needs to be more refined.

5. Application Challenges

Optical application strategies (when to apply, rate of application, frequency of application) are different for different EPN species or strains and targeted pests to maintain high level of control at low cost. Plant morphology and phenology must be considered in predicting whether EPNs are viable candidates for management of insect pests. EPNs should be prepared for field application according to the formulation. If nematodes are in a liquid suspension, the shipment container should be shaken well before mixing with water. If nematodes are in a sponge or in vermiculite or in talc formulation and added directly to water in the application container, it should be agitated in order to keep the nematode in suspension and supplied with water.

Efficacy of EPN applications can be enhanced through improved application equipment or approaches (single versus multiple applications) e.g., optimizing spray systems (e.g., nozzles, pumps, spray distribution) according to the size, for enhanced EPNs survival and dispersion (Shapiro-Ilan et al., 2006a). To minimize desiccation, spraying should be done during the evening hours. It is better to go for supplemental soil wetting prior to application or followed by application. Applicators could also use other methods such as microjet irrigation systems, subsurface injection or baits (Wright et al., 2005; Lara et al., 2008). Duncan et al. (1999) suggested that unequal distribution in some irrigation systems (microjet) can be overcome through addition of extra emitters at the end of the lines.

Application with better strain of EPNs results in good control of pest. Better strain of EPNs have high virulence, host finding behaviour, infection and environmental tolerance and persistence or recycling potential (Lewis et al., 1998; Shapiro-Ilan et al., 2002; Sharifi-Far et al., 2018). Lewis et al. (1992) divided host searching behaviours into cruisers and ambushers and intermediate in both. Sampling pattern and time of sampling, and environmental conditions are considered important aspect in order to select better strain of EPNs. Methods to improve EPNs include strain or species discovery or genetic enhancement via selection, hybridization or molecular manipulation (Gaugler and Campbell, 1991; Mukuka et al., 2010; Salame et al., 2010; Anbesse et al., 2013; Askary and Abd-Elgawad, 2021). The centers of origin for *Steinernema* and *Heterorhabditis* should be determined so that ecological investigations can be carried out in the habitat in which the nematode species presumably evolved.

Enhanced efficacy in EPN applications can be facilitated through improved formulation. For above-ground application, leaf flooding or mixing EPNs with a surfactant and polymer, sprayable gel, chitosan formulation, wood flour foam (Schroeder and Sieblirth,1997; Williams and Walters, 2000; Head et al., 2004; Schroer and Ehlers, 2005; Liacer et al., 2009; Shapiro-Ilan et al., 2010, Lacey et al., 2010; van Niekerk and Malan, 2015) will help in increasing leaf coverage.

Promotion of EPNs also includes employment of novel approaches for applying the symbiotic bacteria of EPNs or their metabolites or byproducts for arthropod pests and plant pathogens (El-Sadawy et al., 2018).

Increased knowledge about the ecological, biological, evolutionary and computational sciences that influence EPN populations will lead to significant impact in biological control of insect pests (Shapiro-Ilan et al., 2018). The diversity of EPNs and their interactions with other organisms can be altered by management practices in the field (Shapiro-Ilan et al., 2002; Barbercheck and Hoy, 2005; Stuart et al., 2006). Millar and Barbercheck (2002) observed EPN abundance along with diversity of hosts where the soil environment was associated with relatively high levels of crop residue in conservation tillage regimes.

Surface residues could benefit nematode persistence through protection from desiccation or ultraviolet light, and increase insect pest suppression by EPNs (Shapiro-Ilan et al., 1999) or enhance nematode movement (Jabbour and Barbercheck, 2008). EPNs can be included in IPM programmes in ways that make them complementary or superior to other agricultural inputs (Griffin, 2015; Laznik and Trdan, 2017; Bajc et al., 2017; Jaffuel et al., 2019). Co-application with entomopathogenic fungi, arbuscular mycorrhizal soil fungi, and entomopathogenic bacteria are some promising approaches for sustainable agriculture, that once optimized, could save costs and time to growers (Shapiro-Ilan et al., 2004; Molina-Acevedo et al., 2007; Ansari et al., 2008; Imperiali et al., 2017; Jaffuel et al., 2019).

Conclusion

A growing interest in alternatives to synthetic insecticides and organic agriculture requires biocontrol agents like EPNs, but they will need to be further improved with respect to efficacy through improved application technologies (Lacey and Georgis, 2012; Shapiro Ilan et al., 2019; Koppenhofer et al., 2020). Regional programs on IPM can promote local EPN production that persist in the environment for several growing seasons, once established. Effective and efficient delivery of entomopathogenic nematodes can only be achieved with careful consideration of moisture, temperature, aeration, soil type, salinity, organic content of soil, exposure to UV light as well as foraging strategy of EPNs and pathogenicity for the targeted insect pests. Inappropriate application methods, dose of application, use of poor quality products, suboptimal application conditions or time may result in poor efficacy of EPN against the target pests. Therefore, an improvement in application technology of EPNs will help dealing with the crop insect pests in a tactful manner.

Acknowledgements

The author is thankful to her institute for the support provided during the preparation of this chapter.

References

Abd-Elgawad, M. M. M., Askary, T. H. and Coupland, J. (2017). *Biocontrol Agents: Entomopathogenic and Slug Parasitic Nematodes*. CAB International, Wallingford, UK.

Akhurst, R. and Smith, K. (2002). Regulation and safety. In: *Entomopathogenic Nematology* (Eds. Gaugler, R.). CAB International, Wallingford, UK, pp. 311-332.

Alumai, A. and Grewal, P. S. (2004). Tank-mix compatibility of the entomopathogenic nematodes, *Heterorhabditis bacteriophora* and *Steinernema carpocapsae*, with selected chemical pesticides used in turfgrass. *Biocontrol Science and Technology*, 14: 725-730.

Anbesse, S., Sumaya, H. N., Dorfler, V. A., Strauch, O., and Ehlers, R. U. (2013). Stabilization of heat tolerance traits in *Heterorhabditis bacteriophora* through selective breeding and creation of inbred lines in liquid culture. *BioControl,* 58: 85-93.

Anbesse, S. A., Adge, B. J. and Gebru, W M. (2008). Laboratory screening for virulent entomopathogenic nematodes (*Heterorhabditis bacteriophora* and *Steinernema yirgalemense*) and fungi (*Metarhizium anisopliae* and *Beauveria bassiana*) and assessment of possible synergistic effects of combined use against grubs of the barley chafer *Coptognathus curtipennis*. *Nematology*, 10: 701-709.

Ansari, M. A., Shah, F. A. and Moens, T. M. (2006). Field trials against *Hoplia philanthus* (Coleoptera:Scarabaeidae) with a combination of an entomopathogenic nematode and the fungus *Metarhizium anisopliae* CLO53. *Biological Control*, 39: 453-459.

Ansari, M. A., Shah, F. A. and Butt, T. M. (2008). Combined use of entomopathogenic nematodes and *Metarhizium anisopliae* as a new approach for black vine weevil, *Otiorhynchus sulcatus*, control. *Entomologia Experimentalis et Applicata*, 129: 340-347.

Ansari, M. A., Tirry, L. and Moens, M. (2004). Interaction between *Metarhizium anisopliae* CLO53 and entomopathogenic nematodes for the control of *Hoplia philanthus*. *Biological Control*, 31: 172-180.

Antwi, F. B. and Reddy, G. V. P. (2016). Efficacy of entomopathogenic nematodes and sprayable polymer gel against crucifer flea beetle (Coleoptera: Chrysomelidae) on canola. *Journal of Economic Entomology*, 109: 1706-1712.

Arthurs, S., Heinz, K. M. and Prasifka, J. R. (2004). An analysis of using entomopathogenic nematodes against above-ground pests. *Bulletin of Entomological Research*, 94: 297-306.

Askary, T. H. (2010). Nematodes as biocontrol agents. In: *Sociology, Organic Farming, Climate Change and Soil Science* (Ed. Lichtfouse, E.). Springer, The Netherlands, pp. 347-378.

Askary, T. H. and Abd-Elgawad, M. M. M. (2021). Opportunities and challenges of entomopathogenic nematodes as biocontrol agents in their tripartite interactions. *Egyptian Journal of Biological Pest Control*, 31: 42. https://doi.org/10.1186/s41938-021-00391-9.

Askary, T. H. and Ahmad, M. J. (2017). Entomopathogenic nematodes: mass production, formulation and application. In: *Biocontrol Agents: Entomopathogenic and Slug Parasitic Nematodes* (Eds. Abd-Elgawad, M. M. M., Askary, T. H. and Coupland, J.). CAB International, Wallingford, UK, pp. 261-286.

Askary, T. H. and Ahmad, M. J. (2020). Efficacy of entomopathogenic nematodes against the cabbage butterfly (*Pieris brassicae* (L.) (Lepidoptera: Pieridae) infesting cabbage under field conditions. *Egyptian Journal of Biological Pest Control*, 30: 39. https://doi.org/10.1186/s41938-020-00243-y.

Askary, T. H. and Ahmad, M. J. (2021a). Survival and virulence capacity of native strain of entomopathogenic nematode, *Steinernema cholashanense* in different formulations. *Agricultural Research*, 10: 390-397.

Askary, T. H. and Ahmad, M. J. (2021b). Biocidal efficacy of some native isolates of entomopathogenic nematodes against oriental armyworm, *Mythimna separata* Walker (Lepidoptera: Noctuidae). *Indian Journal of Nematology*, 51: 67 – 73.

Askary, T. H., Ahmad, M. J., Wani, A. R., Mohiddin, S. and Sofi, M. A. (2018). Behavioural ecology of entomopathogenic nematodes, *Steinernema* and *Heterorhabditis* for insect biocontrol. In: *Sustainable Agriculture Reviews 31* (Ed. Lichtfouse, E.). Springer, The Netherlands, pp. 425-441.

Askary, T. H., Bhat, A. H., Ahmad, M. J., Chaubey, A. K. and Spiridonov, S. E. (2020). *Steinernema feltiae* (Rhabditida: Steinernematidae) from hilly areas of Kashmir valley, India with a note on its geographical distribution. *Russian Journal of Nematology*, 28: 99 – 106.

Askary, T. H., Khan, A. A., Waliullah, M. I. S., Banday, S. A., Iqbal, U. and Mir, M. M. (2012). Slug pest management through nematodes in agricultural and horticultural crops. In: *Nematodes: Morphology, Functions and Management Strategies*. (Eds. Boeri, F. and Chung, J. A.). Nova Publishers, New York, pp.197-211.

Askary, T. H. and M. M. M. Abd-Elgawad (2017) Beneficial nematodes in agroecosystem: A global perspective. In: *Biocontrol Agents: Entomopathogenic and Slug Parasitic Nematodes* (eds. Abd-Elgawad, M. M. M., Askary, T. H. and Coupland, J.). CAB International, Wallingford, UK, pp. 3-25.

Askary, T. H., Nermut, J., Ahmad, M. J. and Ganai, M. A. (2017). Future thrusts in expanding the use of entomopathogenic and slug parasitic nematodes in agriculture. In: *Biocontrol Agents: Entomopathogenic and Slug Parasitic Nematodes* (Eds. Abd-Elgawad, M. M. M., Askary, T. H. and Coupland, J.). CAB International, Wallingford, UK, pp. 620-627.

Bajc, N., Drzaj, U., Trdan, S. and Laznik, Z. (2017). Compatibility of acaricides with entomopathogenic nematodes (*Steinernema* and *Heterorhabditis*). *Nematology,* 19: 891-898.

Barbercheck, M. E. (1992). Effect of soil physical factors on biological control agents of soil insect pests. *The Florida Entomologist*, 75: 539-548.

Barbercheck, M. E. and Hoy, C. W. (2005). A systems approach to conservation of nematodes. In: *Nematodes as Biocontrol Agents* (Eds. Grewal, P. S., Ehlers, R. U. and Shapiro–Ilan, D. I.). CAB International, Wallingford, UK, pp. 331-347.

Barbercheck, M. E. and Kaya, H. K. (1990). Interaction between *Beauveria bassiana* and the entomogenous nematodes *Steinernema feltiae* and *Heterorhabditis heliothidis*. *Journal of Invertebrate Pathology*, 55: 225-234.

Batista, E. S. de P. and Auad, A. M. (2010). Application method of entomopathogenic nematodes for control of *Mahanarva spectabilis* (Hemiptera: Cercopidae). *Biocontrol Science and Technology*, 20: 1079-1085.

Baur, M. E., Kaya, H. K. and Thurston, G. S. (1995). Factors affecting entomopathogenic nematode infection of *Plutella xylostella* on a leaf surface. *Entomologia Experimentalis et Applicata*, 77: 239-250.

Baur, M. E., Kaya, H. K., Gaugler, R. and Tabashnik B. (1997a). Effects of adjuvants on entomopathogenic nematode persistence and efficacy against *Plutella xylostella*. *Biocontrol Science and Technology*, 7: 513-526.

Baur, M. E., Kaya, H. K. and Tabashnik B. (1997b). Efficacy of a dehydrated steinernematid nematode against black cutworm (Lepidoptera:Noctuidae) and diamond backmoth (Lepidoptera:Plutellidae). *Journal of Economic Entomology* 90:1200-1206.

Bednarek, A. and Gaugler, R. (1997). Compatibility of soil amendments with entomopathogenic nematodes. *Journal of Nematology*, 29: 220-227.

Begley, J. W. (1990). Efficacy against insects in habitats other than soil. In: *Entomopathogenic Nematodes in Biological Control* (Eds. Gaugler, R. and Kaya, H. K.). CRC Press, Boca Raton, Forida, pp. 215-232.

Bender, G. S., Bates, L. M., Bethke, J. A., Lewis, Ed., Tanizaki, G. and Morse, J. G., Godfrey, K. E. (2014). Evaluation of insecticides, entomopathogenic nematodes, and physical soil barriers for control of *Diaprepes abbreviates* (Coleoptera: Curculionidae) in citrus. *Journal of Economic Entomology*, 107: 2137-2146.

Benz, G. (1971). Synergism of micro-organisms and chemical insecticides. *In: Microbial Control of Insects and Mites* (Eds. Burgess, H. D. and Hussey, N. W.). Academic Press, London, pp. 327-355.

Bhat, A. H., Askary, T. H., Ahmad, M. J., Suman, Aasha and Chaubey, A. K. (2019). Description of *Heterorhabditis bacteriophora* (Nematoda: Heterorhabditidae) isolated from hilly areas of Kashmir valley. *Egyptian Journal of Biological Pest Control*, 29: 96. https://doi.org/10.1186/s41938-019-0197-6.

Brinkman, M. A. and Gardner, W. A. (2000). Possible antagonistic activity of two entomopathogens infecting workers of the red imported fire ant (Hymenoptera: Formicidae). *Journal of Entomological Science*, 35: 205-207.

Broadbent, A. B. and Olthof, T. H. A. (1995). Foliar application of *Steinernema carpocapsae* (Rhabditida: Steinernematidae) to control *Liriomyza trifolii* (Diptera: Agromyzidae) larvae in chrysanthemums. *Environmental Entomology*, 24: 431-435.

Bruck, D. J., Shapiro-Ilan, D. I. and Lewis, E. E. (2005). Evaluation of application technologies of entomopathogenic nematodes for control of the black vine weevil. *Journal of Economic Entomology*, 98: 1884-1889.

Brusselman, E., Beck, B., Pollet, S., Temmerman, F., Spanoghe, P., Moens, M. and Nuyttens, D. (2012). Effect of spray volume on the deposition, viability and infectivity of entomopathogenic nematodes in a foliar spray on vegetables. *Pest Management Science*, 68: 1413–418.

Burnell, A. M. and Stock, P. (2000). *Heterorhabditis*, *Steinernema* and their bacterial symbionts:lethal pathogens of insects. *Nematology*, 2: 31-42.

Cabanillas, H. E. and Raulston, J. R. (1996). Evaluation of *Steinernema riobravis, S.carpocapsae,* and irrigation timing for the control of corn earworm, *Helicoverpa zea*. *Journal of Nematology*, 28: 75-82.

Campbell, J. and Gaugler, R. (1993). Nictation behavior and its ecological implications in the host search strategies of entomopathogenic nematodes (Heterorhabditidae and Steinernematidae). *Behaviour*, 126: 154-169.

Chang, F. N. and Gehret, M. J. (1992). Stabilizer insect nematode compositions. *WIPO Patent No. WO 92/10170.*

de Altube, M. D. M., Strauch, O., de Castro, G. F. and Pena, A. M. (2008). Control of the flat-headed root borer *Capnodis tenebrionis* (Linne) (Coleoptera: Buprestidae) with the entomopathogenic nematode *Steinernema carpocapsae* (Weiser) (Nematoda: Steinernematidae) in a chitosan formulation in apricot orchards. *BioControl*, 53: 531-539.

Del Valle, E. E., Dolinksi, C., Barreto, E. L. S. and Souza, R. M. (2008). Dispersal of *Heterorhabditis baujardi* LPP7 (Nematoda:Rhabditida) applied to the soil as infected host cadavers. *International Journal of Pest Management*, 54: 115-122.

Del Valle, E. E., Dolinksi, C., Barreto, E. L. S. and Souza, R. M. (2009). Effect of cadaver coatings on emergence and infectivity of the entomopathogenic nematode *Heterorhabditis baujardi* LPP7(Rhabditida: Heterorhabditidae) and the removal of cadavers by ants. *Biological Control*, 50: 21-24.

Del Valle E. E., Dolinski, C., Barreto, E. L. S. and Souza, R. M. (2011). Time period between infection by *Heterorhabditis baujardi* LPP7 (Nematoda:Rhabditida) and soil application of *Galleria mellonella* (Lepidoptera:Pyralidae) cadavers on the emergence of infective juveniles. *Nematologia Brasileira*, 37: 26-28.

Dembilio, O., Llacer, E., de Altube, M. M. and Jacas, J. A. (2010). Field efficacy of imidacloprid and *Steinernema carpocapsae* in a chitosan formulation against the red palm weevil *Rhynchophorus ferrugineus* (Coleoptera: Curculionidae) in *Phoenix canariensis*. *Pest Management Science*, 66: 365-370.

Deol, Y. S., Jagdale, G. B., Canas, L. and Grewal, P. S. (2011). Delivery of entomopathogenic nematodes directly through commercial growing media via the inclusion of infected host cadavers: A novel approach. *Biological Control*, 58: 60-67.

Dolinski, C., Shapiro-Ilan, D. I. and Lewis, E. E. (2015). Insect cadaver applications: pros and cons. In: *Nematode Pathogenesis of Insects and Other Pests: Ecology and Applied Technologies for Sustainable Plant and Crop Protection* (Ed. Campos-Herrera, R.). Springer, Switzerland, pp. 207-230.

Duncan, L. W., Graham, J. H., Dunn, D. C., Zellers, J., McCoy, C. W. and Nguyen, K. (2003). Incidence of endemic entomopathogenic nematodes following application of *Steinernema riobrave* for control of *Diaprepes abbreviates*. *Journal of Nematology*, 35: 178-186.

Duncan, L. W. and McCoy, C. W. (1996). Vertical distribution in soil, persistence, and efficacy against citrus root weevil (Coleoptera: Curculionidae) of two species of entomogenous nematodes (Rhabditida: Steinernematidae; Heterorhabditidae). *Environmental Entomology*, 25: 174-178.

Duncan, L. W., Shapiro-Ilan, D. I., McCoy, C. W. and Graham, J. H. (1999). Entomopathogenic nematodes as a component of citrus root weevil IPM. *Proceedings of Workshop on Optimal Use of Insecticida Nematodes in Pest Management.* (Ed. Polvarapu, S.). Rutgers University, New Jersey, pp. 69-78.

El-Borai, F., Killiny, N., Duncan, L. W. (2016). Concilience in entomopathogenic nematode responses to water potential and their geospatial patterns in Florida. *Frontiers in Microbiology*, 7: 356. http://doi.org/10.3389/fmicb.2016.00356.

El-Borai, F. E., Campos-Herrera, R., Stuart, R. J. and Duncan, L. W. (2011). Substrate modulation, group effects and the behavioral responses of entomopathogenic nematodes to nematophagous fungi. *Journal of Invertebrate Pathology*, 106: 347-356.

El-Sadawy, H. A., Elnamaky, A. H., Elsayed, E. H., Abdelmaguid, B. B., Ayaad, T. H. and Ahmed, A. M. (2018). Silver nanoparticles enhance the larvicidal toxicity of *Photorhabdus* and *Xenorhabdus* bacterial toxins: an approach to control the filarial vector, *Culex pipiens*. *Tropical Biomedicine*, 35: 392-407.

Fife, J. P., Ozkan, H. E., Derksen, R. C., Grewal, P. S. and Krause, C. R. (2005). Viability of a biological pest control agent through hydraulic nozzles. *Transactions of the ASAE*, 48: 45-54.

Fitters, P. F. L., Dunne, R. and Griffin, C. T. (2001). Improved control of *Otiorhynchus sulcatus* at 9°C by cold-stored *Heterorhabditis megidis* UK211. *Biocontrol Science and Technology*, 11: 483-492.

García, L. C., Raetano, C. G. and Leite, L. G. (2008). Application technology for the entomopathogenic nematodes *Heterorhabditis indica* and *Steinernema* sp. (Rhabditida: Heterorhabditidae and Steinernematidae) to control *Spodoptera frugiperda* (Smith) (Lepidoptera: Noctuidae) in corn. *Neotropical Entomology*, 37: 305-311.

Gassmann, A. J., Stock, S. P., Sisterson, M. S., Carriere, Y. and Tabashnik, B. E. (2008). Synergism between entomopathogenic nematode and *Bacillus thuringiensis* crops: integrating biological control, and resistance management. *Journal of Applied Ecology*, 45: 957-966.

Gaugler, R. (1999). Matching Nematode and insect to achieve optimal field performance. *Proceedings of Workshop on Optimal Use of Insecticida Nematodes in Pest Management.* (Ed. Polvarapu, S.). Rutgers University, New Jersey, pp. 9-14.

Gaugler, R. and Boush, G. M. (1978). Effects of ultraviolet radiation and sunlight on the entomogenous nematode, *Neoaplectana carpocapsae*. *Journal of Invertebrate Pathology*, 32: 291-296.

Gaugler, R. and Campbell, J. F. (1991). Selection for enhanced host finding of scarab larvae (Coleoptera, Scarabaeidae) in an entomopathogenic nematode. *Environmental Entomology*, 20: 700-706.

Gaugler, R., Campbell, J. F., Selvan, S. and Lewis, E. E. (1992). Large-scale inoculative releases of the entomopathogenic nematode *Steinernema glaseri*: Assessment 50 years later. *Biological Control,* 2: 181-187.

Georgis, R. (1990). Formulation and application technology. In: *Entomopathogenic Nematodes in Biological Control* (Eds. Gaugler, R. and Kaya, H. K.). CRC Press, Boca Raton, Florida. pp. 173-194.

Georgis, R. and Hague, N. G. M. (1991). Nematodes as biological insecticides. *Pesticide Outlook,* 3: 29-32.

Georgis, R. and Poinar, G. O., Jr. (1983). Effect of soil texture on the distribution and infectivity of *Neoaplectana carpocapsae* (Nematoda: Steinernematidae). *Journal of Nematology*, 15: 308-311.

Glazer, I., Klein, M., Navon, A. and Nakache, Y. (1992). Comparison of efficacy of entomopathogenic nematodes combined with antidesiccants applied by canopy sprays against three cotton pests (Lepidoptera: Noctuidae). *Journal of Economic Entomology*, 85: 1636-1641.

Glazer, I. and Navon, A. (1990). Activity and persistence of entomopathogenic nematodes used against *Heliothis armigera* (Lepidoptera: Noctuidae). *Journal of Economic Entomology*, 83: 1795-1800.

Grewal, P. S. (2002). Formulation and application technology. In: *Entomopathogenic Nematology* (Ed. Gaugler, R.). CAB International, Wallingford, UK, pp. 265-288.

Grewal, P. S., Ehlers, R. U. and Shapiro-Ilan, D. (2005). Critical issues and research needs for expanding the use of nematodes in biocontrol. In: *Nematodes as Biocontrol Agents* (Eds. Grewal, P. S., Ehlers, R. U. and Shapiro-Ilan, D. I.). CAB International, Wallingford, UK, pp. 479-498.

Grewal, P. S., Selvan, S. and Gaugler, R. (1994). Thermal adaptation of entomopatho-genic nematodes-niche breadth for infection, establishment and reproduction. *Journal of Thermal Biology*, 19: 245-253.

Griffin, C. T. (2015). Behaviour and population dynamics of entomopathogenic nematodes following application. In: *Nematode Pathogenesis of Insects and Other Pests: Ecology and Applied Technologies for Sustainable Plant and Crop Protection* (Ed. Campos-Herrera, R.). Springer, Switzerland, pp. 57-95.

Gulzar, S., Usman, M., Wakil, W., Gulcu, B., Hazir, C., Karagoz, M., Hazir, S., Shapiro Ilan, D. I. (2020). Environmental tolerance of entomopathogenic nematodes differs among nematodes arising from host cadavers versus aqueous suspension. *Journal of Invertebrate Pathology*, 175: 107452. https://doi.org/10.1016/j.jip.2020.107452.

Gumus, A., Karagoz, M., Shapiro-Ilan, D. I. and Hazir, S. (2015). A novel approach to biocontrol: release of live insect hosts pre-infected with entomopathogenic nematodes. *Journal of Invertebrate Pathology*, 130: 56-60.

Hayes, A. E., Fitzpatrick, S. M. and Webster, J. M. (1999). Infectivity, distribution, and persistence of the entomopathogenic nematode *Steinernema carpocapsae* All strain (Rhabditida: Steinernematidae) applied by sprinklers or boom sprayer to dry-pick cranberries. *Journal of Economic Entomology*, 92: 539-546.

Head, J., Lawrence, A. J. and Walters, K. F. A. (2004). Effficacy of the entomopatho-genic nematode, *Steinernema feltiae* against *Bemisia tabaci* in relation to plant species. *Journal of Applied Entomology*, 128: 543-547.

Hiltpold, I. (2015). Prospects in the application technology and formulation of entomopathogenic nematodes for biological control of insect pests. In: *Nematode Pathogenesis of Insects and Other Pests*: *Ecology and Applied Technologies for Sustainable Plant and Crop Protection* (Ed. Campos-Herrera, R.). Springer, Switzerland, pp. 187-205.

Hiltpold, I., Hibbard, B. E., French, B. W. and Turlings, T. C. J. (2012). Capsules containing entomopathogenic nematodes as a Trojan horse approach to control the western corn rootworm. *Plant and Soil*, 358: 11-25.

Imperiali, N., Chiriboga, X., Schlaeppi, K., Fesselet, M., Villacres, D., Jaffuel, G., Franz Bender, S., Dennert, F., Blanco-Perez, R., van der Heijden, M. G., Maurhofer, M., Mascher, F., Turlings, T. C. J., Keel, C. J., Campos-Herrera, R. (2017). Combined field inoculations of *Pseudomonas* bacteria, arbuscular mycorrhizal fungi, and entomopathogenic nematodes and their effects on wheat performance. *Frontiers in Plant Science*. 8: 1809. http://doi.org/10.3389/fpls.2017.01809.

Ishibashi, N. and Kondo, E. (1990). Behavior of infective nematodes. In: *Entomopathogenic Nematodes in Biological Control* (Eds. Gaugler, R. and Kaya, H. K.). CRC Press, Boca Raton, Florida, pp.139-150.

Jabbour, R. and Barbercheck, M. E. (2008). Soil and habitat complexity effects on movement of the entomopathogenic nematode *Steinernema carpocapsae* in maize. *Biological Control*, 47: 235-243.

Jaffuel, G., Imperiali, N., Shelby, K., Campos-Herrera, R., Geisert, R. W., Maurhofer, M., Loper, J., Keel, C., Turlings, T. and Hibbard, B. E. (2019). Protecting maize from root worm damage with the combined application of arbuscular mycorrhizal fungi, Pseudomonas bacteria and entomopathogenic nematodes. *Scientific Reports*, 9 (2019). http://doi.org/10.1038/s41598-019-39753-7.

Janson, R. K. and Lecrone, S. H. (1994). Application methods for entomopathogenic nematodes (Rhabditida: Heterorhabditidae): aqueous suspensions versus infected cadavers. *Florida Entomologist*, 77: 281-284.

Kaya, H. K. (1990). Soil ecology. In: Entomopathogenic Nematodes in Biological Control (Ed. Gaugler, R.). CRC Press, Boca Raton, Forida, pp. 93-116.

Kaya, H. K. and Gaugler, R. (1993). Entomopathogenic nematodes. *Annual Review of Entomology*, 38: 181-206.

Kaya, H. K., Joos, J. L., Falcon, L. A. and Berlowitz, A. (1984). Suppression of the codling moth (Lepidoptera:Olethreutidae) with the entomogenous nematode, *Steinernema feltiae* (Rhabditida:Steinernematidae). *Journal of Economic Entomology*, 77: 1240-1244.

Kaya, H. K. and Nelson, C. E. (1985). Encapsulation of Steinernematid and Heterorhabditid nematodes with calcium alginate: A new approach for insect control and other applications. *Environmental Entomology*, 14: 572-574.

Khan, M. R., Khan, U., Askary, T. H., Mohiddin, F. A. and Khan, M. M. (2007). Pathogenicity and host suitability for *in vivo* mass production of *Steinernema masoodi* AMU EPN-1. *International Journal of Nematology*, 17: 151-157.

Klein, M. G. (1990). Efficacy against soil-inhabiting insect pests. In: *Entomopathogenic Nematodes in Biological Control* (Eds. Gaugler, R. and Kaya, H. K.). CRC Press, Boca Raton, Florida, pp. 195-214.

Klein, M. G. and Georgis, R. (1994). Application techniques for entomopathogenic nematodes, *Proceedings of the VIth International Colloquium on Invertebrate Pathology and Microbial Control*, Montpellier, France, pp. 483-484.

Klein, M. G. and Georgis, R. (1992). Persistence of control of Japanese beetle (Coleoptera: Scarabaeidae) larvae with steinernematid and heterorhabditid nematodes. *Journal of Economic Entomology*, 85: 727-730.

Koppenhofer, A. M. and Fuzy, E. M. (2008). Early timing and new combinations to increase the efficacy of neonicotinoid-entomopathogenic nematode (Rhabditida: Heterorhabditidae) combinations against white grubs (Coleoptera: Scarabaeidae). *Pest Management Science*, 64: 725-735.

Koppenhofer, A. M. and Grewal, P. S. (2005). Compatibility and interactions with agrochemicals and other biocontrol agents. In: *Nematodes as Biological Control Agents* (Eds. Grewal, P. S., Ehlers, R. U. and Shapiro-Ilan, D. I.). CAB International, Wallingford, UK, pp. 363-381.

Koppenhofer, A. M. and Kaya, H. K. (1997). Additive and synergistic interactions between entomopathogenic nematodes and *Bacillus thuringiensis* for scarab grub control. *Biological Control*, 8: 131-137.

Koppenhofer, A. M. and Kaya, H. K. (1998). Synergism of imidacloprid and an entomopathogenic nematode: a novel approach to white grub (Coleoptera: Scarabaeidae) control in turfgrass. *Journal of Economic Entomology*, 91: 618-623.

Koppenhofer, A. M., Kaya, H. K. and Taormino, S. P. (1995). Infectivity of entomopathogenic nematodes (Rhabditida: Steinernematidae) at different soil depths and moistures. *Journal of Invertebrate Pathology*, 65: 193-199.

Koppenhofer, A. M., Shapiro-Ilan, D. I. and Hiltpold, I. (2020). Advances in the use of entomopathogenic nematodes (EPNs) as biopesticides in suppressing crop insect pests. In: *Biopesticides for Sustainable Agriculture* (Eds. Birch, N. and Glare, T.). Burleigh Dodds Science Publishing, Cambridge, UK, 1-39.

Kung, S., Gaugler, R. and Kaya, H. K. (1990). Soil type and entomopathogenic nematode persistence. *Journal of Invertebrate Pathology*, 55: 401-406.

Lacey, L. A., Grzywacz, D., Shapiro-Ilan, D. I., Frutos, R., Brownbridge, M. and Goettel, M. S. (2015). Insect pathogens as biological control agents: Back to the future. *Journal of Invertebrate Pathology*, 132: 1-41.

Lacey, L. A., Shapiro-Ilan, D. I. and Glenn G. M. (2010). Post-application of anti-desiccant agents improves efficacy of entomopathogenic nematodes in formulated host cadavers or aqueous suspension against diapausing codling moth larvae (Lepidoptera: Tortricidae). *Biocontrol Science and Technology*, 20: 909-921.

Lara, J. C., Dolinski, C., Fernandes de Sousa, E. and Figueiredo, D. E. (2008). Effect of mini-sprinkler irrigation system on *Heterorhabditis baujardi* LPP7 (Nematoda: Heterorhabditidae) infective juvenile. *Scientia Agricola*, 65: 433-437.

Laznik, Z. and Trdan, S. (2017). The influence of herbicides on the viability of entompathogenic nematodes (Rhabditida: Steinernematidae and Heterorhabditidae). *International Journal of Pest Management*, 63: 105-111.

Lello, E. R., Patel, M. N., Matthews, G. A. and Wright, D. J. (1996). Application technology for entomopathogenic nematodes against foliar pests. *Crop Protection*, 15: 567-574.

Lewis, E. E., Campbell, J. F. and Gaugler, R. (1998). A conservation approach to using entomopathogenic nematodes in turf and landscapes. In: *Conservation Biological Control* (Ed. Barbosa, P.). Academic Press, San Diego, USA, pp. 235-254.

Lewis, E. E., Gaugler, R. and Harrison, R., (1992). Entomopathogenic nematode host finding: response to host contact cues by cruise and ambush foragers. *Parasitology*, 105: 309-315.

Liacer, E., Martýnez de Altube, M. M. and Jacas, J. A. (2009). Evaluation of the efficacy of *Steinernema carpocapsae* in a chitosan formulation against the red palm weevil, *Rhynchophorus ferrugineus*, in *Phoenix canariensis*. *Biocontrol*, 54: 559-565.

MacVean, C. M., Brewer, J. W. and Capinera, J. L. (1982). Field test of antidesiccants to extend the infection period of an entomogenous nematode, *Neoaplectana carpocapsae*, against the Colorado potato beetle. *Journal of Economic Entomology*, 75: 97-101.

Mason, J. M., Matthews, G. A. and Wright, D. J. (1998). Screening and selection of adjuvants for the spray application of entomopathogenic nematodes against a foliar pest. *Crop Protection,* 17: 463-470.

Mason, J. M., Matthews, G. A. and Wright, D. J. (1999). Evaluation of spinning disc technology for the application of entomopathogenic nematodes against a foliar pest. *Journal of Invertebrate Pathology*, 73: 282-288.

Millar, L. C. and Barbercheck, M. E. (2002). Effects of tillage practices on entomopathogenic nematodes in a corn agroecosystem. *Biological Control,* 25: 1-11.

Modic, S., Zigon, P., Kolmanic, A., Trdan, S. and Razinger, J. (2020). Evaluation of the field efficacy of *Heterorhabditis bacteriophora* Poinar (Rhabditida: Heterorhabditidae) and synthetic insecticides for the control of western corn rootworm larvae. *Insects*, 11: 202.

Molina-Acevedo, J. P., Samuels, R., Machado, I. and Dolinski, C. (2007). Interactions between isolates of the entomopathogenic fungus *Metarhizium anisopliae* and the entomopathogenic nematode *Heterorhabditis bacteriophora* JPM4 during infection of the sugarcane borer *Diatraea saccharalis* (Lepidoptera: Pyralidae). *Journal of Invertebrate Pathology*, 96: 187-192.

Molyneux, A. S. and Bedding, R. A. (1984). Influence of soil texture and moisture on the infectivity of *Heterorhabditis* sp. D1 and *Steinernema glaseri* for larvae of the sheep blowfly *Lucilia cuprina*. *Nematologica*, 30: 358-365.

Morton, A. and Garcıa Del Pino, F. (2008). Field efficacy of the entomopathogenic nematode *Steinernema feltiae* against the Mediterranean flat-headed root borer *Capnodis tenebrionis*. *Journal of Applied Entomology*, 132: 632-637.

Mukuka, J., Strauch, O., Hoppe, C. and Ehlers, R. U. (2010). Improvement of heat and desiccation tolerance in *Heterorhabditis bacteriophora* through cross-breeding of tolerant strains and successive genetic selection. *Biocontrol*, 55: 511-521.

Nachtigall, S. and Dickler, E. (1992). Experiences with field applications of entomoparasitic nematodes for biological control of cryptic living insects in orchards. *Acta Phytopathologica et Entomologica Hungarica*, 27: 485-490.

Navon, A., Nagalakshmi, V. K., Levski, S., Salame, L. and Glazer, I. (2002). Effectiveness of entomopathogenic nematodes in an alginate gel formulation against lepidopteran pests. *Biocontrol Science and Technology*, 12: 737-746.

Nickle, W. R. and Shapiro, M. (1994). Effects of eight brighteners as solar radiation protectants for *Steinernema carpocapsae*, All Strain. *Journal of Nematology*, 26: 782-784.

Nilson, U. and Gripwall, E. (1999). Influence of application technique on the viability of the biological control agents *Verticillium lecanii* and *Steinernema feltiae*. *Crop Protection*, 18: 53-59.

Nishimatsu, T. and Jackson, J. J. (1998). Interaction of insecticides, entomopathogenic nematodes and larvae of the western corn rootworm (Coleptera: Chrysomelidae). *Journal of Economic Entomology*, 91: 410-418.

Oliveira-Hofman, C., Kaplan, F., Stevens, G., Lewis, E., Wu, S., Alborn, H. T., Perret-Gentil A., Shapiro-Ilan, D. I. (2019). Pheromone extracts act as boosters for entomopathogenic nematodes efficacy. *Journal of Invertebrate Pathology*, 164: 38-42.

Parkman, J. P., Frank, J. H., Nguyen, K. B. and Smart G. C. Jr. (1994). Inoculative release of *Steinernema scapterisci* (Rhabditida: Steinernemaidae) to suppress pest mole crickets (Orthoptera: Gryllotapidae) on golf courses. *Environmental Entomology*, 23: 1331-1337.

Perez, E. E., Lewis, E. E. and Shapiro-Ilan, D. I. (2003). Impact of host cadaver on survival and infectivity of entomopathogenic nematodes (Rhabditida: Steinernematidae and Heterorhabditidae) under desiccating conditions. *Journal of Invertebrate Pathology*, 82: 111-118.

Raja, R. K., Hazir, C., Gumus, A., Asan, C., Karagos, M. and Hazir, S. (2015). Efficacy of the entomopathogenic nematode *Heterorhabditis bacteriophora* using different application methods in the presence or absence of a natural enemy. *Turkish Journal of Agriculture and Forestry*, 39: 277-285.

Salame, L. and Glazer, I. (2000). Osmotic survival of the entomopathogenic nematode *Steinernema carpocapsae*, *Biological Control*, 18: 251-257.

Salame, L., Glazer, I., Chubinishvilli, M. T. and Chkhubianishvili, T. (2010). Genetic improvement of the desiccation tolerance and host-seeking ability of the entomopathogenic nematode *Steinernema feltiae*. *Phytoparasitica*, 38: 359-368.

Sankaranarayanan, C. and Askary, T. H. (2017) Status of Entomopathogenic Nematodes in Integrated Pest Management Strategies in India. In: *Biocontrol Agents: Entomopathogenic and Slug Parasitic Nematodes* (eds. M. M. M. Abd-Elgawad, T. H. Askary and James Coupland). CAB International, Wallingford, UK, pp. 362-382.

Schroeder, W. J. and Sieblirth, P. J. (1997). Impact of surfactants on control of the root weevil Diaprepes abbreviates larvae with *Steinernema riobravis*. *Journal of Nematology*, 29: 216-219.

Schroer, S. and Ehlers, R. U. (2005). Foliar application of the entomopathogenic nematode *Steinernema carpocapsae* for biological control of diamondback moth larvae (*Plutella xylostella*). *Biological Control*, 33: 81-86.

Shapiro-Ilan, D. I., and Brown, I. (2013). Earthworms as phoretic hosts for *Steinernema carpocapsae* and *Beauveria bassiana*: implications for enhanced biological control. *Biological Control*, 66: 41-48.

Shapiro-Ilan, D. I., Cottrell, T. E., Mizell, R. F., Horton, D. L. and Zaid, A. (2015). Field suppression of the peachtree borer, *Synanthedon exitiosa*, using *Steinernema carpocapsa*e: effects of irrigation, a sprayable gel and application method. *Biological Control*, 82: 7-12.

Shapiro-Ilan, D. I. and Dolinski, C. (2015). Entomopathogenic nematode application technology. In: *Nematode Pathogenesis of Insects and Other Pests*: *Ecology and Applied Technologies for Sustainable Plant and Crop Protection* (Ed. Campos-Herrera, R.). Springer, Switzerland, pp. 231–254.

Shapiro-Ilan, D. I. and Glazer, I. (1996). Comparison of entomopathogenic nematode dispersal from infected hosts versus aqueous suspension. *Environmental Entomology*, 25: 1455-1461.

Shapiro-Ilan, D. I., Gouge, D. H. and Koppenhofer, A. M. (2002). Factors affecting commercial success: Case studies in cotton, turf and citrus. In: *Entomopathogenic Nematology* (Eds. Gaugler, R.). CAB International Wallingford, UK, pp. 333-356.

Shapiro-Ilan, D. I., Gouge, D. H., Piggott, S. J., Fife, P. J. (2006a). Application technology and environmental considerations for use of entomopathogenic nematodes in biological control. *Biological Control*, 38: 124-133.

Shapiro-Ilan, D. I., Hazir, S. and Glazer, I. (2017). Basic and applied research: Entomopathogenic nematodes. In: *Microbial Agents for Control of Insect Pests from Discovery to Commercial Development and Use* (Ed. Lacey, L. A.). Academic Press, Amsterdam, pp. 91-105.

Shapiro-Ilan, D. I., Hazir, S. and Lete, L. (2015). Viability and virulence of entomopathogenic nematodes exposed to ultraviolet radiation. *Journal of Nematology*, 47: 184-189.

Shapiro-Ilan, D. I., Hiltpold, I. and Lewis, E. E. (2018). Ecology of invertebrate pathogens: nematodes. In: *Ecology of Invertebrate Diseases* (Eds. Hajek, A. E. and Shapiro-Ilan, D. I.). John Wiley and Sons, Hoboken, New Jersey, pp. 415-440.

Shapiro-Ilan, D. I., Jackson, M., Reilly, C. C. and Hotchkiss, M. W. (2004). Effects of combining an entomopathogenic nematodes on mortality of *Curculio caryae* (Coleoptera: Curculionidae). *Biological Control*, 30: 119-126.

Shapiro-Ilan, D. I., Kaplan, F., Oliveira-Hofman, C., Schliekelman, P., Alborn, H. T., and Lewis, E. E. (2019). Conspecific pheromone extracts enhance entomopathogenic infectivity. *Journal of Nematolology*, 51: 1-5. https://doi.org/10.21307/jofnem-2019-082.

Sharma, S., Kooner, R. and Arora, R. (2017). Insect pests and crop losses. In: *Breeding Insect Resistant Crops for Sustainable Agriculture* (Eds. Arora, R. and Sandhu, S.). Springer, Singapore, pp. 45-66.

Shapiro-Ilan, D. I. and Lewis, E. E. (1999). Comparison of entomopathogenic nematode infectivity from infected hosts verses aqueous suspension. *Environmenatal Entomology*, 28: 907-911.

Shapiro-Ilan, D. I., Lewis, E. E., Behle, R. W. and McGuire, M. R. (2001). Formulation of entomopathogenic nematode infected cadavers. *Journal of Invertebrate Pathology*, 78: 17-23.

Shapiro-Ilan, D. I., Morales-Ramos, J. A., Rojas, M. G. and Tedders, W. L. (2010). Effects of a novel entomopathogenic nematode infected host formulation on cadaver integrity, nematode yield, and suppression of *Diaprepes abbreviates* and *Aethina tumida* under controlled conditions. *Journal of Invertebrate Pathology*, 103: 103-108.

Shapiro-Ilan, D. I., Obrycki, J. J., Lewis, L. C. and Jackson, J. J. (1999). The effects of crop residue on the persistence of *Steinernema carpocapsae*. *Journal of Nematology*, 31: 517-519.

Sharifi-Far, S., Shapiro-Ilan, D. I., Brownbridge, M., and Hallett, R. H. (2018). The combined approach of strain discovery and the inbred line technique for improving control of *Delia radicum* with *Heterorhabditis bacteriophora*. *Biological Control*, 118: 37-43.

Shapiro-Ilan, D. I., Stuart, R. J., and McCoy, C. W. (2006b). A comparison of entomopathogenic nematode longevity in soil under laboratory conditions. *Journal of Nematology*, 38: 119-129.

Shapiro-Ilan, D. I., Tylka, G. L. and Lewis, L. C. (1996). Effects of fertilizers on virulence of *Steinernema carpocapsae*. *Applied Soil Ecology*, 3: 27-34.

Shetlar, D. J., Power, K. T. and Niemczyk, H. D. (1993). Effect of a subsurface application of entomopathogenic nematodes on non-target arthropods in turfgrass. *International Turfgrass Society Research Journal*, 7: 385-390.

Shetlar, D. J., Suleman, P. E. and Georgis, R. (1988). Irrigation and use of entomogenous nematodes, *Neoaplectana* spp. and *Heterorhabditis heliothidis* (Rhabditida: Steinernematidae and Heterorhabditidae), for control of Japanese beetle (Coleoptera: Scarabaeidae) grubs in turfgrass. *Journal of Economic Entomology*, 81: 1318-1322.

Shields, E. J., Testa, A., Miller, J. M., and Flanders, K. L. (1999). Field efficacy and persistence of the entomopathogenic nematodes *Heterorhabditis bacteriophora* 'Oswego' and *H. bacteriophora* 'NC' on Alfalfa snout beetle larvae (Coleoptera: Curculionidae). *Environmental Entomology*, 28: 128-136.

Smits, P. H. (1996). Post-application persistence of entomopathogenic nematodes. *Biocontrol Science and Technology*, 6: 379-388.

Stuart, R. J., Barbercheck, M. E. and Grewal, P. S. (2015). Entomopathogenic nematodes in the soil environment: distribution, interactions and the influence of biotic and abiotic factors. In: *Nematode Pathogenesis of Insects and Other Pests: Ecology and Applied Technologies for Sustainable Plant and Crop Protection* (Ed. Campos-Herrera, R.). Springer, Switzerland, pp. 97-138.

Stuart, R. J., Barbercheck, M. E., Grewal, P. S., Taylor, R. A. J. and Hoy, C. W. (2006). Population biology of entomopathogenic nematodes: Concepts, issues, and models. *Biological Control*, 38: 80-102.

Thurston, G. S., Kaya, H. K. and Gaugler, R. (1994). Characterizing the enhanced susceptibility of milky disease-infected scarabaeid grubs to entomopathogenic nematodes. *Biological Control*, 4: 67-73.

Toepfer, S., Kurtz, B. and Kuhlmann, U. (2010). Influence of soil on the efficacy of entomopathogenic nematodes in reducing *Diabrotica virgifera virgifera* in maize. *Journal of Pest Science*, 83: 257-264.

Unruh, T. R. and Lacey, L. A. (2001). Control of codling moth, *Cydia pomonella* (Lepidoptera: Tortricidae), with *Steinernema carpocapsae*: effects of supplemental wetting and pupation site on infection rate. *Biological Control*, 20: 48-56.

Usman, M., Gulzar, S., Wakil, W., Pinero, J. C., Leskey, T. C., Nixon, L. J., Hofman, C. O., Wu, S., and Shapiro-Ilan, D. I. (2020). Potential of entomopathogenic nematodes against the pupal stage of the apple maggot *Rhagoletis pomonella* (Walsh) (Diptera: Tephritidae). *Journal of Nematology*, 52: 1-9.

Van Niekerk, S. and Malan, A. P. (2015). Adjuvants to improve aerial control of the citrus mealybug *Planococcus citri* (Hemiptera: Pseudococcidae) using entomopathogenic nematodes. *Journal of Helminthology*, 89: 189-195.

Welch, H. E. and Briand, L. J. (1960). Field experiment on the use of a nematode for the control of vegetable crop insects. *Proceedings of the Entomological Society of Ontario* 91: 197-202.

Willett, D. S., Alborn, H. T., Stelinski, L. L. and Shapiro-Ilan, D. I. (2018). Risk taking of educated nematodes. *PLoS ONE* 13:e0205804. http://doi.org/10.1371/journal.pone.0205804.

Williams, E. C. and Walters, K. F. A. (2000). Foliar application of the entomopathogenic nematode *Steinernema feltiae* against leafminers on vegetables. *Biocontrol Science and Technology*, 10: 61-70.

Womersley, C. Z. (1990). Dehydration survival and anhydrolic potential. In: *Entomopathogenic Nematodes in Biological Control* (Gaugler, R. and Kaya, H. K.). CRC Press, Boca Raton, Florida, pp.117-138.

Wright, D. J., Peters, A., Schroer, S., and Fife, J. P. (2005). Application technology. In: *Nematodes as Biocontrol Agents* (Eds. Grewal, P. S., Ehlers, R. U. and Shapiro-Ilan, D. I.). CAB International, Wallingford, UK, pp. 91-106.

Wu, S., Kaplan, F., Lewis, E., Alborn, H. T., and Shapiro-Ilan, D. I. (2018). Infected host macerate enhances entomopathogenic nematode movement towards hosts and infectivity in a soil profile. *Journal of Invertebrate Pathology*, 159: 141-144.

Zhu, H., Grewal, P. S. and Reding, M. E. (2011). Development of a desiccated cadaver delivery system to apply entomopathogenic nematodes for control of soil pests. *Applied Engineering in Agriculture,* 27: 317-324.

Chapter 9

Application of Fungal Biocontrol Agents for the Management of Plant Parasitic Nematodes

K. Sankari Meena[1,*], S. Lakshmi Prasanna[1], Tarique Hassan Askary[2], R. D. Prasad[1] and M. Sujatha[1]

[1]ICAR-Indian Institute of Oilseeds Research,
Hyderabad, Telangana, India
[2]Division of Entomology,
Faculty of Agriculture,
Sher-e-Kashmir University of Agricultural Sciences
and Technology of Kashmir, Jammu and Kashmir, India

Abstract

Plant parasitic nematodes are one of the major threats to quantitative and qualitative production of several economic crops growing around the world. Nematodes are expected to cause annual global loss of $358 billion to agricultural crops. Growers generally depend primarily on chemical nematicides to solve the nematode problem but the negative impact of chemicals on the environment and the resistance developed by these plant parasitic nematodes against the nematicides demand for an alternative approach that should be eco-friendly and cost effective. Biocontrol agents are important components of eco-friendly pest management systems. Among the biocontrol agents, fungi have been found the most effective that reduce nematode populations. Fungal

* Corresponding Author's Email: meenaa5@rediffmail.com.

In: Pest Management: Methods, Applications and Challenges
Editor: Tarique Hassan Askary
ISBN: 979-8-88697-268-9

antagonists of nematodes consist of a wide range of organisms belonging to divergent orders and families, which include nematode trapping fungi, endoparasitic fungi, parasite of nematode eggs and cysts and symbiotic fungi. They reduce nematode population directly through parasitism and antibiosis or by providing higher nutrient and water uptake to the plant or by modifying the root morphology. Besides, they can also induce resistance against nematodes by activating hormone mediated plant defense mechanisms. In order to maximize the benefits, fungal biocontrol agents need to be included in integrated nematode management (INM) programs. This chapter provides an overview of the potential of various fungal biocontrol agents against plant parasitic nematodes. Besides, cost effective technologies for the mass production of these biocontrol agents have been also discussed.

Keywords: plant parasitic nematodes, biocontrol agents, fungi, management, eco-friendly

1. Introduction

Plant parasitic nematodes (PPNs) are one of the major threats to global agricultural production. Existence of nematodes is widespread in almost all the soil types and their distribution is determined by the degree of temperature, moisture, soil particle size and the presence of acceptable food supply. Infection of plant roots by the nematodes affects the water and nutrients intake by the plants which alters their mineral and photosynthates translocation and ultimately modify the shoot to root ratio of the plants (Saikia et al., 2013). Besides, nematode infestation exposes the plants to the secondary infection by fungal pathogens by establishing nematode - fungus disease complex which produce negative impact on the crop production (Askary and Ali, 2012; Ramyabharathi et al., 2014; Meena et al., 2015). These nematode-fungus interactions alter the chlorophyll and total sugar content in the plants (Meena et al., 2016).

Controlling of PPNs is essential for the effective crop production. Currently, application of the most potential chemical nematicides (Abd-Elgaward and Kabeil, 2010) are limited due to their socio-economic and environmental pollution issues, which demand for the alternative method to reduce content of chemical nematicides (Abd-Elgaward, 2008; Askary, 2015a,b).

Application of biocontrol agents is an efficient and environmentally safe approach to combat the PPNs (Abdelnabby et al., 2011; Abd-Elgawad and Askary, 2020) and to create a productive agricultural system. Many bacterial and fungal biocontrol agents have been used to manage the nematode populations (Meena et al., 2011; Devapriyanka et al., 2011; Jonathan et al., 2012; Askary and Martinelli, 2015) and their related disease complexes (Meena et al., 2016; Askary, 2017; Ramyabharathi et al., 2018, 2020) in various crops. The two groups have established a wider range of interrelationships. Moreover, fungi have been found to continuously kill nematodes in nearly all the soils and therefore fungal protection of nematodes is of high significance. Several compounds produced from the group of ascomycetes and basidiomycetes fungal taxa such as alkaloids, quinones, isoepoxydon, pyran, terpenoid, furan, peptide, macrolide, fatty acids, diketopiperazine, aphthalene and simple aromatic compounds have been recorded to possess nematicidal properties (Li and Zhang, 2014), which have made them the promising biocontrol agents for nematode management.

Fungal biocontrol agents use various mechanisms to regulate the nematode population, such as parasitism, predation, antagonism or competition (Davies and Spiegel, 2011) which cannot be considered as entirely independent of one another because biocontrol is likely a result of combination of various mechanisms (Cameron et al., 2013). The relative importance of each mechanism varies, depending on how biocontrol agents communicate with the nematodes. Hence, a better understanding of mode of action of each fungal biocontrol agent would aid in increasing the efficacy of the organisms. Understanding the possible mechanisms of biocontrol agents is also essential for the selection of active isolates against nematodes and for the development of improved application methods.

The present chapter aims to provide the significance of fungal biocontrol agents against PPNs. A brief description has also been provided for making a cost-effective production of these beneficial fungi.

2. Fungi as Nematode Control Agents

PPNs that move in the soil face a challenging situation teeming with fungal foes. In general, PPNs feed through the stylet, which has an aperture that is too narrow to admit the spores produced by certain fungal species. In this condition, nematode won't interactwith such fungi and those fungi cannot play any role in the biological control of PPNs. *Harposporium* and related genera are included in this category. Some fungi, on the other hand, produce spores that are adapted for adhesion or penetration of the nematode cuticle, which makes them the potential antagonists of PPNs. Fungal biocontrol agents of nematodes include a wide range of species classified as nematophagous fungi and nematode antagonistic fungi. It is noteworthy that these groups of fungi belong to highly disparate orders and families and plays a crucial role in reducing the PPNs (Jatala, 1986).

2.1. Nematophagous Fungi

Nematophagous fungi are the natural rivals of nematodes present in the soil. They are classified into three major group *viz*., i) predacious fungi or nematode trapping fungi, ii) endozoic or endoparasitic fungi and iii) egg parasitic fungi.

2.1.1. Predacious Fungi/Nematode Trapping Fungi

Nematode trapping fungi are distinctive category of carnivorous microorganisms that use specialized trapping structures to capture and digest the nematodes. The specialized trapping structures include adhesive hyphae, adhesive knobs, adhesive networks, constricting rings and non-constricting rings (Askary, 2015a). Adhesive knob is thought to be the ancestral type of all trapping devices, from which constricting rings and adhesive networks evolved via two pathways. Adhesive knobs retained their adhesive material and formed two- and three-dimensional networks in the first pathway. Adhesive knobs lost their adhesive materials in the second pathway, resulting in the formation of non-constricting and constricting rings (Li et al., 2005).

Nematode trapping fungi produce certain chemo attractant substances which attract the nematodes towards the fungal mycelia that help the trapping structures of the fungus to capture the nematodes. Ability of the nematophagous fungi to kill the nematodes is also linked to the production of carbohydrate binding protein (lectin) and an extracellular serine protease (Ahman et al., 2002; Yang et al., 2007) which play an important role in degrading the structural components of the nematodes. Here, proteases play a significant role in the immobilization of nematodes captured by the fungi (Tunlid and Jansson 1991).

More than 50 species of predacious fungi have been identified against nematodes in the soil (Siddiqui and Mahmood, 1996). These fungi have been evaluated as a source of biological nematicides to reduce the substantial damage caused by the PPNs. These predatory hypomycetes fungi mainly include four genera *viz*., *Dactylellina, Gamsylella, Arthrobotrys* and *Drechslerella* (Hagedorn and Scholler, 1999). *Dactylellina* is characterized by the non-constricting rings and stalked adhesive knobs, *Gamsylella* is distinguished by the presence of adhesive branches and unstalked knobs. *Arthrobotrys* is characterized by its adhesive networks, while *Drechslerella* is characterized by the constricting rings.

2.1.1.1. Arthrobotrys spp.

Among the four genera, *Arthrobotrys* spp. exhibit effective control mechanism against PPNs owing to the presence of various trapping structures which functions as nematode capturing device. The fungi completely digest the nematodes, after capturing them (Alkader, 2008; Singh et al., 2012). Besides the formation of trapping structures through the induction of systemic resistance (ISR), the fungus provides resistance against the nematodes (Singh et al., 2012). ISR is a well-established phenomenon in the plant growth promoting rhizobacteria that boosts the defense mechanism of the plants through the potentiated expression of defense related genes (Kim et al., 2004; Meena et al., 2012), and enables the plant to respond more effectively to nematodes by enhancing the infection induced cellular and biochemical defense responses (Becker and Conrath, 2007). *Arthrobotrys oligospora* is widely recognized as a potential predator of PPNs. It is a saprophytic fungus, which in the presence of nematode juveniles, enters into the parasitic mode of life and forms a complex three dimensional network to trap the juveniles (Yang et al., 2011). Trapping by the fungus triggers a sequence of events that include adhesion, penetration and immobilization of the juveniles (Hertz, 2004). Development of trapping structures by *A. oligospora* against the juveniles of root-knot nematode has been demonstrated by *in vitro* experiments where it has been evidenced that increasing the exposure period contributes to a higher rate of fungus trapping (Bakr et al.,2014) which is primarily dependent on mobility of the nematodes, fungus density and nutrient supply on the plates.

Application of *Arthrobotrys* to various vegetable and medicinal plants such as ashwagandha, tomato was found to reduce the root galls, egg mass and juveniles of root-knot nematode, *M. incognita* (Sharma and Pandey, 2009). Besides reducing nematode population, the fungus has been recorded to increase the plant growth by increasing the nutrient uptake by the plants. Similarly, application of *A. dactyloides* has been shown to decrease the population of root-knot nematodes in tomato by around 75-80 per cent (Stirling et al., 1998).

2.1.1.2. Dactylaria spp.

Formation of constricting rings is the unique phenomenon in nematode trapping fungi. These fungi develop three-dimensional ring structures, which aids in capturing and killing the nematodes. *Dactylaria brochopaga,* the nematode trapping fungus is well known for the production of constricting rings that trap and kill the saprophytic and parasitic nematodes (Kumar et al., 2010). It is a common fungus found in the agricultural soils and decaying substrates and its effectiveness in reducing *M. graminicola* and *M. incognita* populations has been well documented by other researchers (Kumar and Singh, 2010). An increase in the population of saprophytic nematodes in the soil leads to an increase in such type of fungal population and their impact against PPNs. Thus, presence of saprophytic nematodes in the field indirectly leads to the decline of PPN population.

Application of fungal spores of *D. brochopaga* has been reported to cause substantial reduction of *M. graminicola* population in rice and wheat (Kumar and Singh, 2011) as well as *M. incognita* population in brinjal (Kumar and Singh, 2010) and improved the plant growth and yield parameters. The efficacy of fungus increased when applied in conjunction with cow dung manure. This may be due to the sufficient proliferation of fungi in the soil, owing to the presence of cow dung manure (Kumar and Singh, 2011).

2.1.1.3. Monacrosporium spp.

Nematode trapping fungi, *Monacrosporium lysipagum, M. ellipsosporium* and *M. cionopagum* were recorded to be highly effective against root-knot nematodes, *M. incognita* and *M. javanica* (Jaffee and Muldoon, 1995). Besides their individual application, combination of these fungi with other biocontrol agents resulted in the substantial reduction of nematode population in the soil. Combined application of *M. lysipagum* with *Purpureocillium lilacinum* reduced the population of *M. javanica* in tomato (Khan et al., 1995), burrowing nematode, *Radopholus similis* in banana, and cereal cyst nematode, *Heterodera avenae* in barley (Wollenwebber, 1924).

Khan et al. (1995) reported that when *M. lysipagum* and *P. lilacinum* were used together, *P. lilacinum* destroyed the eggs of root-knot nematode that were located on the plant roots, while *M. lysipagum* killed the nematode juveniles present in the soil. As a result, for the subsequent generation, only fewer nematode juveniles invaded the plant roots and formed only fewer root galls. Similarly, combined application of these two fungi decreased the population of *R. similis* in banana. Even though, *R. similis* is a migratory endoparasite, their juveniles and adults, that wandered out into the soil to find a fresh root were captured by *M. lysipagumisis* (Rubner, 1996). *P. lilacinum* infects the egg stage, which is regarded as the stationary stage of the nematode (Khan et al., 2006).

2.1.2. Endoparasitic Fungi

Endoparasitic fungi produce special spores known as zoospores. These spores after being ingested by the nematodes, easily reach into the nematode body. The spores may get attach with the cuticle of nematode, germinate and penetrate into the nematode body, where they secrete toxins to immobilize the nematodes. Mycelia that emerge from the spore infest the internal content of the nematodes. Such fungi are highly host specific and are difficult to culture artificially in the absence of their host. Some of the major groups of endoparasitic fungi of PPNs include *Catenaria anguillulae, Haptocillium, Harposporium* or *Drechmeria*, *Hohenbuehelia* (*Nematoctonus*), etc.

2.1.2.1. Catenaria spp.

Catenaria anguillulae, one of the promising endoparasitic fungi, is a facultative parasite of nematodes that produce motile spores (zoospores) to kill the nematodes (Askary, 1996; Askary, 2015a). Juveniles of wheat seed gall nematode, *Anguina tritici*, dagger nematode, *Xiphinema basiri*, females of root-knot nematode, *M. javanica* and pigeon pea cyst nematode, *Heterodera cajani* were recorded to be highly susceptible to the infection by *C. anguillulae*. Initially fungal infection paralyze the invaded portion of the nematode body and cause restricted endobiotic growth as observed in the case of PPNs such as stunt nematode, *Tylenchorhynchus brassicae*, sheathoid nematode, *Hemicriconemoides mangiferae*, lance nematode, *Hoplolaimus indicus*, fungivorous nematode, *Aphelenchus avenae*, spiral nematode, *Helicotylenchus dihystera* and British spiral nematode, *Scutellonema brachyurum.* After the death of nematode, the fungus produces sporangium inside its body (Singh et al., 1996).

Juveniles of *A. tritici* are highly susceptible to *C. anguillulae*. It has been reported that infestation with the fungal zoospores of *C. anguillulae* resulted in the mortality of about 82 per cent juveniles of *A. tritici* (Singh et al., 2012). Despite the fact that this fungi has a higher infection rate in nematode eggs than in the juveniles, degree of parasitism by the fungi differs with the severity of the nematode disease present in the plants. Plants recorded with higher root gall index were found to have more fungal infection in their eggs too (Singh et al., 2007).

2.1.2.2. Hirsutella spp.

Very few *Hirsutella* species have been found to be effective against PPNs. When tested against the juveniles of strawberry crimp nematode, *Aphelenchoides fragariae*, stem and bulb nematode, *Ditylenchus dipsaci* and root-knot nematode, *M. incognita*, the three fungal species *Hirsutella thompsoni, H. satumaensis*, and *H. subulata* failed to adhere to the cuticle of these nematodes. Similarly, *H. vermicola* was also found to be nonpathogenic or weakly pathogenic to the PPNs. It might be due to their shorter conidiogenous cells with a swollen base, and wider conidia, which might find difficult to enter into the nematodes (Xiang et al., 2006).

H. rhossiliensis, on the other hand, was found to be effective against some PPNs, killing 100 percent of *D. dipsaci* and *M. incognita* and 45 to 65 percent of *A. fragariae* within four days after application (Cayrol et al., 1986). The fungus, initially attach its conidia to the surface of nematode cuticle, which after germination, penetrate into the nematode cuticle and produce fungal hyphae, which feed on the internal content of nematodes. When food supply is exhausted, conidia emerge out from the nematode and infect the other nematodes in their immediate vicinity (Devi, 2018).

2.1.2.3. Drechmeria spp.

Drechmeria coniospora, an obligate fungal pathogen infects the nematodes *via* adhesion of specialized spores to the host cuticle. Conidia of this fungus initially bind to the sensory organs of the nematode followed by penetration into the nematode cuticle which lead to subsequent infection process. Adhesion is generally mediated by the protein(s) present in the adhesive part of the conidia which bind with the protein(s) excreted from the sensory organs of the nematodes. Hydrophobicity and electrical charges as well as carbohydrates, have no role to play in the conidial adhesion process (Jansson, 1993). *Drechmeria* sp. is reported to kill its host within 24 hours of infection and produce around 5,000 to 10,000 conidia per infected nematode (Gams and Zare, 2003). The fungus has also been found effective against the juveniles of *M. incognita* present in the soil (Jansson et al., 1985).

2.1.2.4. Baeuveria spp.

Beauveria bassiana is a well-known biopesticide for insect control, however few studies have shown its efficacy on PPNs such as *M. incognita*, potato rot nematode, *Ditylenchus destructor* and soybean cyst nematode, *Heterodera glycines* (Bekanayake and Jayasundar, 1994; Liu et al., 2008). *B. bassiana* grows saprophytically within the host and produce metabolites such as beauvericin, bassianin, beauverolides, bassianolide and oosporein that suppress the nematodes (Butt et al., 2001). Drenching of soil with the spores or fermented products of *B. bassiana* resulted in the decline of *M. hapla* with a concomitant reduction in plant damage. Contrarily, application of *B. bassiana* increased the population of PPNs such as *D. destructor* and *D. dipsaci* with corresponding increased damage to potato tubers (Mwaura et al., 2017). Only few strains of this fungus secrete nematicidal compounds (Zhao et al., 2013) and the application of inefficient strains results in increasing the population of nematodes. Thus, attention needs to be paid to the selection and application of this fungus.

2.1.2.5. Metarhizium spp.

Metarhizium spp. are green muscardine fungus that have the potential to minimize nematode infestations to some degree. It is a facultative parasite with a well-known microbial control mechanism (Hoe et al., 2009). This fungus produces certain cyclic peptides such as destruxins (Hsiao and Ko, 2001), which may be responsible for its pathogenicity (Kershaw et al., 1999). Conidia of *Metarhizium brunneum* emit certain volatile compounds which in turn attract the root-knot nematode, *M. hapla*. Their efficacy varies depending upon the fungal strains and the application dose of the fungus (Hussain et al., 2010; Bojke et al., 2018). Volatile compounds such as 1-octen-3-ol and 3-octanone, produced from the fungi, attracted the second stage juveniles of *M. hapla* at their low concentrations, while the same compounds at their relatively higher concentrations seemed to be toxic to the nematodes. Direct exposure of the nematodes to these volatiles was found fatal. In addition to root-knot nematodes, effect of *Metarhizium anisopliae* has been reported on sugarcane nematodes (Zorilla, 2001), *H. avenae* (Ghayedi and Abdollahi, 2013) and *R. reniformis* in tomato (Tribhuvaneshwar et al., 2008).

Till now, very less works have been carried out on the efficacy of *Metarhizium* spp. against nematodes. The exact mode of action of this fungi on nematodes is still not clearly understood, but they are likely similar to other fungi producing sticky spores. The sticky spores can stick to the nematode body at any point and mainly cause the infection in the head region of nematode. The fungal spores germinate and penetrate into the nematode body through the cuticle and produce infective hyphae within the body cavity of nematode.

2.1.3. Egg Parasitic Fungi

These fungi mainly attack the egg stages of nematodes, especially of root-knot and cyst nematodes (Abd-Elgawad and Askary, 2018). Several egg parasitic fungi have been identified against PPNs, some of them are mentioned below under the following subheadings:

2.1.3.1. Purpureocillium Lilacinum (Thom) Luangsa-ard, Houbraken, Hywel-Jones & Samson 2011

P. lilacinum is the only permitted bionematicide in the countries like Switzerland and one of the three identified biological nematicides in the European Union (EU) pesticides database. It is filamentous ascomycete fungus that controls phytoparasitic nematodes such as *Meloidogyne* spp. (Mukhtar et al., 2013; Udo et al., 2014), *Radopholus similis* (Mendoza et al., 2007), *Heterodera* spp*.*, *Globodera* spp. and *Pratylenchus* spp. (Askary and Martinelli, 2015a). The fungi use their hyphal tips to attack nematode eggs and cysts. By direct hyphal penetration it colonizes the root surface and parasitizes all the life stages of root-knot nematodes *viz*., eggs, juveniles and females. However, it is mainly hazardous to sedentary stages of root-knot nematodes (Hertz et al., 2006; Yang and Zhang, 2014).

Direct interaction of *P. lilacinum* on the eggs and females of root-knot nematodes were demonstrated through *in vitro* studies (Holland et al*.*, 1999). After penetration into the nematode eggs, the fungal hypha grows profusely inside and over them thereby completely inhibits the juvenile growth. Infected eggs swell and buckle. As penetration progresses, the vitelline layer of the egg splits into three bands and a large number of vacuoles, lipid layer vanishes at this stage. The rapidly growing hyphae kill the developing juveniles within the egg. After infecting the eggs completely, the fungal hypha moves to the adjacent eggs to infect them.

Besides direct penetration, the fungi release compounds that play a significant role in reducing the nematode population. Production of antibiotics such as leucinostatin, lilacin and enzymes like protease and chitinase, lipases and acetic acid by the fungi is linked to the infection process of nematodes (Wang et al., 2010) which aid in the degradation of egg shell during the colonization of nematode eggs. Protease causes degradation of the nematode eggshell and inhibits egg hatching. Chitinase breaks down the eggshell and makes the route for the fungus to pass through. Decomposition of chitin present in nematode cuticle by the fungus releases ammonia, which is toxic to second stage juveniles of root-knot nematodes. Fungal hypha also penetrates and infests the vulva and anus of root-knot nematode females.

Undiluted crude filtrate of *P. lilacinum* (UdeA0106) is reported to significantly retard the hatching of eggs and the mobility of juveniles of *M. incognita* and *M. javanica* which has been observed through *in vitro* studies (Bustos et al., 2014). In tomato, the fungus has been reported to be the excellent remedy for *M. incognita*, which resulted in the reduction of about 88.23 per cent galls and 76.94 per cent egg mass of root-knot nematodes (Khalil et al., 2012).

It is a well-known fact that *P. lilacinum* has a significant impact in reducing eggs of sedentary nematodes. Besides egg parasitism, it has the ability to reduce the nematode juveniles in the soil. However, it takes at least 25 - 28 days to reduce the juveniles. As a result, effect of the fungi on nematode juveniles become slower than that of egg masses of root-knot nematodes. However, the impact on juveniles is significant, because the fungi can have an effect as long as the juveniles are exposed to the soil. But the suppression of root-knot nematode juveniles in the soil is still surprising as *P. lilacinum* is best recognized as a nematode egg parasite and only a small amount of research has been done on its ability to regulate root-knot nematode juveniles in the soil (Paul et al., 2019).

P. lilacinum at the time of planting has also been found to be much effective against nematodes than post planting application (Kalele et al., 2010). Its application immediately after planting resulted in the greater juvenile suppression, implying that early application of this fungus could prevent nematode establishment, provided periodic soil sampling for nematode diagnosis should be carried out for the early application of bioagents to combat the nematodes. Application of *P. lilacinum* in the field tracks the nematode population for the whole crop season (Gine and Sorribas, 2017). Since, efficacy of this fungus primarily depends on the time of application and the extent of nematode establishment in the soil, abundance of nematodes in the plant is a major factor in determining the success of fungi. This has been proved in a relationship study between *P. lilacinum* and *M. arenaria* in tomato, where under optimal colonization of egg masses by the fungi, only 50 per cent of the nematode eggs were parasitized (Carneiro and Cayrol, 1991) and a high number of nematode inoculum were left by the fungi for the next reproductive cycle. This may be due to high nematode pressure during planting which reduced the ability of the fungi to combat the nematodes (Kiewnick and Sikora, 2006).

2.1.3.2. Pochonia chlamydosporium (Verticillium chlamydosporium) (Goddard, 1913) Zare and Gams, 2001

It is a facultative parasite of root-knot and cyst nematodes. Moreover, it also parasitize citrus, burrowing and reniform nematodes. It thrives as saprophyte in the soil, but in the presence of nematodes, it may switch to a parasitic mode of life (Llorca et al., 2002; Tobin et al., 2008). Generally, *Pochonia chlamydosporium* parasitizes the eggs and adult females of PPNs.

P. chlamydosporium enters into the nematode cysts either through natural openings or by penetrating the cyst wall directly, which later forms a branched mycelial network within it. The fungus produces an appressorium that adheres to the eggshell by mucigens, from which an infection peg develops and penetrates into the eggshell. Penetration also occurs from the lateral branches of the mycelium. This results in disintegration of vitelline layer of the egg shell and also partial dissolution of the chitin and lipid layers. Egg hatching is inhibited due to the secretion of enzymes and toxins produced by the fungus (Askary and Martinelli, 2015).

2.1.3.3. Aspergillus niger (van Tieghem, 1867)

Aspergillus niger is saprophytic fungus that generally grows on dead and decaying plant material, stored grain and compost piles. The fungus produces nematicidal substance. Moreover, they colonize the nematode eggs and cysts and prevent the formation of juveniles and in case if juveniles are formed the eggs become less vulnerable (Askary, 2015a). The culture filtrates of *A. niger* has been reported to cause 93.3 per cent juvenile mortality on the eggs of *Meloidogyne* spp. (Dahiya and Singh, 1985). Singh et al. (1991) reported lesser damage to tomato by *M. javanica* in the presence of *A. niger*. The secondary metabolites of *A. niger* was found to suppress nematode populations and root-knot index and promoted the plant growth of tomato (Li et al., 2011). The fungus has also been recorded to induce systemic resistance against the nematodes (Askary and Martinelli, 2015).

2.1.3.4. Mortierella Globalpina

M. globalpina is reported to parasitize eggs, larvae and adults of PPNs. DiLegge et al. (2019) developed a technique for the identification of new fungal bioagents, wherein they first used a free living nematode, *Caenorhabditis elegans* to assess effectiveness of the beneficial fungus, and then developed a targeted approach to PPNs. Since PPNs and free living nematodes share the same soil environment, they were exposed to similar stresses, and the fungus that affected *C. elegans* infested PPNs as well. Using the free-living nematode, C. *elegans* as a guide, this new method pave a way for quickly identifying generalist nematode pathogens.

M. globalpina was found to be pathogenic against *M. chitwoodi* under *in vitro* condition. The fungus trapped the nematode by hyphal adhesion and penetration into the cuticle layer, followed by the digestion of nematode cellular contents. The fungus has also been recorded to reduce root-knot nematode symptoms in tomato crop by significantly reducing the galls on root.

Only few studies on the effectiveness of *Mortierella* against nematodes are available. The hyphae of *Mortierella* spp. have been found within the gelatinous matrix of egg masses of root-knot nematode (Orion et al., 2001) as well as in hatched root-knot nematode juveniles, indicating that this fungus is nematophagous in nature. Under greenhouse conditions, the fungus has also been found effective against *M. javanica* and *M. alpina* when co-inoculated with *T. longibrachiatum* (AL-Shammari et al., 2013).

Extracellular enzymes produced by *M. globalpina* can degrade the cuticle of both *C. elegans* and *M. chitwoodi*. Since the cuticle of most nematodes is primarily made up of collagen (Davies and Curtis, 2011), *M. globalpina* is likely to secrete collagenases. In addition, application of *M. globalpina* spores alone has been recorded to increase root and shoot biomass in tomato (var.Rutgers) plants. This indicates that the fungus can also promote plant growth. *M. antarctica* and *M. verticillata* is reported to synthesize gibberellic acid and indoleacetic acid, which may justify the growth promoting effects by *M. globalpina* (Ozimek et al., 2018).

2.1.3.5. Paraboeremia taiwanensis and Samsoniella sp.

P. taiwanensis and *Samsoniella* sp. has also been found effective against PPNs. The efficacy of two newly identified nematophagous fungi, *P. taiwanensis* and *Samsoniella* sp., against *M. enterolobii* was evaluated on Chinese cabbage in greenhouse trails. Both the fungi were effective in reducing the nematode disease severity by 29-63 per cent in the crop. Similarly, under *in vitro* condition, *P. taiwanensis* and *Samsoniella* sp. infected *M. enterolobii* eggs by 47.83 and 47.50 per cent, respectively. After 6 days of infection, the fungal hyphae continued to develop inside the nematode eggs without causing damage to the outer egg shell and the overall suppression efficacy of these fungi was comparable to that of the commercial *P. lilacinum* strain, PL251 (Liang et al., 2020).Various enzymes secreted by the nematophagous fungi are thought to be an important criteria in the degradation of nematode body structure (Bird and McClure, 1976), which illustrate the significance of these fungi in the management of PPNs.

2.2. Antagonistic Fungi

2.2.1. Trichoderma Harzianum Rifai 1969 (Teleomorph: Hypocrea Albofulva Berk. and Br. 1873)

Trichoderma spp. has been recognized as the effective nematode antagonistic fungi. They grow very fastly and rapidly colonize the substrates and effectively control the nematodes by using various mechanisms (Vinale et al., 2008). *Trichoderma* secrete many lytic enzymes such as chitinases, glucanases and proteases as well as antibiotics such as trichodermin, dermadin, trichoviridin, and sesquiterpene heptalic acid which kill the nematodes by impairing their viability (Degenkolb and Vilcinskas, 2016; Sharapova, 2019). Among the different compounds produced by the fungi, only few fungal secondary metabolites have been evaluated for nematicidal activities to date. Hence, further research is required in this field to confirm the properties of the metabolites produced by the fungi.

Trichoderma exhibit mechanisms such as antibiosis, plant growth promotion, induced systemic resistance, inactivation of nematode enzymes through enzymatic hydrolysis to regulate PPNs. Primary mechanism of *Trichoderma* against root-knot nematodes is the direct parasitization and antagonistic activity through the production of secondary metabolites. Several authors have reported the efficacy of *Trichoderma* in the regulation of PPNs (Rao et al., 1996; Ozbay and Newman, 2004; Medina et al., 2017; Sonkar et al., 2018).

T. viride, T. asperellum and T. harzianum, T. longibrachiatum, T. koningii and *T. hamatum* have been identified as the potential fungi against root-knot, cyst, reniform, spiral and yam nematodes (Saikia et al., 2013; Medina et al., 2017). Among the different species, T. harzianum has been identified to exhibit strongest egg parasitic potential. Culture filtrate of *T. harzianum* was recorded to be more effective on root-knot nematode eggs than on the juveniles The chitinolytic enzyme produced by the fungus play a significant role during the egg parasitism. Transcriptional activity of *chit 42* and *chit 33* of *T. harzianum* (SZMC 1647) has been characterized during the egg parasitism. *T. harzianum* directly penetrates into the females of potato cyst nematode, *Globodera rostochiensis* (Saifullah and Thomas, 1996), and their eggs inside the cysts, and cause the death of nematodes. Similarly proven efficacy of *T.viride* (Sonkar et al., 2018) and *T. citrinoviride* (Fan et al., 2020) in the management of root-knot nematode, *M. incognita* has been well evidenced in tomato crop. Besides decreasing the nematode population, increase in the production of plant biomass were observed with the application of these fungi, which might be due to the production of plant growth promoters that indirectly stimulated the nutrient uptake by the plants.

2.3. Symbiotic Fungi - Arbuscular Mycorrhizal *Fungi*

Arbuscular Mycorrhizal *Fungi (*AMF) are obligate root symbionts that protect the plants from several biotic stress factors including PPNs. AMF employs variety of mechanisms to combat the nematodes. It entails increased plant tolerance, direct competition for nutrients and space, induced systemic resistance and altered rhizosphere interactions. AMF encourages plant growth by increasing water and mineral nutrient uptake such as phosphate and nitrogen (Parniske, 2008), as well as micro elements like zinc (Smith and Smith, 2011). In exchange of photosynthetic carbon from their host (Smith et al., 2010), they minimize the plant stress caused due to the infestation of PPNs (Singh et al., 2011). While increased phosphate absorption has been suggested as the mechanism for the AMF mediated biocontrol, additional phosphate supply even seemed to increase the nematode incidence.

Competition for space could also be involved in AMF-nematode interactions, since they both share the same niche in the root system (Jung et al., 2012). But some endoparasitic nematodes escape from the competition such as cyst nematode, whose feeding cells are confined inside the endodermis of the roots, and hence, these nematodes are less affected by AMF. However, as mycorrhizal arbuscules are short-lived structures (Javot et al., 2011), it is difficult to distinguish whether AMF or nematode colonized the same root part or not. Competition for space also implies that a higher degree of AMF colonization leads to a higher-level biocontrol, provided the presence of arbuscules are prerequisite for biocontrol (Khaosaad et al., 2007).

AMF has also been recorded to have a systemic biocontrol effect on *R. similis* and *P. coffeae* in banana (Elsen et al., 2008) and *X. index* in grapevine (Hao et al., 2012). They should be inoculated into the plants prior to nematode infection to achieve an effective nematode control (DosAnjos et al., 2010), while their co inoculation has no effect on nematode control, except in a few cases such as *M. exigua* in coffee plants. (Alban et al., 2013). This indicates that nematode infection can also affect AMF (Hol and Cook, 2005), which is influenced by variety of factors such as species of AMF, nematodes and the plant host.

3. Production and Multiplication of Fungal Biocontrol Agents

In order to make the fungal biocontrol agents readily available to the farmers and to expand their application against PPNs, it becomes essential to produce and multiply them on a commercial scale. Diverse non-traditional substrates were being evaluated for their ability to support the development of biocontrol agents. Non edible deoiled cakes, such as neem (*Azadirachta indica*), castor (*Ricinus communis*), karanja (*Pongamia pinnata*), jatropha (*Jatropha curcas*) and mahua (*Madhuca indica*) are rich in micro and macronutrients and considered to be the suitable substrates for the growth of *P. lilacinum* (Sharma et al., 2014). The substrates, such as barley (Moity and Shatala, 1981), Sorghum (Upadhyay and Mukopadhyay, 1986), tapioca rind (Kousalya and Jeyarajan, 1988) and cow dung gas slurry (Jacob and Sivaprakasam, 1993) are being used for the mass production of *Trichoderma* species.

Several soil fungi produce nematicidal volatile compounds that inhibit nematode hatching and infectivity. Each microbe produces different nematicidal compounds indicates which that these characteristics developed independently to protect the host plant from nematode attack (Haraguchi and Yoshiga, 2020). The prospect of developing bioformulations containing bioactive compounds of the fungi would be exciting if the volatile metabolites responsible for the nematicidal activity could be identified. Volatile compounds such as aminoacids and leucinostatins have recently been identified as possible factors to destroy the nematodes (Huang et al., 2020). These volatile compounds draw the nematodes towards them (Niu et al., 2010). Polarity of the volatile plays a major role in determining the nematicidal properties (Siddiqui et al., 2000). Ethyl acetate extract of *P. lilacinum* has been recorded more lethal to the juveniles of *M. javanica* than hexane extract due to the variation in their polarity. Analytical study of ethyl acetate extract of the fungal filtrate by FT-IR spectroscopy and GC-MS demonstrated the presence of phenolic compounds *viz*., 2-ethyl butyric acid, phenyl ethyl alcohol, benzoic acid, benzene acetic acid and 3,5-Di-t-butylphenol (Breinig et al., 2000), implying that the phenyl moiety can play a major role as potential nematicidal compounds.

4. Mode of Application of Fungal Biocontrol Agents

Fungal biocontrol agents are generally applied in two ways a. seed treatment and b. soil application.

4.1. Seed Treatment

Application of fungal biocontrol agents is done either individually or in combination with the seeds to minimize the nematode infestation. Several species of *Trichoderma*, *Purpureocillium*, *Pochonia* and *Aspergillus* were found to reduce nematode population when used as seed treatment (Askary, 2015a). Once applied as seed treatment, these fungi reproduce quickly and establish in the rhizosphere within 4-5 days (Pathak and Kumar, 2003), thereby play a major role in the reduction of nematodes (Karthikeyan et al., 2001). Mycoparasitism, antibiosis or competition with nematodes are all possible modes of action for the fungi in nematode control.

Seed dressing with *T. harzianum* and *A. niger* was found to reduce the population of *M. javanica* in sunflower and okra. The germination and growth parameters of both okra and sunflower plants were significantly improved when the coating content concentration was increased (Dawar et al., 2008). In tomato, *A. niger* alone and in combination with other bacterial inoculants prevented the formation of root-knot nematode galling (Siddiqui et al., 2003). Similarly, combining *Trichoderma* spp. with a series of other biocontrol agents reduced the nematode population and increased the growth parameters of the plants. Combination of *Trichoderma* spp. (*T. harzianum* and *T. viride*) with other biocontrol agents such as *P. chlamydosporia, P. lilacinum* and *Psudomonas fluorescens* has been found effective in increasing growth and yield of okra and decreasing the population of *M. incognita* (Kumar et al., 2012). Seed treatment with 2% (w/w) *T. harzianum* with *P. lilacinum,* and *P. fluorescens* against maize cyst nematode, *Heterodera zeae* registered an increase in plant growth parameters and substantial reduction in nematode population (Mehta et al., 2016). Seed treatment of fungal biocontrol agent, *P. lilacinum* with *Pochonia chlamydosporia* and *Glomus fasiculatum* at the rate of 12 g/ kg seed was reported to be most effective in enhancing the plant growth parameters and reducing the infection of *M. incognita* in bitter gourd (Bhati et al., 2019).

4.2. Soil Application

Application of conidial suspension of fungal antagonists as soil amendment has been found to decrease the root invasion as well as soil densities of PPNs (Thakur and Devi, 2007). *T. harzianum* @ 3g/ kg soil was found to enhance plant growth and reduce *M. incognita* infestation on brinjal (Sen et al., 2016). When *T. harzianum* (10 g/ plant) was combined with *P. fluorescens* (10 g/ plant), the root-knot nematode disease was effectively controlled in brinjal plants (Pavithra et al., 2014). *Trichoderma* sp. in combination with *P. chlamydosporia* reduced the population of *Pratylenchus brachyurus* by 43.7 - 58.7 per cent on soybean and corn (Pacheco et al., 2020).

Some of the improved application strategies for nematode control include spraying the biocontrol agents in plant rhizospheres, placing them in tea bags for gradual release and dipping seedling roots in biocontrol agent suspension (Askary, 2017; Abd-Elgawad, 2019). These methods are cost-effective as lesser amount of biocontrol agents are used in comparison to other methods.

5. Factors Affecting the Efficacy of Fungal Biocontrol Agents

Soil texture and structure have a significant impact on the growth and spread of microorganisms as well as their nematode hosts. Soil moisture does not restrict the growth of fungi but affects their spore dispersal, especially zoospores. Temperature, yet another important factor influences the efficacy and persistence of biocontrol agents in the soil. For most of the crop growing season, soil temperatures vary from 10 - 15°C in temperate regions and 20 - 25 °C in warm climates at 10 cm depth of the soil. Such ranges have a direct impact on the growth and sporulation of the biocontrol agents, as well as on the nematode production (Kerry, 2000). *P. lilacinum* needs high soil temperature which limits its application in the field, despite its beneficial effects such as easy *in vitro* production, rhizosphere competition and attack on nematode eggs (Askary and Martinelli, 2015; Helmberger et al.,2017). In some cases, the added fungal biocontrol agents may compete with indigenous organisms present in the soil for the same resources such as for the prey or even other hosts (Abd-Elgawad and Vagelas, 2015), which will decrease efficacy of the biocontrol agents. Major factors that hinder the efficacy of biocontrol agents should be considered during the formulation of biocontrol agent and their application time. Adequate sampling to analyze the nematode genera and their related fauna and flora in the soil should be a preconsideration before the application of biocontrol agents in the soil (Abd-Elgawad, 2016).

Due to the complexity of soil ecology and biology, myriad interacting biotic and abiotic factors impact the biocontrol agents. Although, biocontrol agents are applied in high numbers, only a small fraction of these succeed to suppress PPN population. Hence, to enhance their efficacy, adequate environmental and biological factors are to be considered. Based on the microbial loads, application rates should be optimized for different soil types. Biocontrol agents that are successful at low doses can have the potential to control the nematodes in their natural environment. This would further reduce the application rates of biopesticides in the field. Higher microbial loads combined with improved controlled release strategies may be used to repel the nematodes, allowing crops to grow and establish natural resistance to the nematodes.

Environmental and biological factors often interact with one another. The most important ones are soil texture, moisture, temperature, predation, competition and starvation. It is often difficult to control these factors in the field. Hence, the complex network of interactions among biotic and abiotic factors in intimate contact with the biocontrol agents should be focussed to maximize the benefits of biocontrol agents via safe and skillful application and advanced technology. In recent days, many technologies are being used to protect the biocontrol agents from the biotic and abiotic stresses. Addition of adjuvants to the biocontrol agents has been found effective in enhancing their efficacy. Addition of avermectin as adjuvant with *P. lilacinum*, was recorded to enhance its efficacy against the nematodes (Kiewnick and Sikora, 2006). Another technically advanced alternative is the use of biocontrol agent filled capsules. Encapsulation of the fungi in nutrient amended hollow beads is a promising approach to combat the stresses. *Hirsutella rhossiliensis* encapsulated in the beads provide a wide protection against PPNs (Patel et al., 2002). Besides protecting the biocontrol agents from adverse environmental and biological factors, encapsulate formulations have several advantages like easy handling, long shelf life, adequate release of the biocontrol agents into the soil based on the environmental conditions and capsule materials. Amending capsules with adjuvants such as water retaining substances, nutrients, and fillers will further increase the efficacy of biocontrol agents in terms of their dispersal, rhizosphere competition and compatibility with the sustainable environment. Furthermore, encapsulated bio formulations in micro or nano forms are found as suitable alternative to traditional nematicides. Many nano formulated agricultural inputs have been applied against *M.*

incognita in tomato (El-Sherif et al., 2019), which should be exploited further in order to increase their efficacy against other nematodes too.

6. Fungal Biocontrol Agents in Integrated Nematode Management (INM)

Despite the above mentioned application strategies, utilization of biocontrol agents as a sole management option would not provide the level of economic nematode control but when used in conjunction with other management strategies as part of an integrated nematode management program, they will provide a long term protection to the plants against nematode infestation. In general, biocontrol agents are living systems that are responsive to both biotic and abiotic factors resulting from other inputs, especially in the soil rhizosphere. Hence, to optimize the benefits of biocontrol agents, it should be consistent with other inputs. Biocontrol agents can act synergistically or additively with other agricultural inputs which indicate their priorities in sustainable agriculture practices.

Application of manures in the field reduces nematode population due to the release of antimicrobial compounds during their decomposition, which in turn help in the multiplication of beneficial microorganisms in the soil (Oka, 2010). Hence, application of enriched biocontrol agents from the initial stages of crops has been suggested as better management practice in INM (Rao et al., 2017). Addition of oil cake or farmyard manure to fungal biocontrol agent increases their antagonistic activity against nematodes (Parvatha Reddy et al., 1996).

Application of *T. harzianum* with neem cake amendment has been recorded to reduce *M. incognita* infestation in tomato and this combination significantly enhanced growth of the plant (Rao et al., 1997). Addition of karanj (*Pongamia pinnata*), and castor (*Ricinus communis*) oil cakes to *T. harzianum* reduced the population of citrus nematode, *Tylenchulus semipenetrans* and enhanced the growth of acid lime (*Citrus aurantifolia*) (Parvatha Reddy et al., 1996). Combined application of *T. harzianum* with chemical nematicide, carbofuran was found to reduce the population of root-knot nematodes and root galling on french bean and brinjal plants (Gogoi and Mahantra, 2013; Devi et al., 2016). Rao (2017) reported enhanced efficacy of *P. lilacinum* against PPNs when they were enriched in vermicompost and farm yard manure (FYM). In general, organic amendments had a greater influence on the density and trapping behaviour of nematode trapping fungi. Dry alfalfa leaves, when applied in a very small quantity (360kg/ha) had a substantial impact on the population density of nematophagous fungi, *Dactylellina haptotyla,* Application of greater quantity of leaves (9000 kg/ha) had a negligible effect on its population. Contrarily, inclusion of higher amount of alfalfa amendment increased the population density of *A. oligospora,* but the trapping effectiveness of the fungi was decreased regardless of the amendment, confirming that the trapping efficacy and population density of the fungi vary with the fungal genera (Jaffee, 2004).

In organic debris, presence of tannins and phenolic compounds, free ammonium gases, nitrate, sulfur gases and organic acids play a significant role in exhibiting the antimicrobial properties against PPNs. In addition, the decayed debris contains significant quantities of organic carbon, total nitrogen, available phosphorus, absorbable potassium and micronutrients like iron, manganese, and zinc. Significant decrease in pH, calcium carbonate and copper has also been observed in decayed debris. Addition of organic amendments to the soil, alter the physical and chemical properties of soil and subsequently provide a favorable environment to enhance the efficacy of fungal biocontrol agents. Thus, application of fungal biocontrol agents added with organic amendments has a major impact in reducing of nematode population.

In recent years, use of binary and multiple mixtures of beneficial microorganisms (Bashan and Bashan 2005; Kavitha et al., 2010), as well as the use of microbial species with multiple functions (Singh, 2007) has become increasingly important in INM. For example, *P. chlamydosporia, when used in conjunction with other biocontrol agents and compost amendments resulted in better control of PPNs* (Zakaria et al., 2013). *P. chlamydosporia in combination with Trichoderma sp. was found effective in the management of Pratylenchus brachyurus in soybean and corn* (Pacheco et al., 2020).

In order to determine the effective species or combination of biocontrol agents for the management of PPNs, different species of biocontrol agents are usually compared singly or in combination with other agricultural inputs (Timper, 2014; Dawabah et al., 2019). In such cases, the biocontrol processes should be carefully designed to achieve the best possible match between the various attributes of both nematodes and biocontrol agents. For example, egg parasitic fungi can easily kill the eggs of root-knot nematodes but not the cyst nematodes. This is because, root-knot nematode eggs are usually exposed outside on the plant roots, but in case of cyst nematodes, their eggs are protected within the cysts. On the contrary, non-parasitic biocontrol agents release nematicidal compounds and kill the nematodes present in the soil. Similarly, endophytic fungi compete with a wide range of nematodes found inside the plant roots. Hence, effective native species should be selected after a detailed review of the effect of biocontrol agents since they are supposed to be adapted to local environmental conditions and attack the nematode directly. Otherwise, effective biocontrol programmes should consider identifying and deploying highly effective fungal strain(s) against nematodes in specific ecological and biological conditions (Shehata et al., 2019). During the selection of native biocontrol strains, laboratory and greenhouse screening steps are most important in rejecting low virulent biocontrol agent species and strains. Since many factors other than virulence are essential for field performance, it is necessary to carry more than one virulent strain to field testing.

Conclusion

Biocontrol strategies for PPNs provide a viable alternative to chemical nematicides, and the decline of most important chemical nematicides created urgency in this search area. Application of fungi is a promising biocontrol strategy against PPNs due to their wide diversity and mechanism of action. The efficacy of various fungal biocontrol agents in the soil is generally dependent on the dynamics of agricultural ecosystem, making their performance difficult to predict (Askary, 2015b). Furthermore, their activities are slow because they may take time to build up their population to the level needed for nematode control. A feasible solution to this problem is to use stabilized approaches, such as applying chemical nematicides initially to the fields, then introducing biocontrol agents along with complementary control measures such as host plant resistance, with the aim to gradually reduce the reliance on chemical nematode management.

Due to the wide spectrum behavior of all the chemical nematicides, a large number of nematodes are likely to be killed within a short period of time. Contrarily, application of fungal biocontrol agents take longer time to control nematodes, the higher economic threshold levels of nematode damage can be offset by the long term benefits of lower bionematicides cost and environment pollution problems. Hence, an integrated approach of the effective fungal biocontrol agents with other methods is needed to optimize the benefits without interfering with the efficacy of other practices. Since, fungal biocontrol agents interact synergistically with a variety of chemicals and other biocontrol agents. The combination of multiple tactics can be used to achieve satisfactory results. However, in terms of application technologies, further advancements are needed to make the fungal biocontrol agents competitive with the chemical nematicides. Some important issues that need attention on priority basis are improved shelf life of fungal biocontrol agents, transportation logistics and commercialization.

Acknowledgments

Authors are thankful to all those who have provided information required for this chapter.

References

Abd-Elgawad, M. M. M. (2008). The current status of phytonematode management in Egypt with special reference to applicable nematicides. *Egyptian Journal of Agronematology*, 6: 33-46.

Abd-Elgawad, M. M. M. (2016). Use of Taylor's power law parameters in nematode sampling. *International Journal Pharmtech Research*, 9: 999-1004.

Abd-Elgawad, M. M. M. (2019). Plant-parasitic nematodes of strawberry in Egypt: a review. *Bulletin of the National Research Centre*, 43: 7. https://doi.org/10.1186/s42269-019-0049-2.

Abd-Elgawad, M.M.M. and Askary, T.H. (2018). Fungal and bacterial nematicides in integrated nematode management strategies. *Egyptian Journal of Biological Pest Control,* 28:74. https://doi.org/10.1186/s41938-018-0080-x.

Abd-Elgawad, M. M. M. and Askary, T. H. (2020). Factors affecting success of biological agents used in controlling the plant-parasitic nematodes. *Egyptian Journal of Biological Pest Control*, 30:17. https://doi.org/10.1186/s41938-020-00215-2.

Abd-Elgawad, M. M. M. and Kabeil, S. S. A. (2010). Management of the root-knot nematode, *Meloidogyne incognita* on tomato in Egypt. *Journal of American Science*, 6: 256–262.

Abd-Elgawad, M. M. M. and Vagelas, I. K. (2015). Nematophagous bacteria: field application and commercialization. In: *Biocontrol Agents of Phytonematodes* (Eds. Askary, T. H. and Martinelli, P. R. P.). CAB International, Wallingford, UK, pp. 276–309.

Abdelnabby, H., Mohamed H. and Aly, H. A. (2011). Nematode-antagonistic compounds from certain bacterial species. *Egyptian Journal of Biological Pest Control*, 21: 209–217.

Ahman, J., Johansson, T., Olsson, M., Punt, P. J., Hondel, C. and Tunlid, A. (2002). Improving the pathogenicity of a nematode trapping fungus by genetic engineering of a subtilisin with nematotoxic activity. *Applied Environmental Microbiology*, 68: 3408-3415.

Alban, R., Guerrero, R. and Toro, M. (2013). Interactions between a root-knot nematode (*Meliodogyne exigua*) and arbuscular mycorrhizae in coffee plant development (*Coffea arabica*). *American Journal of Plant Sciences*, 4: 19-23.

Alkader, M. A. (2008). *In vitro Studies on Nematode Interactions with Their Antagonistic Fungi in the Rhizosphere of Various Plants*. Ph.D. Thesis, Faculty of Forest and Environmental Sciences, Albert Ludwigs University, Germany.

Al-Shammari, T. A., Bahkali, A., Elgorban, A., Elkahky, M. and Al-Sum, B. (2013). The use of *Trichoderma longibrachiatum* and *Mortierella alpina* against root-knot nematode, *Meloidogyne javanica* on tomato. *Journal of Pure and Applied Microbiology*, **7** (S): 199–207.

Askary, T. H. (1996). *Studies on some nematophagous fungi in agriculture soil of Pusa Farm, Samastipur,* Bihar. M.Sc. (Ag.) Thesis, Department of Nematology, Rajendra Agricultural University, Pusa, Bihar, India.

Askary, T. H. (2015a). Nematophagous fungi as biocontrol agents of phytonematodes. In: *Biocontrol Agents of Phytonematodes* (Eds. Askary, T. H. and Martinelli, P. R. P.). CAB International, Wallingford, UK, pp. 81-125.

Askary, T. H. (2015b). Limitations, research needs and future prospects in the biological control of phytonematodes. In: *Biocontrol Agents of Phytonematodes* (Eds. Askary, T. H. and Martinelli, P. R. P.). CAB International, Wallingford, UK, pp. 446-454.

Askary, T. H. (2017). Diversity of plant parasitic nematodes in pulses. In: *Plant Diversity: Present Situation and Future Scenario* (Eds. Abid, A. A., Gill, S. S., Abbas, Z. K. and Naeem, M.). CAB International, Wallingford, UK, pp. 239-274.

Askary, T. H. and Ali, S. S. (2012). Effect of *Meloidogyne javanica* and *Fusarium udum* singly and concomitantly on wilt resistant accessions of pigeon pea. *Indian Journal of Plant Protection*, 40: 167-170.

Askary, T. H. and Martinelli, P. R. P. (2015). *Biocontrol Agents of Phytonematodes*. CAB International, Wallingford, UK, 470 pp.

Bakr, R. A., Mahdy, M. E. and Mousa, E. M. (2014). Biological control of root-knot nematode *Meloidogyne incognita* by *Arthrobotrys oligospora. Egyptian Journal of Crop Protection*, 9: 1-11.

Bashan, Y. and Bashan, D. L. E. (2005). Bacterial plant growth promotion. In: *Encyclopedia of Soils in the Environment* (I) (Ed. Hillel, D). pp. 103-115.

Becker, G. J. and Conrath, U. (2007). Priming for stress resistance; from the lab to field. *Current Opinion in Plant Biology*, 10: 425–431.

Bekanayake, H. M. R. K and Jayasundar, N. J. (1994). Effect of *Paecilomyces lilacinus* and *Beauveria bassiana* in controlling *Meloidogyne incognita* on tomato in Sri Lanka. *Nematologia Mediterranea*, 22: 87-88.

Bhati, S. S., Baheti, B. L. and Kumar, R. N. (2019). Bioagents: An effective method for suppression of root-knot nematode, *Meloidogyne incognita* infecting bitter gourd (*Momordica charantia* L.) as seed treatment. *Journal of Entomology and Zoology Studies*, 7: 1197-1201.

Bird, A. F. and McClure, M. A. (1976). The tylenchid (Nematoda) eggshell: Structure, composition and permeability. *Parasitology*, 72: 19-28.

Bojke, A., Tkaczuk, C., Stepnowski, P. and Gołębiowski, M. (2018). Comparison of volatile compounds released by entomopathogenic fungi. *Microbiological Research*, 214: 129-136.

Breinig, S., Schiltz, E. and Fuchs, G. (2000). Genes involved in anaerobic metabolism of phenol in the bacterium *Thauera aromatica*. *Journal of Bacteriology*, 182: 5849-5863.

Bustos, N. L. C., Pavas. H. and Fernandez, E. P. (2014). Effect of the raw filtrate of *Purpureocillium* sp. (UdeA0106 strain) on the eclosion of eggs and the motility of *Meloidogyne incognita javanica* juveniles. *Revista Colombiana de Biotecnologia*, 16: 37-44.

Butt T. M., Jackson, C. and Magan N. (2001). Introduction-fungal biological control agents: progress, problems and potential. In: *Fungi as Biocontrol Agents: Progress, Problems and Potential* (Eds. Butt, T. M., Jackson, C. and Magan, N.). CAB International, Wallingford, UK, pp.1-8.

Cameron, D., Neal, A., VanWees, S. and Ton, J. (2013). Mycorrhiza induced resistance: more than the sum of its parts? *Trends in Plant Science*, 18: 539-545.

Carneiro, R. and Cayrol, J. C. (1991). Relationship between inoculum density of the nematophagous fungus *Paecilomyces lilacinus* and control of *Meloidogyne arenaria* on tomato. *Revue de Nematology*, 14: 629-634.

Dahiya, J. S. and Singh, D. P. (1985). Inhibitory effects of *Aspergillus niger* culture filtrate on mortality and hatching of larvae of *Meloidogyne* spp. *Plant and Soil*, 86: 145–146.

Davies, K. G. and Curtis, R. H. (2011). Cuticle surface coat of plant parasitic nematodes. *Plant Pathology and Microbiology*, 49: 135-156.

Davies, K. G. and Spiegel, Y. (2011). Root patho-systems nematology and biological control. In: *Biological Control of Plant Parasitic Nematodes: Building Coherence between Microbial Ecology and Molecular Mechanisms.* Springer, Dordrecht, pp. 291-303.

Dawabah, A. A. M., Al-Yahya, F. A. and Lafi, H. A. (2019). Integrated management of plant parasitic nematodes on guava and fig trees under tropical field conditions. *Egyptian Journal of Biological Pest Control,* 29: 29. https://doi.org/10.1186/s41938-019-0133-9.

Dawar, S., Sattar, A. and Zaki, M. J. (2008). Seed dressing with biocontrol agents and nematicides for the control of root-knot nematode on sunflower and okra. *Pakistan Journal of Botany*, 40: 2683-2691.

Degenkolb, T. and Vilcinskas, A. (2016). Metabolites from nematophagous fungi and nematicidal natural products from fungi as an alternative for biological control. Part I: Metabolites from nematophagous ascomycetes. *Applied Microbiology and Biotechnology*, 100: 3799-3812.

Devapriyanka, R., Jonathan, E. I., Sankari Meena, K. and Kavitha, P. G. (2011). Defense relate enzymatic activities mediated by *Pseudomonas* and *Bacillus* isolates against root-knot nematode, *Meloidogyne incognita* in black pepper cv. Panniyur 1. *Indian Journal of Nematology*, 41: 150-155.

Devi, G. (2018). Utilization of nematode destroying fungi for management of plant parasitic nematodes-A review. *Biosciences Biotechnology Research Asia*, 15: 377-396.

DiLegge, J. M., Manter, D. K. and Vivanco, J. M. (2019). A novel approach to determine generalist nematophagous microbes reveals *Mortierella globalpina* as a new biocontrol agent against *Meloidogyne* spp. nematodes. *Scientific Reports*, 9: 7521. https://doi.org/10.1038/s41598-019-44010-y.

Elsen, A., Gervacio, D., Swennen, R. and De Waele, D. (2008). AMF-induced biocontrol against plant parasitic nematodes in *Musa* sp.: a systemic effect. *Mycorrhiza*, 18: 251-256.

ElSherif, A. G., Gad, S. B., Megahed, A. and Sergany, M. I. (2019). Induction of tomato plants resistance to *Meloidogyne incognita* infection by mineral and nano-fertilizer. *Journal of Entomology and Nematology*, 11: 21-26.

Fan, H., Yao, M., Wang, H., Zhao, D., Zhu, X., Wang, Y., Liu, X., Duan, Y. and Chen, L. (2020). Isolation and effect of *Trichoderma citrinoviride Snef* 1910 for the biological control of root-knot nematode, *Meloidogyne incognita. BMC Microbiology*, 20: 299. https://doi.org/10.1186/s12866-020-01984-4.

Gams, W. and Zare, R. (2003). A taxonomic review of the clavicipitaceous anamorphs parasitizing nematodes and other micro invertebrates. In: *Clavicipitalean Fungi: Evolutionary Biology, Chemistry, and Cultural Impacts* (Eds. White, J. F. Jr., Bacon, C. W., Hywel-Jones, N. L. and Spatafora, J. W.). Marcel Dekker, New York.

Ghayedi, S. and Abdollahi, M. (2013). Biocontrol potential of *Metarhizium anisopliae* (Hypocreales: Clavicipitaceae), isolated from suppressive soils of Boyer-Ahmad region, Iran, against J2s of *Heterodera avenae*. *Journal of Plant Protection Research*, 53: 165- 171.

Gine, A. and Sorribas, F. J. (2017). Effect of plant resistance and BioAct WG (*Purpureocillium lilacinum* strain 251) on *Meloidogyne incognita* in a tomato-cucumber rotation in a greenhouse. *Pest Management Science,* 73: 880–887.

Hagedorn, G. and Scholler, M. (1999). Reevaluation of predatory orbiliaceous fungi. I. Phylogenetic analysis using rDNA sequence data. *Sydowia,* 51: 27-48.

Hao, Z., Fayolle, L., VanTuinen, D., Chatagnier, O., Li, X., Gianinazzi, S. and Pearson, V.G. (2012). Local and systemic mycorrhiza-induced protection against the ecto parasitic nematode *Xiphinema index* involves priming of defence gene responses in grape vine. *Journal of Experimental Botany*, 63: 3657–3672.

Haraguchi, S. and Yoshiga, T. (2020). Potential of the fungal feeding nematode *Aphelenchus avenae* to control fungi and the plant parasitic nematode *Ditylenchus destructor* associated with garlic. *Biological Control*, 143: 1-7.

Helmberger, M. S., Shields, E. J. and Wickings, K. G. (2017). Ecology of belowground biological control: entomopathogenic nematode interactions with soil biota. *Applied Soil Ecology*, 121: 201-213.

Hertz, N. B. (2004). Morphogenesis in the nematode trapping fungus, *Arthrobotrys oligospora* -an extensive plasticity of infection structures. *Mycologist*, 18: 125-133.

Hertz, N. B., Jansson, H. B. and Tunlid, A. (2006). Nematophagous Fungi. In: *Encyclopedia of Life Sciences.* John Wiley and Sons Ltd., pp. 1-11. http://doi.org/10.1038/npg.els.0004293.

Hoe, P. K., Bong, C. F. J., Jugah, K. and Rajan, A. (2009). Evaluation of *Metarhizium anisopliae* var. anisopliae (Deuteromycotina: Hyphomycete) isolates and their effects on subterranean termite *Coptotermes curvignathus* (Isoptera: Rhinotermitidae). *American Journal of Agricultural and Biological Sciences*, 4: 289-297.

Holland, R. J., Williams, K. L. and Khan, A. (1999). Infection of *Meloidogyne javanica* by *Paecilomyces lilacinus*. *Nematology*, 1: 131-139.

Hsiao, Y. M. and Ko, J. L. (2001). Determination of destruxins, cyclic peptide toxins produced by different strains of *Metarhizium anisopliae* and their mutants induced by ethyl methane sulfonate and ultraviolet using HPLC method. *Toxicon*, 39: 837-841.

Huang, D., Yu, C., Shao, Z., Cai, M., Li, G., Zheng, L., Yu, Z. and Zhang, J. (2020). Identification and characterization of nematicidal volatile organic compounds from deep-sea *Virgibacillus dokdonensis* MCCC 1A00493. *Molecules*, 25: 744. http://doi.org/10.3390/molecules25030744.

Hussain, A., Tian, M. Y., He, Y. R. and Lei. Y. Y. (2010). Differential fluctuation in virulence and VOC profiles among different cultures of entomopathogenic fungi. *Journal of Invertebrate Pathology*, 104: 166-171.

Jacob, C. K. and Sivaprakasam, K. (1993). Development of a cheap system for mass multiplication and delivery of *Trichoderma harzianum* and *T. viride*. In: *Crop Diseases Innovative Techniques and Management* (Eds. Sivaprakasam, K. and Seetharaman, K.). Kalyani Publishers, New Delhi, pp.143-155.

Jaffee, B. A. (2004). Do organic amendments enhance the nematode trapping fungi *Dactylellina haptotyla* and *Arthrobotrys oligospora*. *Journal of Nematology*, 36: 267-275.

Jaffee, B. A. and Muldoon, A. E. (1995). Numerical responses of the nematophagous fungi *Hirsutella rhossiliensis, Monacrosporium cionopagum* and *M. ellipsosporum*. *Mycologia*, 87: 643-650.

Jansson, H. B. (1993). Adhesion to nematodes of conidia from the nematophagous fungus *Drechmeria coniospora*. *Microbiology*, 139: 1899-1906.

Jansson, H. B., Jeyaprakash, A. and Zuckerman, B. M. (1985). Control of root knot nematodes on tomato by the endoparasitic fungus *Meria coniospora*. *Journal of Nematology*, 17: 327-329.

Jatala, P. (1986). Biological control of plant parasitic nematodes. *Annual Review of Phytopathology*, 24: 453-489.

Javot, H., Penmetsa, R. V., Breuillin, F., Bhattarai, K. K., Noar, R. D., Gomez, S. K., Zhang, Q., Cook, D. R. and Harrison, M. J. (2011). *Medicago truncatula* mtpt4 mutants reveal a role for nitrogen in the regulation of arbuscule degeneration in arbuscular mycorrhizal symbiosis. *The Plant Journal*, 68: 954–965.

Jonathan, E. I., Raguchander, T., Meena, K. S. and Kavitha, P. G. (2012). Plant growth promoting rhizobacteria in the management of *Radopholus similis* and *Meloidogyne incognita* in black pepper. *Madras Agricultural Journal*, 99: 356-358.

Jung, S. C., Martinez Medina, A., Lopez Raez, J. A. and Pozo, M. J. (2012). Mycorrhiza-induced resistance and priming of plant defenses. *Journal of Chemical Ecology*, 38: 651-664.

Kalele, D. N., Affokpon, A., Coosemans, J. and Kimenju, J. W. (2010). Suppression of root-knot nematodes in tomato and cucumber using biological control agents. *African Journal of Horticultural Science*, 3: 17-23.

Karthikeyan, G., Doraisamy, S., Sivakumar, C. V. and Doraisamy, S. (2001). Biological control of *Pythium aphanidermatum-Meloidogyne incognita* disease complex in brinjal with organic amendments. *Madras Agricultural Journal*, 88: 40-42.

Kavitha, P. G., Jonathan, E. I. and Sankari Meena, K. (2010). Antagonistic potential of *Pseudomonas fluorescens* against *Meloidogyne incognita* in tomato. *Madras Agriculture Journal,* 97: 399-401.

Kershaw, M. J., Moorhouse, E. R., Bateman, R., Reynolds, S. E. and Charnley A. K. (1999). The role of destruxins in the pathogenicity of *Metarhizium anisopliae* for three species of insect. *Journal of Invertebrate Pathology*, 74: 213-223.

Khalil, M. S. H., Allam, A. F. G. and Barakat, A. S. T. (2012). Nematicidal activity of some biopesticide agents and microorganisms against root-knot nematode on tomato plants under greenhouse conditions. *Journal of Plant Protection Research*, 52: 47-52.

Khan, A., Williams, K. L. and Nevalainen, H. K. (2006). Infection of plant parasitic nematodes by *Paecilomyces lilacinus* and *Monacrosporium lysipagum. Biological control*, 51: 659-678.

Khan, M. M. A., Mian, I. H. and Zahid, M. I. (1995). Post penetration development of *Meloidogyne graminicola* in rice root. *Thai Journal of Agriculture Science*, 28: 311–319.

Khaosaad, T., Garcia-Garrido, J. M., Steinkellner, S. and Vierheilig, H. (2007). Take-all disease is systemically reduced in roots of mycorrhizal barley plants. *Soil Biology and Biochemistry,* 39: 727-734.

Kiewnick, S. and Sikora, R. A. (2006). Biological control of the root-knot nematode *Meloidogyne incognita* by *Paecilomyces lilacinus* strain 251. *Biological Control*, 38: 179-187.

Kim, M. S., Kim, Y. C. and Cho, B. H. (2004). Gene expression analysis in cucumber leaves primed by root colonization with *Pseudomonas chlororaphis* O6 upon challenge-inoculation with *Corynespora cassiicola. Plant Biology*, 6: 105-108.

Kousalya, G. and Jayarajan, R. (1988). Techniques for mass multiplication of Trichoderma viride and T. harzianum. *Abstracts of papers presented in National Seminar on "Management of Crop Diseases with plant products and biological agents" held during January 10- 12, Agricultural College. Madurai.* pp. 32-33.

Kumar, N. and Singh, K. (2011). Use of *Dactylaria brochopaga*, a predacious fungus, for managing root-knot disease of wheat (*Triticum aestivum*) caused by *Meloidogyne graminicola. Mycobiology*, 39: 113-117.

Kumar, N. and Singh, K. P. (2010). Effect of mass culture and spore suspension of *Dactylaria brochopaga* on growth of egg plant and population of the root-knot nematode, *Meloidogyne incognita. Indian Journal of Scientific Research*, 2: 9-14.

Kumar, N., Singh, R. K. and Singh, K. P. (2010). Occurrence and colonization of nematophagous fungi in different substrates, agricultural soils and root-knots. *Archieves of Phytopathology and Plant Protection*, 44: 1182-1195.

Kumar, V., Singh, A. U. and Jain, R. K. (2012). Comparative efficacy of bioagents as seed treatment for the management of *Meloidogyne incognita* infecting okra. *Nematologia Mediterranea*, 40: 209-2011.

Li, G. H. and Zhang, K. Q. (2014). Nematode-toxic fungi and their nematicidal metabolites. In: *Nematode-Trapping Fungi* (Eds. Zhang, K. Q. & Hyde, K.). Springer, Dordrecht, pp. 313-375.

Li, S., Duan, Y. X., Zhu, X. F., Chen, L. J. and Wang, Y. Y. (2011). The effects of adding secondary metabolites of *Aspergillus niger* on disease resistance to root-knot nematode of tomato. *China Vegetables*, 1: 44–49.

Li, Y., Hyde, K. D., Jeewon, R., Cai, L., Vijaykrishna, D. and Zhang, K. (2005). Phylogenetics and evolution of nematode-trapping fungi (Orbiliales) estimated from nuclear and protein coding genes. *Mycologia*, 97: 1034-1046.

Liu, T., Wang, L., Duan, Y. X. and Wang, X. (2008). Nematicidal activity of culture filtrates of *Beauveria bassiana* against *Meloidogyne hapla*. *World Journal of Microbiology and Biotechnology*, 24: 113-118.

Llorca, L. L. V., Bordallo, J. J., Salinas, J., Mon-fort, E. and Lopez Serna, M. L. (2002). Use of a light and scanning electron microscopy to examine colonization of barley rhizosphere by the nematophagous fungus *Verticillium chlamydosporium*. *Micron*, 33: 61-67.

Medina, M. A, Fernandez, I., Lok, G. B., Pozo, M. J., Pieterse, C. M. and Van Wees, S. C. (2017). Shifting from priming of salicylic acid to jasmonic acid regulated defences by *Trichoderma* protects tomato against the root knot nematode, *Meloidogyne incognita*. *New Phytologist*, 213: 1363-1377.

Meena, K. S., Ramyabharathi, S. A. and Raguchander, T. (2016). Biomanagement of nematode-fungus disease complex in tuberose using plant growth promoting rhizobacteria. *International Journal of Science and Nature*, 7: 557-565.

Meena, K. S., Jonathan, E. I., Devrajan, K. and Raguchander, T. (2012). *Pseudomonas fluorescens* induced systemic resistance in tomato against *Meloidogyne incognita*. *Indian Journal of Nematology*, 42: 5-10.

Meena, K. S., Ramyabharathi, S. A., Raguchander, T. and Jonathan, E. I. (2015). *Meloidogyne incognita* and *Fusarium oxysporum* interaction in Gerbera. *African Journal of Microbiological Research*, 9: 1281-1285.

Mehta, S. K., Baheti, B. L., Nama, C. P. and Rathore, B. S. (2016). Farmer friendly integrated management strategy for maize cyst nematode, *Heterodera zeae* on maize (*Zea mays* L.). *Progressive Research*, 11: 407-410.

Meena, S. K., Jonathan, E. I. and Kavitha, P.G. (2011). Systemic resistance in tomato against *Meloidogyne incognita* induced by plant growth promoting rhizobacteria, *Pseudomonas fluorescens*. *Madras Agricultural Journal*, 98: 255-257.

Mendoza, A. R., Sikora, R. A. and Kiewnick, S. (2007). Influence of *Paecilomyces lilacinus* 251 on the biological control of the burrowing nematode *Radopholus similis* in banana. *Nematropica,* 37(II): 203-215.

Moiety, A. E. T. H. and Shatla, M. N. (1981). Biological control of white rot disease of onion *(Sclerotium cepivorum)* by *Trichoderma harzianum*. *Phytopathology*, 100: 29-35.

Mukhtar, T., Arshad, M. A. and Zameer Kayani, M. (2013). Biocontrol potential of *Pasteuria penetrans, Pochonia chlamydosporia, Paecilomyces lilacinus* and *Trichoderma harzianum* against *Meloidogyne incognita* in okra. *Phytopathologia Mediterranea*, 52: 66-76.

Mwaura, P., Niere, B. and Vidal, S. (2017). Application of an entomopathogenic fungus (*Beauveria bassiana*) increases potato nematodes reproduction and potato tubers damage caused by *Ditylenchus destructor* and *D. dipsaci*. *Biological Control*, 115: 23-29.

Niu, Q., Huang, X., Zhang, L., Xu, J., Yang, D., Wei, K., Niu, X., An, Z., Bennett, J. W. and Zou, C. (2010). A trojan horse mechanism of bacterial pathogenesis against nematodes. *Proceedings of the National Academy of Sciences,* 107: 16631-16636.

Oka, Y. (2010). Mechanisms of nematode suppression by organic soil amendments-a review. *Applied Soil Ecology*, 44: 101-115.

Orion, D., Kritzman, G., Meyer, S. L., Erbe, E. F. and Chitwood, D. J. (2001). A role of the gelatinous matrix in the resistance of root knot nematode (*Meloidogyne* spp.) eggs to microorganisms. *Journal of Nematology*, 33: 203-207.

Ozbay, N. and Newman, S. E. (2004). Biological control with *Trichoderma* spp. with emphasis on *T. harzianum*. *Pakistan Journal of Biological Sciences*, 7: 478–484.

Ozimek, E., Jaroszuk-Scisel, J., Bohacz, J., Kowalska, K. T., Tyskiewicz, R., Slomka, A., Nowak, A. and Hanaka, A. (2018). Synthesis of indole acetic acid, gibberellic acid and ACC-Deaminase by *Mortierella* strains promote winter wheat seedlings growth under different conditions. *International Journal of Molecular Sciences, 19:*3218. http://doi.org/10.3390/ijms19103218.

Pacheco, P. V. M., Monteiro, T. S. A., Coutinho, R. R., Balbino, H. M. and Freitas, L. D. G. (2020). Fungal biocontrol reduces the populations of the lesion nematode, *Pratylenchus brachyurus*, in soybean and corn. *Nematology*, 23: 1-8. http://doi.org/10.1163/15685411-bja10064.

Parniske, M. (2008). Arbuscular mycorrhiza: the mother of plant root endosymbioses. *Natural Reviews Microbiology*, 6: 763-775.

Parvatha Reddy, P., Rao, M. S., and Nagesh, M. (1996). Management of citrus nematode, *Tylenchulus semipenetrans* by integration of *Trichoderma harzianum* with oil cakes. *Nematolgia Mediterranea*, 24: 265-267.

Patel, A. V., Rose, T. and Vorlop, K. D. (2002). Encapsulation of *Hirsutella rhossiliensis* in hollow beads based on sulfo ethylcellulose to control plant parasitic nematodes. *Landbauforschung Volkenrode*, 241:145-150.

Pathak, K. N. and Kumar, B. (2003). Effects of culture filtrates of *Gliocladium virens* and *Trichoderma harzianum* on the penetration of rice roots by *Meloidogyne graminicola*. *Indian Journal of Nematology*, 33: 149-151.

Paul, D., Eder, R., Consoli, E., Krauss, J. and Kiewnick, S. (2019). Integrated control of *Meloidogyne incognita* in tomatoes using fluopyram and *Purpureocillium lilacinum* strain 251. *Crop Protection*, 124: 104874. https://doi.org/10.1016/j.cropro.2019.104874.

Pavithra, S., Suresh, S. R. and Jayashree, M. (2014). Effect of *Pseudomonas fluorescens* and *Trichoderma viride* against root knot nematode in brinjal. *Environment and Ecology*, 32: 1380-1382.

Ramyabharathi, S. A., Sankari Meena, K., Rajendran, L., Raguchander, T. and Jonathan, E. I. (2020). Potential of a rhizobacterium, *Bacillus subtilis* (Bbv 57) on *Fusarium oxysporum* f. sp. *gerberae* and *Meloidogyne incognita* infecting Gerbera grown in protected cultivation. *European Journal of Plant Protection*, 158: 615-632.

Ramyabharathi, S. A., Sankari Meena, K., Rajendran, L., Karthikeyan, G., Jonathan, E. I. and Raguchander, T. (2018). Biocontrol of wilt-nematode complex infecting gerbera by *Bacillus subtilis* under protected cultivation. *Egyptian Journal of Biological pest Control,* 28:21. http://doi.org/10.1186/s41938-018-0027-2.

Ramyabharathi, S. A., Sankari Meena, K., Raguchander, T. and Jonathan, E. I. (2014). Incidence and prevalence of Gerbera *Fusarium* wilt in association with nematodes in Tamil Nadu. *International Journal of Tropical. Agriculture*, 32: 97-102.

Rao, M. S., Kamalnath, M., Umamaheswari, R., Rajinikanth, R., Prabu, P., Priti, K., Grace, G. N., Chaya. M. K. and Gopalakrishnan, C. (2017). *Bacillus subtilis* IIHR BS-2 enriched vermicompost controls root-knot nematode and soft rot disease complex in carrot. *Scientia Horticulturae*, 218: 56-62.

Rao, M. S., Reddy, P. P. and Hagesh, M. (1996). Evaluation of plant based formulations of *Trichoderma harzianum* for the management of *Meloidogyne incognita* on eggplant. *Nematologia Mediterranea*, 26: 59-62.

Rubner, A. (1996). Revision of predacious hyphomycetes in the *Dactylella Monacrosporium* complex. *Studies in Mycology*, 39:1-134.

Saifullah, and Thomas, B. J. (1996). Studies on the parasitism of *Globodera rostochiensis* by *Trichoderma harzianum* using low temperature scanning electron microscopy. *Afro Asian Journal of Nematology*, 6: 117- 122.

Saikia, K. S., Tiwari, S. and Pandey, R. (2013). Rhizospheric biological weapons for growth enhancement and *Meloidogyne incognita* management in *Withania somnifera* cv. *Poshita. Biological Control*, 65: 225-234.

Sen, K., Sharma, H. K., Bhati, S. S., Nama, C. P. and Pushpa (2016). Effect of bio-agents as soil application against root knot nematode, *Meloidogyne incognita* infecting brinjal. *International Journal of Agriculture Sciences*, 8: 1687-1689.

Sharapova, I. E. (2019). Prospects of using entomopathogenic fungus in development of a biopesticide product with nematicidal activity. *Biocatalysis and Agricultural Biotechnology*, 19: 101098. https://doi.org/10.1016/j.bcab.2019.101098.

Sharma, A., Sharma, S., Mittal, A. and Naik, S. N. (2014). Statistical optimization of growth media for *Paecilomyces lilacinus* 6029 using non edible oil cakes. *Annals of Microbiology*, 64: 515-520.

Sharma, P. and Pandey, R. (2009). Biological control of root-knot nematode; *Meloidogyne incognita* in the medicinal plant; *Withania somnifera* and the effect of biocontrol agents on plant growth. *African Journal of Agricultural Research*, 4: 564-567.

Shehata, I. E., Hammam, M. M. A., El-Borai, F. E., Duncan, L. W. and Abd-Elgawad, M. M. M. (2019). Comparison of virulence, reproductive potential, and persistence among local *Heterorhabditis indica* populations for the control of *Temnorhynchus baal* (Reiche and Saulcy) (Coleoptera: Scarabaeidae) in Egypt. *Egyptian Journal of Biological Pest Control*, 29: 32. https://doi.org/10.1186/s41938-019-0137-5.

Siddiqui, I. A., Shaukat, S. S. and Khan, A. (2003). Differential impact of some *Aspergillus* species on *Meloidogyne javanica* biocontrol by *Pseudomonas fluorescens* strain CHA0. *Letters in Applied Microbiology,* 39: 74-83.

Siddiqui, I. A., Qureshi, S. A., Sultana, V., Ehteshamul-Haque, S. and Ghaar, A. (2000). Biological control of root rot-root knot disease complex of tomato. *Plant and Soil*, 227: 163-169.

Siddiqui, Z. A. and Mahmood, I. (1996). Biological control of plant parasitic nematodes by fungi: a review. *Bioresource Technology*, 58: 229-239.

Singh, K. P., Jaiswal, R. K., Kumar, N. and Kumar, D. (2007). Nematophagous fungi Associated with root galls of rice caused by *Meloidogyne graminicola* and its control by *Arthrobotrys dactyloides* and *Dactylaria brochopaga. Journal of Phytopathology*, 155: 193-197.

Singh, K. P., Niranjan Kumar, S. S. and Minakshi Kumari, K. D. (2012*). Catenaria anguillulae* as an efficient biological control agent of *Anguina tritici in vitro. Biological Control*, 61: 185-193.

Singh, K. P., Stephen, R. A. and S. S. Vaish (1996). Pathogenicity and development of *Catenaria anguillulae* on some nematodes, *Mycological Research*, 100: 1204-1206.

Singh, S. M., Azam, M. F., Khan, A. M. and Saxena, S. K. (1991). Effect of *Aspergillus niger* and *Rhizoctonia solani* on development of *Meloidogyne incognita* on tomato. *Current Nematology*, 2: 163–166.

Smith, F. A. and Smith, S. E. (2011). What is the significance of the arbuscular mycorrhizal colonization of many economically important crop plants? *Plant and Soil*, 348: 63-79.

Smith, S. E., Facelli, E., Pope, S. and Smith, F. A. (2010). Plant performance in stressful environments: interpreting new and established knowledge of the roles of arbuscular mycorrhizas. *Plant and Soil*. 326: 3–20.

Sonkar, S. S, Bhatt, J., Meher, J. and Kashyap, P. (2018). Bioefficacy of *Trichoderma viride* against the root-knot nematode (*Meloidogyne incognita*) in tomato plant. *Journal of Pharmacognosy and Phytochemistry*, 7: 2010-2014.

Stirling, G. R., Smith, L. J., Licastro, K. A. and Eden, L. M. (1998). Control of root-knot nematode with formulations of the nematode-trapping fungus *Arthrobotrys dactyloides*. *Biological Control*, 11: 224-230.

Thakur, N. S. and Devi, G. (2007). Management of *Meloidogyne incognita* attacking okra by nematophagous fungi, *Arthrobotrys oligospora*and *Paecilomyces lillacinus. Agricultural Science Digest*, 27: 50-52.

Timper, P. (2014). Conserving and enhancing biological control of nematodes. *Journal of Nematology*, 46: 75-89.

Tobin, J. D., Haydock, P. P. J., Hare, M. C., Woods, S. R. and Crump, D. H. (2008). Effect of fungus *Pochonia chlamydosporia* and fosthiazate on the multiplication rate of potato cyst nematodes (*Globodera pallida* and *G. rostochiensis*) in potato crops grown under UK field conditions. *Biological Control*, 46: 194-201.

Tribhuvaneshwar, Sharma, M. K. and Bhargava, S. (2008). Efficacy of green muscardine fungi, *Metarhizium anisopliae* against reniform nematode, *Rotylenchulus reniformis* on tomato. *Indian Journal of Nematology*, 38: 242-244.

Tunlid, A. and Jansson, S. (1991). Proteases and their involvement in the infection and immobilization of nematodes by the nematophagous fungus, *Arthrobotrys oligospora. Applied and Environmental Microbiology*, 57: 2868- 2872.

Udo, I. A., Osai, E. O. and. Ukeh, D. A. (2014). Management of root-knot disease on tomato with bioformulated *Paecilomyces lilacinus* and leaf extract of *Lantana camara*. *Brazilian Archives of Biology and Technology*, 57: 486-492.

Upadhyay, J. P. and Mukopadhyay, A. N. (1986). Biological control of *Sclerotium rolfsi* by *Trichodenlla harziallum* in sugar beet. *Tropical Pest Management*, 32: 215-220.

Vinale, F., Sivasithamparam, K., Ghisalberti, E. L., Marra, R, Woo, S. L. and Lorito M. (2008). *Trichoderma*-plant-pathogen interactions. *Soil Biology and Biochemistry*, 40: 1-10.

Wang, J., Wang, J., Liu, F., Pan, C. (2010). Enhancing the virulence of *Paecilomyces lilacinus* against *Meloidogyne incognita* eggs by over expression of a serine protease. *Biotechnology Letters,* 32: 1159-1166.

Wollenweber, H. W. (1924). Zur Kenntnis der Kartoffel-Heteroderen [*To the knowledge of potato heteroders*]. *Illustrierte Landwirtschaftliche Zeitung*, 44: 100-101.

Xiang, M. C., Yang, E. C., Xiao, Q. M., Liu, X. Z. and Chen, S. Y. (2006). *Hirsutella vermicola* sp. nov., a new species parasitizing bacteria-feeding nematodes. *Fungal Diversity Research Series*, 22: 255-266.

Yang, J., Tian, B., Liang, L. and Zhang, K. Q. (2007). Extracellular enzymes and the pathogenesis of nematophagous fungi. *Applied Microbiology and Biotechnology*, 75: 21–31.

Yang, J. and Zhang, K. Q. (2014). Biological control of plant parasitic nematodes by nematophagous fungi. In: *Nematode-Trapping Fungi* (Eds. Zhang, K. Q. & Hyde, K.). Springer, Dordrecht, pp. 231-262.

Yang, J., Wang, L., Ji, X., Feng, Y., Li, X., Zou, C., Xu, J., Ren, Y., Mi, Q., Wu, J., Liu, S., Liu, Y., Huang, X., Wang, H., Niu, X., Li, J., Liang, L., Luo, Y., Ji, K., Zhou, W., Yu, Z., Li, G., Liu, Y., Li, L., Qiao, M., Feng, L. and Zhang, K. Q. (2011). Genomic and proteomic analyses of the fungus *Arthrobotrys oligospora* provide insights into nematode trap formation. *PLoS Pathogens*, 7: e1002179. https://doi.org/10.1371/journal.ppat.1002179.

Zakaria, H. M., Kassab, A. S., Shamseldean, M. M., Oraby, M. M. and El Mourshedy, M. M. F. (2013). Controlling the root-knot nematode, *Meloidogyne incognita* in cucumber plants using some soil bioagents and some amendments under simulated field conditions. *Annals of Agricultural Sciences*, 58: 77-82.

Zhao, D., Liu, B., Wang, Y., Zhu, X., Duan, Y. and Chen, L. (2013). Screening for nematicidal activities of *Beauveria bassiana* and associated fungus using culture filtrate. *African Journal of Microbiology Research*, 7: 974-978.

Zorilla, R. A. (2001). *Monitoring and Management of Sugarcane Diseases Due to Nematode.* National Crop Protection Center. University of the Philippines at Los Banos College, Laguna 4031, Philippines, pp. 19.

Chapter 10

Field Applications of Bacteria for Plant-Parasitic Nematode Management

Fernando S. Rocha[1,*], Vicente P. Campos[2], Maria de Fátima S. Muniz[3] and Willian C. Terra[2]

[1]Phytopathology Research Laboratory,
Federal University of Minas Gerais, Brazil
[2]Department of Plant Pathology,
Federal University of Lavras, Brazil
[3]Agricultural Science Center,
Federal University of Alagoas, Brazil

Abstract

Nematophagous bacteria are an important group that survive in the rhizosphere of plants and exhibit different modes of action against nematodes. This group of bacteria are produced on a mass scale by several companies worldwide, aiming at application in the field for control of plant-parasitic nematodes. In this chapter, we have addressed the bacterial biopesticides (*Pasteuria* spp., Rhizobacteria-*Bacillus*, *Pseudomonas*, actinomycetes, Bacteria VOCs; Symbiotic bacteria-*Rhizobium*) and their modes of action against plant-parasitic nematodes. In addition, we have also described the production process of products based on nematophagous bacteria and their field application tactics.

Keywords: nematophagous bacteria, rhizosphere, field application, bionematicides, biopesticides

---

* Corresponding Author's Email: rochafs@ufmg.br.

In: Pest Management: Methods, Applications and Challenges
Editor: Tarique Hassan Askary
ISBN: 979-8-88697-268-9

1. Introduction

Plant-parasitic nematodes (PPNs) constitute about 4100 species within the phylum Nematoda which cause high impact on economically important crops (Ferraz and Brown, 2016). Globally, these organisms cause an estimated yield loss of 12.3 percent which in monetary terms is $157 billion (Askary and Martinellli, 2015). In the field, PPNs not only cause single parasite infections but configure disease-complexes with other micro-organisms increasing huge crop losses (Singh et al., 2004; Askary and Ali, 2012; Askary 2017). Furthermore, the symptoms are non-specific and resemble other diseases and abiotic stresses such as water and mineral deficiency (Singh et al., 2015). The combination of chemical control, crop rotation, and use of resistant cultivars is the most reliable practice for the management of nematodes. However, with the exclusion of many nematicides from the market, more sustainable approaches which reduce environmental impact, based on bio-control agents, have been stimulated (Sikora and Roberts, 2018).

Based on recent research, nematode antagonistic bacteria are divided into six groups which include parasitic bacteria (nematophagous bacteria), opportunistic parasitic bacteria, rhizobacteria, endophytic bacteria, symbionts of entomopathogenic nematodes and cry protein-forming bacteria (Tian et al., 2007). Nematodes can be affected by bacteria in different ways, such as direct suppression, promotion of plant growth and facilitation of rhizosphere colonization. Based on the effect of soil inhabiting bacteria on nematodes, they can be classified as toxin producing, antibiotic producing and enzyme producing bacteria as well as plant growth promoting bacteria. The efficiency of nematophagous bacteria to control PPNs including root-knot nematode (*Meloidogyne* spp.), cyst nematode (*Heterodera* spp.), golden cyst nematode (*Globodera* spp.), root lesion nematode (*Pratylenchus* spp.), sting nematode (*Benololaimus* spp.), sheath nematode (*Hemicycliophora* spp.), spiral nematode (*Helicotylenchus* spp.) and stunt nematode (*Tylenchorhynchys* spp.) has been well documented (Abd-Elgawad and Vagelas, 2015; Abd-Elgawad and Askary, 2018; Abd-Elgawad, 2020; Migunova and Sasanelli, 2021).

In the last two decades, the global bionematicide market has witnessed an increase in the development of products called eco-friendly, based on antagonistic bacteria and/or in combination with fungal biocontrol agents, chemical nematicides, fertilizers, biostimulants or root enhancers with the potential to suppress PPNs. The technological advances from companies operating in the global nematicide market in partnership with the seed industry and agricultural equipment companies have led to the emergence of new biological products and greater precision in the application and integrated management of nematodes in different crops. However, at the level of applicability of bionematicides under field conditions, the essential factors playing significant role in nematode management are the mechanism of action of antagonist bacteria, the target nematode, type of formulation, mode of application and the environmental conditions.

The present chapter focusses on the bacterial biopesticides used as bionematicides and their mechanism of action against PPNs. In addition, we also described the production process of nematophagous bacteria based products and their field application methods for successful nematode management.

2. Bacterial Biopesticides

Bacteria produce an arsenal of bioactive compounds that are able to control PPNs in the field (Huang et al., 2010; Subedi et al., 2020). Several companies have largely focused upon the bacterial group called plant growth-promoting rhizobacteria (PGPR) to control PPNs. Besides their efficacy as biological control agents (BCAs), PGPR are also known as biostimulants since they facilitate plant growth and productivity as well as adaptation or acclimatization towards various abiotic stresses (Backer et al., 2018). PGPR can reduce PPNs in infested plants through direct and/or indirect antagonistic mechanisms. Direct antagonism occurs by predation, release of nematicidal compounds and semiochemicals. A common bacterial strategy against PPNs has been the production of toxic metabolites. It can be divided into two groups, (1) short-chain molecules called volatiles compounds and (2) long-chain molecules. On the other hand, indirect antagonism may occur by induced systemic resistance (ISR), which primes the whole plant to better react against biotic and abiotic constraints (Subedi et al., 2020). One of the advantages of using PGPR to control PPNs is that a single bacteria strain can use a number of strategies to control nematodes in the field (Hu et al., 2020). Whole genome sequencing of *Bacillus firmus* I-1582, used as seed treatment nematicide (Poncho®/Votivo®), revealed a wide range of possible virulence factors in the bacterial genome, which include 18 homologs nematode-virulent proteases and 28 secondary metabolites, projected to encode antibiotics, small peptides, toxins and siderophores (Susic et al., 2020).

2.1. *Pasteuria* spp.

Pasteuria spp. is a group of gram-positive bacteria which are hyperparasites of PPNs and water fleas and produce highly resistant endospores. All economically important genera of PPNs are parasitized by *Pasteuria* spp. The five species of *Pasteuria* which differ in their host range and pathogenicity are (i) *P. penetrans* parasitic on *Meloidogyne* spp., (ii) *P. thornei* parasitic on *Pratylenchus brachyurus*, (iii) *P. nishizawae* parasitic on *Heterodera* sp., (iv) *P. usage* parasitic on *Belonolaimus longicaudatus* and (v) *P. hartismeri* parasitic on *M. ardensis* (Gowen et al., 2008).

All species within the genera *Pasteuria* produce large endospores (3–6 µm wide), which attach to the nematode cuticle, easily visible under a microscope (150–200x magnifications). The initial infection process of *Pasteuria* spp. is determined by the ability of non-motile endospores that adhere to the cuticle of migrating nematodes. Endospore germinates and produces the peg of infection that penetrates the cuticle toward the pseudocoelom cavity, where the branched hyphae proliferation may be observed. Sporogenesis begins with the production of microcolonies. These mycelial-like structures are subsequently fragmented to give rise to endospores (Davies, 2009). The parasitized nematode may produce a few to 2 million spores. The utilization of *Pasteuria* spp. as a biological control agent is currently limited by its restricted host range and its tough *in vitro* cultivation (Orr et al., 2018).

Generally, isolates of *Pasteuria* exhibit a narrow host range (Channer and Gowen, 1992; Campos et al., 1998; Davies, 2009). Endospore attachment is the primary and, arguably, the most fundamental step in the infection process. However, *Pasteuria* spp. exhibits a fastidious attachment profile that include species and population specificity. Currently, the Velcro-like mechanism has been proposed to explain the bacterial attachment wherein the bacterial collagen-like fibres bind to the microvilli that constitutes part of the nematode's surface coat (Davies, 2009). Environmental factors, such as root exudates, also affect endospore attachment of *Pasteuria* (Mohan et al., 2020).

In vitro Pasteuria spp. is only produced by the multinational Corporation Syngenta®, which currently supply the bioproduct Clariva™ registered in the USA, Canada and Brazil. Yet, the obligate nature of the bacterium life cycle has proved difficult to overcome. It seems that the life cycle involves additional factors besides complex nutrient availability (Duponnois et al., 1999).

In vivo production of *Pasteuria* spp. demands time and space (greenhouse), for obtaining moderate amounts of endospores. In this technique a plant host is inoculated with spore-encumbered nematode. The root systems, containing infected (spore-filled) female nematodes, are harvested after an appropriate period of time and then dried as fine powder (Gowen et al., 2008).

2.2. Rhizobacteria

The use of rhizobacteria to control plant parasitic nematodes has been widely studied and a variety of strains, particularly of *Bacillus,* are currently commercialized as biopesticides (Subedi et al., 2020). Rhizosphere bacteria control PPNs by mechanisms such as: direct antagonism through the production of toxins, enzymes or other secondary metabolites; interference with PPNs recognition; competition for nutrients; plant growth promotion; and induced systemic resistance (ISR). The preference for the use of *Bacillus* spp. in the formulation of biological products is related to several factors such as: they are well-studied organisms which facilitates their rational use; they produce spores which are extremely resistant; dormancy forms capable to withstand high temperatures, unfavorable pH, nutrients or water stress and therefore relatively easy to produce in industries. Additionally, they are motile bacteria that readily move towards and on the root surface, thus facilitating colonization of new ecological niches (Cawoy et al., 2011).

Actinomycetes are also a powerful group of nematode antagonistic bacteria. The members of this group are recognized producers of bioactive compounds, an example is the compound avermectin, produced by *Streptomyces avermitilis*, which are one of the most potent nematicidal compounds. Avermectin is widely used as a seed treatment to control PPNs. *Streptomyces* spp. are able to reduce the damage caused by PPNs by recycling nutrients and stimulate the production of plant growth hormones. Co-inoculation of chilli (*Capsicum frutescens*) with the *Streptomyces* spp. strains KPS-E004 and KPSA032 increased the shoot and root weight by 56.78 and 141.84 per cent, respectively, as well as suppressed the population of *M. incognita* in the roots (Nimonoi and Ruanpanun, 2020).

Another group of bacteria that has been studied as biocontrol agents of PPNs are the gram-negative *Pseudomonas* spp. However, formulations of these bacteria are discouraged, in part, because they do not form spores.

2.3. Bacterial Metabolites Antagonistic to PPN

Bacteria produce diversified metabolites either volatiles or non-volatiles with nematicidal activities (Gu et al., 2007; Castaneda-Alvarez and Aballay, 2016).

2.3.1. Volatile Organic Compounds

Bacteria produce short-chain molecule characterized as volatile organic compounds (VOCs). VOCs affect PPNs in different ways, like nematicidal activity, attraction or repellence.

Extensive investigations have been conducted over the last years to assess the potential of VOCs released by bacteria for the control of PPNs. Gu et al. (2007) identified isolates of *B. simplex*, *B. subtilis*, *B. weihentephanensis*, *Stenotrophomonas maltophilia*, and *Serratia marcences* as producers of highly toxic VOCs against the nematode, *Bursaphelenchus xylophilus*. Huang et al. (2010) demonstrated that *B. megaterium* YFM3.25 acts against different stages of *M. incognita* by producing several nematicidal volatiles. The nematicidal volatiles produced by the bacterium were characterised as benzeneacetaldehyde, 2-nonanone, decanal, 2-undecanone and dimethyl disulphide, which were active against the nematode juveniles and eggs.

Nematodes recognize environmental stimuli through sensory organs, such as amphids and phasmids which guide them to the food source (Perry and Curtis, 2013). They are attracted not only by sensory cues from plants but also from rhizobacteria. Some bacteria have developed a sophisticated system to control PPNs. First, they release attractive volatile compounds to the nematodes in the environment. When the nematodes approach the emitting source they are killed by other VOCs also emitted by the bacteria. *Paenibacillus polymyxa* KM2501-1 releases the VOCs furfural acetone and 2-decanol that are attractive to second stage juveniles (J_2) of *M. incognita.* Bacteria kill the nematode by contact or fumigation. Two VOCs, 2-nonanone and 2-decanone, as well as strain KM2501-1 itself, destroyed the integrity of the nematode intestine and pharynx (Cheng et al., 2017).

Some rhizobacteria release VOCs that are repellent to PPNs. As these bacteria inhabit the rhizosphere of plants, they prevent the nematodes from getting close to the roots. Tomato roots carrying the endophytic bacterium *B. cereus* BCM2 were repellent to *M. incognita* J_2. The compounds 2,4 di-tert-butylphenol and 3,3-dimethyloctane produced in the colonized roots may be responsible for this behavior (Hu et al., 2017; Li et al., 2019).

2.3.2. Long Chain Compounds

Bacteria are able to produce enzymes such as proteases, collagenases, lipases and toxins that act against PPNs.

The cuticle constitutes the outer layer of nematodes. Collagen and collagen-like proteins constitute the vast majority of the cuticular structural components (Page et al., 2014). The wide distribution of cuticle-degrading proteases in *Bacillus* strains with nematicidal activity suggest that these enzymes are likely to play an important role in bacteria–nematode interactions (Lian et al., 2007). *Bacillus cereus* produces an extracellular collagenolytic/ proteolytic enzyme with a very strong proteolytic activity on intact cuticle of J_2 of *M. javanica* (Sela et al., 1998). The Enterobacteriaceae *Serratia* strain A88copa13, which is able to attach on *B. xylophilus*, produces a serine protease with 70 kDa. The presence of the bacteria on the cuticle leads to the degradation of the nematode (Paiva et al., 2013). The antagonism of *P. fluorescens* CHA0 against *M. incognita* involves extracellular protease AprA. Mutation in the aprA gene resulted in reduced biocontrol activity against *M. incognita* in tomato and soybean infections (Siddiqui et al., 2005).

Other secondary metabolites are also important in the antagonistic action of rhizobacteria against PPNs. For instance, *P. fluorescens* produces hydrogen cyanide (HCN), 2,4-diacetylphloroglucinol that are toxic to several species of PPNs (Siddiqui et al. 2006; Meyer et al. 2009).

2.4. Symbiotic Bacteria

Rhizobium forms an endosymbiotic nitrogen fixing association with roots of legumes and it can also act as a biocontrol agent of root-knot nematodes. Inoculation of *Rhizobium* on lentil (*Lens culinaris*) infested with *M. incognita* increased plant growth characters, using four different PGPRs in nematode-inoculated plants. Additionally, the combined use of *Rhizobium* with other species of PGPR caused higher reduction in the number of galls produced and nematode multiplication than their individual inoculation (Siddiqui et al., 2007). *Rhizobium etli* G12 has been repeatedly demonstrated to be capable of suppressing early infection caused by *M. incognita* (Fabry et al., 2007; Adam et al., 2014).

2.5. Induced Systemic Resistance

Rhizobacteria do not protect plants only through direct antagonism against nematodes, but also by triggering induced resistance. Induced systemic resistance (ISR) might explain how rhizobacteria biologically control PPNs (Siddiqui and Shaukat, 2002). Adam et al. (2014) proposed that ISR is the major control mechanism employed by *B. subtilis* strain Sb4-23, Mc2-Re2, and Mc5-Re2 against *M. incognita*. In split-root systems the three BCAs, significantly, reduced nematode infectivity and fecundity. The number of galls or egg masses was 40 - 51 per cent lower in treatments with the biocontrol strains. The highest reduction was obtained by the strains Mc2-Re2 and Mc5-Re2. *Bacillus cereus* strain BCM2 induced resistance in tomato plants against *M. incognita*, which led to reduction of up to 64.2 per cent in nematode fecundity. Sequencing the transcriptomes of tomato roots colonized by *B. cereus* BCM2 revealed 34 defense or stress-related genes playing a vital role on triggering plant defense against *M. incognita* (Hu et al., 2018). The ability of two rhizobacteria *P. aeruginosa* strain IE-6S^{+} and *P. fluorescens* strain CHA0 stimulate ISR against *M. javanica* was demonstrated in split-root trials (Siddiqui and Shaukat, 2002).

3. Production of Nematophagous Bacterial Products

Among the important natural enemies, nematophagous bacteria including members of the genera *Pasteuria*, *Pseudomonas* and *Bacillus* have shown great potential in the biological control of PPNs (Meyer, 2003). Production of these bacteria to control nematodes contribute to reduce damage and losses of economically important crops. Mass production of antagonistic bacteria and their introduction into soil infested with PPNs is the main strategy to induce suppression of PPNs in agricultural soils. Several studies have been conducted on mass production and the formulation of nematophagous bacteria for the control of PPNs (Abd-Elgawad and Vagelas, 2015; Abd-Elgawad and Askary, 2018; Forghani and Hajihassani, 2020; Migunova and Sasanelli, 2021).

The techniques involved in *in vitro* culture of *Pasteuria* and their commercialization as biopesticide have been described by many workers (Abd-Elgaward and Vagelas, 2015; Migunova and Sasanelli, 2021). In the process, infected female nematodes are washed in the absence of disinfectants or antibiotics. The nematodes are then crushed and introduced into a nutrient medium and grown at room temperature. A standard nutrient broth, optionally supplemented with bovine calf serum and/or with an egg yolk preparation, peanut oil, or other source of lipids is used. After a period of few hours to few days, swimming rod-shaped organisms will appear (Gerber and White, 2005).

Growth media for production of *Bacillus* species have been also screened. For instance, *B. thuringiensis* are produced by the use of commercial laboratory medium and industrial raw materials and liquid swine manure as nutrient sources (Valicente et al., 2010). According to these authors, the seed culture of *B. thuringiensis tolworthi* 344, is produced using shake flasks and grown in Luria Bertani (LB) medium plus salts, incubated in a shaker at 200 rpm (rotation per minute) for 96 hours at 30ºC. Medium 1 is constituted of LB plus salts ($FeSO_4$, $ZnSO_4$, $MnSO_4$, $MgSO_4$) and 0.2 per cent glucose; medium 2 is composed of 1.5 per cent glucose and 0.5 per cent soybean flour plus salts; and medium 3 is composed of liquid swine manure 4 per cent and 0.2 per cent glucose. The media are sterilized and inoculated with *B. thuringiensis tolworthi* (seed culture) at a stirrer speed of 200 rpm, for 96 hours at 30ºC. Media 1 and 2 resulted in the highest number of viable spores, 2.0 x 10^8 cfu/ml.

Production of growth medium for non-spore-forming organisms is more difficult to formulate because these organisms do not produce survival structures such as spores, they have a short life cycle and are readily killed by desiccation (Islam et al., 2003). Mputu and Thonart (2013) produced *P. fluorescens* BTP1 using a bioreactor with different against-pressure values (0.1 and 0.3 bar for bioreactor 1 and 2 respectively) harvested cells during stationary phase (2 h and 4 h for bioreactor 1 and 2, respectively). The authors also observed that the survival rate was better at 4 °C than at room temperature.

In 2020, the Brazilian company Biotrop (Empresa Biotrop Soluções Biológicas e Participações Ltda-Registrant/Tacto Biotecnologia e Agro-pecuária Ltda-Manufacturer and formulator, SP-Brazil) launched the bionematicide Furatrop®, based on *B. subtilis* strain CNPSo 2657, using Hayai technology for bacterial multiplication. This technology employs an ancient Japanese technique of fermentation in a semi-solid medium that provided a high load of metabolites and organic compounds on its composition, accelerating the rapid activation of bacteria in the soil and colonization of the roots, promoting root growth, acting by antibiosis, competition (forming a protective biofilm in the roots) and inducing plant defenses to control *Meloidogyne* spp. in various crops such as soybeans, corn, tomatoes, beans, sugarcane and cotton. The company also launched the products Vult® and Nematrop®, both based on *B. subtilis* strain CNPSo 2657.

Bionematicidal products have been produced by different companies in several countries around the world, mainly with only one bacterial species with different formulations and application methods and recommended for the management of PPNs in different crops (Table 1). Some products have the same bacterial species with different trading names however, there is a tendency to formulate products with a mixture of bacterial strains and antagonistic fungi and/or with compatible agrochemicals in new bio-nematicidal products.

In Brazil, bionematicides registered in the Ministry of Agriculture, Livestock and Food Supply-MAPA, with open consultation of information on agrochemicals and related products approved by MAPA, are available in the system of phytosanitary pesticides-AGROFIT[1]. Products commercialized are based on *B. subtilis*, *B. licheniformis*, *B. firmus*, *B. amyloliquefaciens*, *B. methilotrophicus* and *P. nishizawae*.

According to Migunova and Sasanelli, (2021) the strains of *B. firmus* I-1582 and *B. amyloliquefaciens* FZB42 are the species registered as bionematicides to control *Meloidogyne* spp. in vegetable plantations in Europe. Among the main bionematicides registered and commercialized in the last decades, about 77 per cent have *Bacillus* spp. as an active ingredient, 68 per cent of which are formulated as liquids and 35 per cent as powder, with the largest participants in the world market being Brazil (45%) and the United States of America (27%). The minimum concentration of nematophagous bacteria in the composition of bionematicidal products vary between 10^4 and 10^{12} cfu/ml or gm of the commercial product, with an average between 10^9 cfu/ml or gm (Table 1).

Among the commercialized bio-nematicides, the highest concentrations of the active ingredient are observed in Furatrop® (minimum of 1.9×10^{12} cfu/ml), followed by Quartzo® and Presence® (minimum of 1.0×10^{11} cfu/g), which is an important aspect to consider, both for the formulation of a bionematicide product and for the acquisition of the commercial product. Other relevant factors are the active ingredient (only a bacterium, mixture of bacterial or fungal strains), type of formulation, shelf life, storage conditions and recommendations for application in crops.

---

[1] http://agrofit.agricultura.gov.br/agrofit_cons/principal_agrofit_cons.

Table 1. Bionematicidal products based on bacteria used in different countries to manage plant-parasitic nematodes

Bacteria/(CFU/mL, g)	Target PPN	Name/Manufacturer-Country	Formulation/Method of application	References
P. penetrans/(1.5-3x10^5) *P. usage*/(9.8 x 10^8)	*M. incognita* *M. arenaria* *B. longicaudatus*	Econem/Syngenta, Nematech, Bayer-USA, Japan	WP, FS/Surface application, applied at seeding or soil post-transplant	Migunova and Sasanelli (2021) Abd-Elgawad and Askary (2018); Wilson and Jackson (2013) Kokalis-Burelle (2015) Subedi et al. (2020)
Pasteruia sp. Ph3 *P. usage* Bl1 + *Pasteuria* sp. Ph3	*R. reniformis* *B. longicaudatus*	Naviva ST, NewPro/Syngenta-USA	WP, FS/Foliar spraying, seed treatment, soil drench or spraying application	Migunova and Sasanelli (2021) Subedi et al. (2020) https://www3.epa.gov/pesticides/chem_search/ppls/085004-00011-20130422.pdf.
P. nishizawae Pn1/(1x10^{10})	*H. glycines* *H. schachtii*	Clariva PN, Clariva PN BR/Syngenta-Brazil, USA, Canada	FS/Seed treatment	MAPA (2021) https://www.syngenta-us.com/seed-treatment/clariva-pn.
Bacillus subtilis CNPSo2657/(1.9x10^{12})	*M. javanica*	Furatrop, Vult, Nematrop/Biotrop and Tacto-Brazil	SC/Apply to in-furrow by spraying application	MAPA (2021)
B. subtilis/(1x 10^9)	*M. arenaria*	Stanes Sting/T. Stanes and Company Limited-India, Egypt	SC/Soil drenching application	Abd-El-Khair and El-Nagdi (2014) Ameen et al. (2015)
B. subtilis BV-09/(1x10^8)	*M. incognita* *M. javanica* *M. exigua* *P. zeae*	Biobaci/Vittia- Brazil	SC/Soil drenching application and in-furrow	MAPA (2021) https://vittia.com.br/produto/biobaci/.
B. subtilis UFPEDA 764/(3x10^9)	*M. javanica* *P. brachyurus*	Rizos, Rizos OG/Lallemand- Brazil	FS/Seed treatment	MAPA (2021)

Table 1. (Continued)

Bacteria/(CFU/mL, g)	Target PPN	Name/Manufacturer-Country	Formulation/Method of application	References
B. subtilis ATCC6051 + *B. licheniformis* ATCC12713/(1X10^{10}) + *P. lilacinus* CPQBA 040-11 DRM10/(1X10^{10})	*M. incognita* *P. brachyurus*	Profix, Profix A, Profix B, Profix C/Agrivalle - Brazil	WP/Seed treatment, spraying application	MAPA (2021)
B. subtilis FMCH002 + *B. licheniformis* FMCH001/(1x10^{11})	*M. incognita* *M. javanica* *M. graminicola* *M. exigua* *P. zeae* *R. similis*	Quartzo, Avodigen/Chr. Hansen, FMC- Brazil, Canada, USA	WS/Spraying and drenching application	MAPA (2021) https://www.fmc.com/en/innovation/biologicals.
B. subtilis FMCH002 + *B. licheniformis* FMCH001/(1x10^{11})	*M. incognita* *P. brachyurus*	Presence/FMC- Brazil	WS/Seed treatment	MAPA (2021)
B. firmus I-1582/(4.8x10^{9})	*M. javanica* *P. brachyurus*	Andril Prime/Basf- Brazil	FS/Seed treatment	MAPA (2021)
B. firmus I-1582/(4x10^{9})	*M. javanica* *P. brachyurus*	Oleaje Prime/Basf- Brazil	FS/Seed treatment	MAPA (2021)
B. firmus I-1582/(4x10^{9})	*M. javanica* *P. brachyurus*	Votivo Prime/Basf- Brazil	FS/Seed treatment	MAPA (2021)
B. firmus I-1582/(4.8x10^{9})	*Meloidogyne* spp., *Pratylenchus* spp., *H. glycines*, *H. avenae*, *Belonolaimus* sp., *Hoplolaimus* spp.,	Poncho-Votivo, Votivo, Nortica/Bayer - USA, Mexico, New Zealand	FS, WP/Seed treatment, drenching and spray application	Wilson and Jackson (2013) Migunova and Sasanelli (2021) https://www.greenbook.net/bayer-cropscience/poncho-votivo.

Bacteria/(CFU/mL, g)	Target PPN	Name/Manufacturer-Country	Formulation/Method of application	References
B. firmus GB-126/(1x10^6)	*R. reniformis*	VOTiVO/Bayer- Germany	WP/Seed treatment	Castillo et al. (2013) Xiang et al. (2017b) Philbrick et al. (2020)
B. firmus MCC0086/(1x10^8)	*Meloidogyne* spp., *Heterodera* spp., *Globodera* spp., *Rotylenchulus* spp., *Radopholus similis*, *Tylenchulus semipenetrans*, *Pratylenchus* spp.	BioNemagon/Agri-Life- India	WP/Seed treatment, soil treatment and foliar spraying	http://agrilife.in/products/microbial_bionemagon.php.
B. firmus	*Meloidogyne* spp., *Heterodera* spp., *Helicotylenchus* spp.	BioNem, BioSafe, Chancellor/ Agro Green- Israel, India	WP/Irrigation system application, soil treatment	Keren-Zur et al. 2000; Wang and Liu (2004) Terefe et al. (2009) Abd-Elgawad and Vagelas (2015) Philbrick et al. (2020) https://www.abim.ch/fileadmin/abim/documents/presentations2007/session4/8_keren_zur_abim_2007.pdf.
B. amyloliquefaciens MBI600/(1x10^{10})	*M. incognita* *M. javanica* *R. reniformis* *P. brachyurus*	Trunemco/Sumitomo Chemical Brasil Ind. Química S.A. – Brazil,	FS/Seed treatment	MAPA (2021)
B. amyloliquefaciens BV03/(3x10^9)	*Meloidogyne* spp., *H. glycines*, *P. brachyurus*	Nema-Guard, Nema-Attack, No-Nema/Vittia - Brazil	SC/Seed treatment and drenching application	MAPA (2021)
B. amyloliquefaciens UMAF6614/(1x10^{10})	*M. incognita* *M. javanica* *P. brachyurus*	Chevelle, Boneville, Veraneio/Koppert Biological systems- Brazil	WP/Seed treatment	MAPA (2021)

Table 1. (Continued)

Bacteria/(CFU/mL, g)	Target PPN	Name/Manufacturer-Country	Formulation/Method of application	References
B. amyloliquefaciens FZB42/(2.5x10^{10})	*M. incognita*	RhizoVital/ AbiTep GmbH, Andermatt Biocontrol AG- German, Switzerland	SL/Seed treatment, Spray and drench application, irrigation system or hydroponic systems, root dipping or mixing into soils	Migunova and Sasanelli (2021) Burkett-Cadena et al. (2008) https://www.andermattbiocontrol.com/media/pdf/products/biostimulants-bioinnoculants/rhizovital42/general-information-RhizoVital42.pdf.
B. amyloliquefaciens PTA-4838 (6.1x10^{10})	*H. glycines*, *Rotylenchus* spp., *Pratylenchus* spp, *Meloidogyne* spp. and others PPNs	Aveo EZ, Lumialza/Valent - USA, Brazil	FS/Seed treatment	MAPA (2021) https://www.greenbook.net/valent-usa-llc-agricultural-products/aveo-ez.
B. amyloliquefaciens IN937a *B. subtillis* GB03	*M. incognita*	Bio-YIELD/Gustafson LLC-USA	FS/Apply to seed, drenching or in-furrow	Mayer (2003) Burkett-Cadena et al. (2008) Philbrick et al. (2020)
B. amyloliquefaciens SIMBI BS10/(5x10^{9}) *B. amyloliquefaciens* SIMBI BS 10/(1x10^{10})	*M. javanica* *P. brachyurus*	Nemacontrol, Nemacontrol Super/Simbiose-Brazil	SC/Seed treatment	MAPA (2021) https://simbiose-agro.com.br/product/view/nemacontrol#.
B. amyloliquefaciens BV03/(3x10^{9})	*H. glycines* *M. incognita* *M. javanica* *P. brachyurus*	No-Nema/Vittia- Brazil	SC/Seed treatment, application in the planting furrow or via seedling irrigation	MAPA (2021)

Bacteria/(CFU/mL, g)	Target PPN	Name/Manufacturer-Country	Formulation/Method of application	References
B. megaterium/(25x10^6)	*M. incognita* *M. javanica* *T. semipenetrans*	Bio-Arc/ Organic Biotechnology Company- Egypt	WP/Soil drenching application	Migunova and Sasanelli (2021) Abdel-Sattar et al. (2020) Metwally et al. (2019)
B. methilotrophicus UFPEDA20/(1x10^9)	*M. javanica* *P. brachyurus*	Onix, Onix OG/ Lallemand-Brazil	SC/Seed treatment	MAPA (2021)
B. cereus	*Meloidogyne* spp.	Xian Mie/XinYiZhongKai Agro-Chemical Industry Co., Ltd-China	-	Abd-Elgawad and Askary (2018) Abd-Elgawad and Vagelas (2015)
B. cereus BCE-2 + *Bacillus subtilis* CM-5	*Meloidogyne* spp. and others PPNs	BioStart Defensor/Bio-Cat Microbials- USA	SC/-	Abd-Elgawad and Askary (2018) https://patents.justia.com/patent/2018 0116224
B. chitinosporus CM-1+ *B. laterosporus* CM-3 and CM-33 + *B. licheniformis* HB-2	*Meloidogyne* spp.	BioStart/Bio-Cat, Rincon-Vitova- USA	SC/-	Migunova and Sasanelli (2021) Abd-Elgawad and Askary (2018) https://patents.justia.com/patent/2018 0116224.
B. thuringiensis EX297512/(5x10^4)	PPNs	TWO.O/Bayer-USA	SL/Seed treatment	https://agriculture.basf.us/content/dam /cxm/agriculture/crop-protection/ products/documents/PonchoVOTIVO 2_TWO_O_Label.pdf.
B. thuringiensis	*Meloidogyne* spp. *Belonolaimus* sp. *Criconemella* spp.	Avid 0.15 EC (or abamectin)/Syngenta-USA	EC/Spraying and drenching application	Abd-Elgawad and Askary (2018) https://www.greencastonline.com/tech article.aspx?paid=185134.
B. subtilis, *B. licheniformis*, *B. megaterium*, *B. coagulans*, *P. fluorencens*, *Streptomyces* spp. and *Trichoderma* spp.	*Meloidogyne* spp. *R. reniformis*	Pathway Consortia/Pathway Holding-USA	SL, WP/Drenching application	Philbrick et al. (2020) Abd-Elgawad and Askary (2018) Xiang et al. (2018a) Castillo (2012)

Table 1. (Continued)

Bacteria/(CFU/mL, g)	Target PPN	Name/Manufacturer-Country	Formulation/Method of application	References
B. circulans, *B. thuringiensis*, *Serratia* sp., *Pseudomonas* sp., *Azotobacter* sp. (Mix)/($1x10^9$)	*M. incognita*	Micronema/Agricultural Research Centre-Egypt	SL/Drenching application or incorporating in soil	Abd-Elgawad (2020) Youssef et al. (2017) El-Nagdi et al. (2015) Raddy et al. (2013) Abd-Elgawad et al. (2008)
Heat-killed *Burkholderia rinojenses* A396	*Meloidogyne* spp., *Pratylenchus* spp., *Rotylenchulus* spp. *Tylenchorhynchus* sp. *Xiphinema* spp. *H. glycines* *Belonolaimus* sp.	Majestene, Zelto, BIOst Nematicide 100, BioST/Marrone Bio Innovations, Albaugh- USA	SL/Spraying application, chemigation, soil drench and in-furrow application, and seed treatment	Philbrick et al. (2020) https://marronebio.com/products/majestene/#labels. https://www.albaughllc.com/north-america/seed-treatment/biost-platform/biost-nematicide-100/.
Burkholderia cepacia	*M. incognita*	Deny Blue Circle/Stine Microbial Products-USA	SC/Seed treatment and soil drenching application	Mhatre et al. (2019) Meyer and Roberts (2002)
Serratia sp.	*M. incognita*	Nemafree/Agricultural Research Centre-Egypt	SL/Soil drenching application	Raddy et al. (2013)
Serratia marcescens/($1x10^9$)	*M. incognita*	Nemaless/Agricultural Research Centre-Egypt	SL/drenching application	Raddy et al. (2013)
Pseudomonas fluorencens/($2x10^8$)	*Meloidogyne* spp. Cyst and Citrus nematode	Sheathguard or Sudozone/Agriland Biotech, Agri Life-India	WP/Seed treatment, drenchig and spraying application	Abd-Elgawad and Vagelas (2015); Abd-Elgawad and Askary (2018); https://www.agrilandbiotech.com/products/sudozone.

Bacteria/(CFU/mL, g)	Target PPN	Name/Manufacturer-Country	Formulation/Method of application	References
Rhodotorula pustula/($2x10^3$) + *Serratia entomophila*/($6x10^3$) + *Serratia marcescens*/($9x10^8$) + *Pseudomonas fluorescens*/($3x10^5$) +*P. puitida* ($3x10^3$)	*Meloidogyne* spp. and others PPNs	Fornem x5/Bio Nano Tech for Fertilizers Development-Egypt	SC/Drench application	Abd-El-Fattah et al. (2013) El-Nagdi et al. (2015) https://bionano-egy.com/works/ fornem-x5/#.

PPN: Plant-parasitic nematode; WP: Wettable powders; WS: Water dispersible powder for slurry seed treatment; SC: Suspension concentrate; SL: Suspensionliquid; FS: Flowable concentrate; EC: Emulsifiable concentrate.

In the last 10 years, the biocontrol industries have five times more growth as compared to the chemical pesticide industries, which represents new bionematicide products marketed for the integrated management of PPNs. However, several illegal products such as fertilizers, soil conditioners, compost additives, inoculants and substrates with biological actives (bacteria and fungi antagonistic to PPNs) have been formulated and marketed without the proper toxicological, ecotoxicological and efficacy tests that are required for their registration as bionematicides. In addition, due to the still high cost of certain bacterium-based bionematicides, some producers carry out on farm production of biological assets, with ingredients from products marketed by reputable companies. This has serious consequences for the sector of bionematicides such as lack of quality control, risk of biological contamination, discreditability and competitive disadvantage. It should be noted that the process of production, formulation and commercialization of bionematicides must follow the regulations and legislation in force in the country of origin in a similar way to chemical nematicides.

4. Field Application Methods of Bionematicides

Bionematicides based on nematophagous bacteria protect the plant against nematodes, allowing more efficient nutrients uptake by plant and increase crop vigor and yield. The application of bionematicide aims to introduce antagonist bacteria which are exogenous or are present in low population densities in the soil. Therefore, biotic and abiotic factors can have influences on bionematicides and consequently in their commercialization (Moosavi and Zare, 2015). Nematophagous bacteria are formulated in liquid or powder (dry) form for the application on annual and perennial crops of economic importance. They are applied during the planting and post-planting stages, depending on the culture, with the objective to reduce the population density of nematodes to levels below the threshold level of economic damage, but not to eradicate the nematode in the field. Thus, it allows cultivation in areas even in the presence of nematodes, but with a low population of them. In annual crops, the application of the bionematicide in the planting stage allows to protect the plant in the early stages of its life cycle, the most susceptible to nematode, thus favoring plant development and growth. In perennial crops, the application can be accomplished in the vegetative stage of the crop to protect the plant roots from PPN. Because in the vegetative stage there is emergence of new roots and the soil population of PPNs also increases.

The mode of application depends on the formulation of the bionematicide and the trophic group of the target nematode. The application technology of bionematicides does not require any specialized equipment and they are applied with the same equipment as used for the application of insecticides or fertilizers. The compatibility of the equipment and the application method used interferes with the efficiency of field results and involves the volume, agitation system, pressure and recycling time, spray distribution pattern and environmental conditions (Shetlar, 1999; Moosavi and Zare, 2015). Based on the application method, the largest share in the global bionematicides market is mainly segmented in soil treatment, seed treatment and leaf spraying.

4.1. Soil Treatment

Temperature, soil type, humidity and type of formulation can influence the behaviour of most nematophagous bacteria (Abd-Elgaward and Vagelas, 2015; Forghani and Hajihassani, 2020), especially the survival of those that do not have resistance structures such as endospores. One of the causes for inadequacy of granular formulation of chemical nematicides to control cyst nematodes, *Globodera rostochiensis* and *G. pallida* in potato is the depth of incorporation; efficiency is increased when incorporated between 15 and 20 cm depth rather than application on soil surface or up to 35 cm depth (Woods and Haydock, 2000). This result reflects a similar behavior occurring in different nematodes and biological products based on bacteria. Therefore, application technology of bionematicides depends on the bacterial species, the target nematode, the culture and the formulation of the commercial product. In general, bionematicides should be applied to the soil during planting or immediately after planting in case of annual crops, while for perennial crops it should be applied during planting or post-planting in the rainy season or at the time of irrigation.

4.2. Application on Soil Surface

Endospores of *Pasteuria penetrans* are resistant to high temperatures and soil desiccation (Williams et al., 1989). Due to these characteristics, the use of dry root powder containing endospores of *P. penetrans* was the first approach for mass production of the bacterium and its application in soils infested with *Meloidogyne* spp. (Stirling and Wachtel, 1980; Gowen et al. 1998; Sharma and Stirling, 2002). In 2004, with *in vitro* mass multiplication of *P. usage* by Pasteuria Bioscience LLC, Alachua (Florida, USA), acquired in 2012 by Syngenta, the commercial product Econem™ was released to control sting nematodes (*Belonolaimus* spp.) on golf courses in the USA. This bionematicide is marketed in granular and liquid formulation (Table 1). Econem™ in granular formulation is applied using granular applicators, distributing the granules with *P. usage* over the lawn, followed by irrigation to release the endospores, with three applications of the commercial product (Abd-Elgawad and Vagelas, 2015). Luc et al. (2010) verified that the granular and liquid endospore formulations of *Pasteuria* sp. promote the population reduction of *B. longicaudatus* in sandy soil, however the liquid formulation was more efficient in suppressing the nematode.

In another study, Crow et al. (2011) observed that Econem™ in granular formulation at a dosage of 98 and 293kg/ha in three applications over a four-week interval, was not effective in the management of *B. longicaudatus* in golf course lawns. This result can be explained by the loss of virulence of *Pasteuria* sp. during the *in vitro* culture process or during the formulation, as well as due to the greater percolation of the endospores with the application of the liquid formulation in comparison to the granulated product, which favored the adhesion of endospores and parasitism in *B. longicaudatus*. However, losses of endospores through percolation in sandy soil can reach up to 67.7 per cent (Mateille et al., 1996). Other factors such as the adsorption of endospores by clay particles (clay soils), organic matter, pH, high soil moisture, high and low temperatures, fungal and rhizobacterial filtrates, chemical nature of the nematode cuticle and chloropicrin-based nematicide have an impact on adhesion, parasitism and the life cycle of *Pasteuria* sp. in the host nematode (Davies et al., 1991; Chen and Dickson, 1998; Rocha and Campos, 2000; Freitas et al., 2000; Maximiniano et al., 2001).

Bionematicide products based on *Pasteuria* spp. and specific for some economically important nematodes were produced, such as Naviva® ST to control *Rotylenchulus reniformis*, NewPro® for *B. longicaudatus* and Clariva® PN for *H. glycines*. The first two products have a formulation similar to Econem™, but Naviva® ST and Clariva® PN are used in the treatment of cotton and soybean seeds, respectively. Other bionematicidal products based on *Bacillus* spp. with WP formulation are recommended depending on the target nematode and/or the culture.

4.3. Nematigation

Utilization of irrigation systems (conventional sprinkler, center pivot and drip) for the application of fungicides (fungigation) and nematicides (nematigation) to manage fungi and PPNs, respectively is known for decades in several agricultural countries (Keng and Guilk, 1985; Keng et al., 1985; Johnson et al., 1986). Drip nematigation depends exclusively on the movement of water in the soil, and its efficiency is increased according to the dosages that promote effective use in lower doses (Keng and Guilk, 1985; Keng et al., 1985). In addition, low doses reduce the risk of contamination of the water table. As an advantage of nematigation we can mention the maximization of the irrigation system, uniformity of product application, better coverage of the rhizosphere surface, application of the product at different stages of the crop, reduce contamination risk for the operator and save labour. The irregular distribution of the irrigation blade in the sprinklers, corrosion of the pipes and high dilution of the product are some of the disadvantages. Though it is an efficient and safe method, but in some countries it has been slightly used or adopted, due to lack of an adequate scientific basis for its application. In Brazil, there are no chemical nematicides registered for application via irrigation water, but some producers apply bionematicides to control PPNs in fruits, vegetables and cereals using a drip, micro sprinkler and central pivot systems.

The BioNem®, BioSafe® and Chancellor™ products with WP formulation based on *B. firmus* and developed by the company Agro Green are marketed in India, Israel and Ethiopia as a commercial bionematicide for drip application (Terefe et al., 2009; Abd-Elgawad and Askary, 2018; Migunova and Sasanelli, 2021). BioNem® has been found effective in the management of *M. incognita*, *M. hapla*, *Heterodera* spp., *Tylenchulus semipenetrans*, *Xiphinema index* and *Ditylenchus dipsaci* (Keren-Zur et al., 2000; Terefe et al., 2009; Migunova and Sasanelli, 2021), with satisfactory results in managing root-knot nematodes in vegetables including tomato, cucumber, pepper, garlic and some herbs grown in commercial greenhouse (Giannakou et al., 2004; Giannakou et al., 2007). This bionematicide contains 3.0 per cent lyophilized bacteria spores and 97 per cent non-toxic additives, plant and animal extracts (Giannakou et al., 2007). Under *in vitro* conditions BioNem® @ 2 and 2.5 per cent reduces hatching and motility of *M. incognita* juveniles by 100 per cent, and its incorporation into the pot (8g/1200 cc soil) growing tomato seedlings and inoculated with 1000 J_2 of *M. incognita*, reduced the number of eggs by 45 per cent, the final nematode population by 76 per cent and gall formation by 91 per cent at 50 days after application (Terefe et al.,

2009). In a field experiment, BioNem® in combination with soil solarization for 15 - 30 days provided satisfactory control of root-knot nematode in cucumber (Giannakou et al., 2007).

PGPR, *Bacillus amyloliquefaciens* strain FZB42, releases enzymes and other compounds that stimulate the growth and production of several vegetable crops (lettuce, potatoes, carrots, tomatoes, cucumbers, strawberries, turnip cabbage, etc.), cereals (cotton, maize, rice, grains) and ornamentals. It is prepared by the company Andermatt Biocontrol and marketed as Rhizovital®42. It has suppressive action on *M. incognita* (Burkett-Cadena et al., 2008). Rhizovital® is applied in hydroponic or irrigation systems in combination with fertilizers and fungicides, as well as in seed treatment, root dipping, spraying and drench application in different cultures. New generation of this product, RhizoVital P45 based on *B. velezensis* (syn. *B. amyloliquefaciens*) and RhizoVital CS based on *B. atrophaeus*, in different formulations (SC-liquid suspension concentrate, TB-talcum based dry powder, SB-dextrose based dry powder and WG-wettable granules), has been marketed as a biostimulant/bioinoculant by Andermatt Biocontrol AG in Switzerland[2]). Other commercial products based on *B. firmus* (BioNem, BioSafe, Chancellor) and *Burkholderia rinojenses* strain A396 (Majestene®, Zelto) are also suitable for application via irrigation system. They are effective against root-knot nematode, soybean cyst nematode, reniform nematode, citrus nematode, dagger nematode, stem nematode, ring nematode, stubby root nematode, lance nematode and sting nematode (Keren-Zur et al. 2000; Wang and Liu, 2004; Wilson and Jackson, 2013).

4.4. Spraying and Drenching

The application of liquid formulation products in the agricultural segment varies according to the chemical and biological target. The chemical target depends on the structure of the plant in agricultural crops, while the biological target considers the location of the pathogen. Once the biological target is defined, it is necessary to select the product to be applied. Most bionematicides are diluted and applied in liquid form. Among the different ways of applying bionematicidal products on crops, spraying and drenching are widely used. Spraying promotes breaking of the liquid in drops, while drenching is

---

[2] https://www.andermattbiocontrol.com/sites/products/biostimulants-bioinnoculants/rhizovital.html.

application of products in a metered form in a directed jet. In general, spraying promotes greater product losses due to drift and evaporation, mainly due to poor calibration (droplet size, walking speed, type of projection, volume of application) and equipment used. Fewer losses of product occur via drench, with targeted application in the projection below the canopy of the plant. Drench application is a trend in the agricultural segment due to its high efficiency, product and cost savings. Among the bionematicidal products recommended for application to the soil (53%) approximately 40% are recommended for application via drench in annual and perennial crops (Table 1). The agricultural equipment permits these applications to be performed in dose and automatic mode with speed and application volume control, disc and jet mechanisms designed for incorporation and application in-furrows or in the plant projection with repeatability and precision.

The application method can influence the efficiency of the biological product in the control of nematodes. Frederic et al. (2020) studied the product Nem-Out™ registered as an additive for composting and based on *B. subtilis*, *B. licheniformes*, *Trichoderma longibrachiatum* ($3.75x10^8$ cfu/g) and enzymes (protease, cellulase, xylanase) applied on the soil surface and in planting pit at different concentrations (0, 1.25, 2.5, 5, 10, 20 and 40 gm/l) to control *M. incognita* in tomato. It was noted that the interaction between concentrations and form of application significantly influenced the infectivity and reproduction of the nematode, showing a linear increase in the reproduction factor with the increase in concentration of the biological product applied in the pit; and quadratic increase when applied on the surface. The application of the biological product on the soil surface provided higher mass of the roots, regardless of the concentration. The authors argued factors such as the content of organic matter in the substrate, the form of soil treatment, a single application of the product, the irrigation system and the time of colonization of the roots may have negatively interfered in the performance and effectiveness of the biological control agents.

The yam dry rot nematodes (*Scutellonema bradys*, *P. brachyurus* and *P. coffeae*) constitute one of the main problems for producers in the Northeast of Brazil. These nematodes have wide distribution in the production areas and there are no nematicides recommended for control. Under these circumstances, bionematicides are the only feasible alternatives for crop growers. Ramirez et al. (2021) studied the products, Presence®, Quartzo® and Bio Ax® (plant bio-activator with *Bacillus* spp. and organic carbon, AMK) to manage yam dry rot nematodes and verified that Presence® and Bio Ax® applied via drench to seedlings inoculated with the nematode were more effective to reduce the population density of nematodes, while Quartzo® was unable to cause any effect on nematodes. The bionematicide Quartzo® (*B. licheniformis* and *B. subtilis*) was commercially released in 2018 in Brazil and launched in the USA and Canada under the trade name Avodigen®. Quartzo® is recommended for the management of *M. incognita*, *M. javanica*, *M. exigua*, *M. graminicola*, *P. brachyurus*, *P. zeae* and *Radopholus similis* by applying in-furrow spray or drench. It showed good results by increasing productivity in sugarcane plantations infested by *P. zeae* in Northeast Brazil (Porto et al., 2018), as well as 65 per cent reduction of *M. javanica* in soybean at a dosage of 250 gm ha (Soraes et al., 2018).

The bionematicide Nortica® 5% WP based on *B. firmus* strain I-1582 at a concentration of 3.0 x 10^9 cfu/g have been commercialized by Bayer CropScience in Mexico and USA, and applied via spraying or drenching to manage cyst nematode, lance nematode, lesion nematode, RKNs, sheath nematode, spiral nematode, sting nematode and stunt nematode (Wilson and Jackson, 2013). The best results are obtained by pre-planting applications (2 to 7 days) and immediate irrigation to a minimum of 3-4 inches deep. If the product is applied before planting, the soil must be moistened with daily irrigation until planting. Nortica® is recommended to apply with irrigation after foliar treatment or at the beginning of the irrigation cycle with an aim to allow the irrigation water to incorporate the bionematicide in the entire root zone of the plant.

PGPR are exhibiting beneficial results in plant development and nematode suppression. They are used as bioinoculants in several available commercial products (Equity®, Bio-Yield®, AgBlend®, RhizoVital®), applied via drench, in the seed or in the planting pit to control root-knot nematodes (Meyer, 2003; Burkett-Cadena et al., 2008). The bionematicide Bio-Yield® was developed as commercial transplant mix containing *B. amyloliquefaciens* strain IN937a, *B. subtilis* GB03 and *Paenobacillus macerans*, to manage root-knot nematode besides several other PPNs infesting tomatoes, peppers and strawberries in the USA (Meyer, 2003). Products based on *Bacillus*, *Pseudomonas* and *Serratia* are more promising because they suppress PPNs and also stimulate plant growth and control plant pathogenic microorganisms (Migunova and Sasanelli, 2021). In Egypt, the bionematicidal products Nemafree (*Serratia* sp.) and Nemaless (*S. marcescens*) are marketed in liquid formulation for soil treatment with application via drenching in-furrow or planting pit to control PPNs (EI-Nagdi and Youssef, 2004; Raddy et al., 2013). Nemafree and Nemaless applied at recommended rates via soil drench reduced the reproduction factor of *M. incognita* in tomato by 71.34 and 76.22 per cent, respectively (Raddy et al., 2013). In India, the commercial product Sheathguard™ based on *P. fluorencens* ($2x10^8$ cfu/g) is marketed in various formulations by the companies Agri Life and AgriLand Biotech Limited for the control of RKNs, cyst nematode and citrus nematode (Abd-Elgaward and Vagelas, 2015).

The application of bionematicides by drench in only one side of the plant is more practical, however the application on both sides of the plant using fan-type spray nozzles, which have a higher flow rate (150-200 litre/ha), increases the spectrum of application of the product in the root system and colonization of the bacteria, providing greater control of PPNs.

4.5. Seed Treatment

Most of the cultivated plants are propagated by seeds; therefore it is necessary to protect them preceding sowing in order to avoid the incidence of pathogens. Seed treatment with chemicals or biocontrol agents have several advantages such as reducing the application rates, environmental impacts, production costs, impacts on the microbial population of the soil and concentrating the applied product (chemical/biocontrol agent) around the plant (Sharma et al., 2015). Seeds are mostly treated with fungicides, insecticides and nematicides (Abati et al., 2014; Sharma et al., 2015; Lamichhane et al., 2020). Application via in-furrow treatment mixed with fertilizers or foliar sprays (Lamichhane et al., 2020), does not provide the immediate protection observed with the seed treatment. It is reported that 80 per cent of the sown seeds are not treated in several countries, while in developed countries the rate of treated seeds varies between 93 to 100 per cent of the sown seeds (Sharma et al., 2015; Lamichhane et al., 2020).

The technology for seed treatment involves methods varying from the simple immersion of seeds in an aqueous solution, liquid paste or dry formulation for a predetermined time, to methods with higher technology that can vary from basic dressing to coating and pelletizing (Sharma et al., 2015; Lamichhane et al., 2020). However, seed dressing is the most common method of seed treatment used by producers and industries. Seed coating and pelleting has also been used in the treatment of seeds with bacteria as an industrial method, as it requires machines with high technology that provide a smoother treatment to the seeds, with application of polymers and dyes that act as adhesive and protector (colored film or colorless), ensures greater uniformity and fixation of the active ingredient to the seeds or can change the physical shape of the seed depending on the treatment, increase safety and operational planting activities. The colored seeds are often modified to indicate that the seed has been treated become less attractive to birds, warns that it cannot be used for consumption and differentiates commercial brands (Lamichhane et al., 2020). The most advanced seed priming technology called Solid Matrix Priming (SMP) allows the mixture seeds to a solid material and water, which can be combined with bioprotectors and/or chemical pesticides in SMP (Taylor and Harman, 1990; Sharma et al., 2015). It is worth mentioning that the graphite (carbon based powder and water soluble powder) used by farmers after treating the seeds and drying the chemical or biological product, allows to verify the uniformity of the product in the seeds, but its main function is to lubricate the seeds and the components mechanically (eliminate the friction between the disk and the ring, the seed and the dosing mechanism) and increase the fluidity of the seeds during the sowing process. While planting with a mechanical dosing planter, it becomes important that the disc selection is made after seed treatment according to the size of the seeds.

Seed inoculation (Biopriming) with *Rhizobium* spp., *Frankia* spp., *Azospirillum brasilense*, *A. lipoferum*, *Azotobacter chroococum*, *Pseudomonas* strain 186 is a well-known method for promoting the fixation of atmospheric nitrogen and increasing productivity (Mahmood et al., 2016; Schaefer et al., 2019). The treatment of coated soybean seeds with *Rhizobium japonicum* increased seed germination and reduced the severity of *Fusarium solani* and *Macrophomina phaseolina* under field conditions (AL-Ani et al., 2012). Dawar et al. (2008) reported that okra and sunflower seeds treated with *R. meliloti* and coated with sugar, mollasses, glucose and gum arabic at 1 and 2 per cent significantly reduced *M. javanica* infection and provided greater plant growth and development. In an experiment, *Bradyrhizobium* sp., *R. etli* G12 in combination with *Bacillus* sp., was found antagonistic to *M. javanica*, *M. ingonita* and *G. pallida* (Hallmann et al., 2001; Parveen et al., 2019). *Rhizobium* has nematicidal potential and its efficiency is complex that can vary with the population density of the nematode, the concentration of the inoculated bacteria, type of soil, evaluation period and combination with other biocontrol agents or chemical products.

The commercial product Clariva^{TM} based on *P. nishizawae* Pn1, launched by the company Syngenta in 2013, is marketed in Brazil and Canada for the treatment of soybean seeds to control *H. glycines*, and in the USA for the treatment of sugarbeet seeds to control *H. schachtii*. It has been found that the treatment of soybean seeds with Avicta and Clariva^{TM} inhibited movement of *H. glycines* juveniles and their penetration into seed (3 and 5 cm below) as well as nematode development within the roots of young plants (10 and 20 day-old), however these effects were not detected on nematodes developing in other plants (30 and 60 days old) (Jensen, 2018). Avicta was more efficient in reducing penetration in young plants and had no effect on nematode development, while Clariva™ reduced penetration and also affected development 20 days after planting. Kokalis-Burelle (2015) found that seed treatment with *P. penetrans* and drench application in the soil were effective in reducing the reproduction of *M. incognita* on tomato and cucumber as compared to the untreated control.

Environmental conditions can affect the effectiveness of bacteria. Liu and Timper (2020) compared the effect of *P. penetrans* endospores (3.0 x 10^5) applied in-furrow versus seed (peanut) treatment i.e., spores applied along with the seed at planting and inoculated with J_2 of *M. arenaria.* It was observed that the endospores of *P. penetrans* reduced the number of nematode eggs, regardless the method of application, however the application in the furrow caused a greater reduction in the reproduction of the nematode than the endospores applied together with the seed. These results can be explained by the distinctive application methods, in which application in the furrow promotes greater distribution of endospores in the soil, increasing the chances of adhesion to J_2 and preventing *M. arenaria* infection, while seed treatment occurs greater action of root exudates on J_2 and/or endospores causing reduced adhesion of *P. penetrans* endopores to *M. arenaria* J_2 (Rocha et al., 2004; Liu et al., 2017).

Several companies have commercialized bionematicidal products based on *B. subtilis* (Rizos, Rizos OG), *B. subtilis* and *B. licheniformis* (Profix, Presence), *B. firmus* (Andril Prime, Oleaje Prime, Votivo Prime, Poncho-Votivo, Votivo), *B. amyloliquefaciens* (Trunemco, Nema-Guard, Nema-Attack, No-Nema, Chevelle, Boneville, Summer, RhizoVital, Aveo EZ, Lumialza, Bio-YIELD, Nemacontrol, Nemacontrol Super), *B. methilotrophicus* (Onix, Onix OG), *B. thuringiensis* (Two.O, Avid 0.15 EC), *S. marcescens* (Nemaless), *P. fluorencens* (Sudozone or Sheathguard) and *Burkholderia cepacia* (Deny Blue Circle, Majestene, Zelto, BIOst Nematicide 100, BioST) with liquid formulation or powder dispersed in water for slurry seed treatment to control PPNs. In Brazil, No-Nema is a microbiological nematicide and fungicide recommended for the control of cyst nematode, RKN, lesion nematode and *Fusarium verticillioides*. Most of the products sold are based on *B. amyloliquefaciens*, followed by those on *B. firmus* and *B. subtilis* and *B. cepacia*. In 2018, FMC launched the product Presence® in Brazil for the treatment of soybean seeds to control *H. glycines* and also the product Quartzo® for in-furrow application. Both the products containing the same species and concentration of *Bacillus* (*B. subtilis* FMCH002 and *B. licheniformis* FMCH001, minimum 1.0×10^{11} cfu/g), diverging only in the way the product is applied.

The company Bayer CropScience markets the product VOTiVO™ WP based on *B. firmus* GB-126 (Israeli strain), which is used to treat corn, cotton, sorghum, soybean and sugarbeet seeds for the suppression of various PPNs such as *Meloidogyne* spp., *Pratylenchus* spp., *R. reniformis*, *Heterodera* spp., *Hoplolaimus* spp., *Criconema* spp. and *D. dipsaci* (Castillo et al., 2013; Wilson and Jackson, 2013). Two other bionematicides developed by Bayer based on *B. firmus* I-1582 are Poncho®/Votivo® and Votivo®. Poncho®/Votivo® is combination of 8.10 per cent *B. firmus* I-1582 with 40.30 per cent of the insecticide PONCHO (Clothianidin) in FS formulation for the treatment of corn, sorghum, soy and sugar beet seeds. The development of this product by Bayer with an approach for seed treatment enables an insecticidal and nematicidal effect, in addition to reduction of 1,000 times in the number of bacteria to be applied per unit area (Wilson and Jackson, 2013), and also the reduction of applied insecticide. Poncho®/Votivo® and Votivo are marketed for seed treatment in the USA, Mexico and New Zealand to control *Meloidogyne* spp., *Pratylenchus* spp. *Heterodera* spp., *Belonolaimus* sp. and *Hoplolaimus* spp. (Wilson ad Jackson, 2013; Migonova and Sasanelli, 2021). The application of Poncho®/Votivo® formulations has shown satisfactory results in the suppression of PPNs similar to the chemical nematicides Avicta and Aldicarbe (Castillo et al., 2013; Xiang et al., 2017a; Xiang et al., 2018b). In Brazil, *B. firmus* I-1582 is registered in MAPA by BASF and marketed under the trademark Andril Prime®, Oleaje Prime® and Votivo Prime® for the treatment of seeds to manage *M. javanica* and *P. brachyurus* in soybean, rice, corn, wheat, cotton and beans. In 2020, the company Koppert launched the bionematicides Chevelle, Boneville and Veraneio based on *B. amyloliquefaciens* UMAF6614, designated for seed treatment and application in vegetables (Chevelle), grains, cereals (Summer) and sugarcane (Boneville) for the management of *P. brachyurus* and *Meloidogyne* spp. These products reduced the reproductive factor of nematodes by more than 80 per cent and promoted growth and increased crop productivity.

4.6. Seedling Dipping

Seedling dipping method is the immersion of root system of the seedlings in an aqueous suspension for a predetermined period. This method is generally used to treat vegetable and fruit seedlings. Araújo et al. (2018) found that the immersion of banana plantlets cv. Comprida in the solution of Nemathel® (based on *B. subtilis*) at concentrations of 200 and 250 ml/10 l water for 30 minutes, reduced the reproduction of *R. similis*, *Meloidogyne* spp., *Pratylenchus* spp. and *Helicotylenchus* spp. The bacterial efficiency was almost similar to chemical nematicide, Carbofuran.

Pasteuria has shown variable efficiency in the control of PPNs according to the application method (Luc et al., 2010; Crow et al., 2011; Kokalis-Burelle, 2015; Jensen, 2018; Liu and Timper, 2020). Tzortzakakis et al. (1997) applied a suspension of *P. penetrans* endospores (root powder at a concentration of 84 x 10^4/g endospores) in tomato seedlings transplanted into soil infested with *Meloidgyne* spp. The spores were added to the seedling trays or seedling roots were dipped endospore suspension, but these methods did not resulte in significant nematode control. Munif et al. (2013) evaluated endophytic bacteria (*Pantoea agglomerans* MK-29, *Cedecea davisae* MK-30, *Enterobacter* spp. MK-42 and *Pseudomonas putida* MT-19) with different application methods (seed treatment-seed soaked in the bacterial suspension for 30 minutes; root dipping-roots of three-week-old tomato plants dipped for 30 min into bacterial suspension; soil drench- 3 ml of bacterial suspension pipetted onto the soil around the plants) to control *M. incognita.* Bacterial endophytes significantly reduced the infection of *M. incognita* in tomatoes with seed treatment, root immersion or application of irrigation in the soil, as well as 56 per cent reduction in the penetration of J_2 in tomato roots when applied via root immersion and watering in the soil.

The method of applying bionematicides by root dipping is unusual in practice when compared to other methods of application. However, this method can be promising for treatment of seedlings in vegetable and fruit crops (Burkett-Cadena et al., 2008).

4.7. Foliar Application

Some species of PPNs (*Ditylenchus* spp., *Aphelenchoides* spp., *Bursaphelenchus* spp.) attack stems and leaves causing leaf distortion, necrotic lesions, stunting and plant death. Rice, strawberries, ornamental plants, forages, soybeans, cotton and cowpea are among the most affected crops by *Aphelenchoides besseyi* in Brazil (Favoreto and Meyer, 2019), and yam (Noronha et al., 2020). In soybeans, *A. besseyi* incites the disease called "Soja Louca II" (SL-II), causing miscarriage of flowers and pods, wrinkling and darkening of the leaves during the hot and rainy period in the Brazilian states of Mato Grosso, Pará, Maranhão, Tocantins and Amapá, with losses of up to 100 per cent in soybean productivity (Favoreto and Meyer, 2019). *A. fragariae* and *A. ritzemabosi* attack several plant species including some ornamental crops (Franklin, 1965; Hooper, 1972; Kohl, 2011).

Nematode species which attack the aerial part of plants spend a fraction of their life cycle on the ground. Hence, in order to control the nematode, liquid formulations are generally used as foliar application with hydraulic sprayers. Foliar applications of chemical nematicides as well as bionematicides are used to control nematodes by direct contact and/or during feeding on plant tissue. Among the recommended bionematicides for foliar application is BioNemagon™ WP, based on *B. firmus* strain MCC0086, marketed by Indian company Agri-Life for suppression of RKNs, cyst nematodes, remiform nematode, burrowing nematode, citrus nematode, golden cyst nematode and lesion nematode. BioNemagon™ can also be used for the treatment of seeds (seed dressing) or soil application.

Chemical control, e.g., oxamyl and parathion have been used in the past for the management of leaf nematodes in ornamental plants (Johnson and Grill, 1975), however due to toxicity and environmental issues such products have now been withdrawn (Kohl, 2011). Avermectins are macrocyclic lactones produced by *Streptomyces avermitilis* (gram-positive filamentous bacteria) with insecticidal and nematicidal activity against PPNs (Stretton et al., 1987; Cayrol et al., 1993). Cayrol et al. (1993) studied the effect of foliar and soil application of Abamectin (Vertimec-1.8% w/v emulsifiable concentrate) applied @ 100 mg/l for the management of *M. arenaria.* Foliar application of Abamectin B1 was not effective, while soil application induced a significant reduction in nematode population. Stretton et al. (1987) demonstrated that foliar application with 1000 mg/l solutions of Abamectin reduced the number of root galls on tomato plant. Garebedian and Van Gundy (1983) investigated the efficacy of Abamectin applied in soil as granules and liquid in furrows or by low pressure drip irrigation systems, to protect tomato plants infested with *M. incognita.* It was noted that Abamectin was at least ten times more efficient than the chemicals, Aldicarb and oxamil.

Bacillus thuringiensis (Bt) has been formulated in water dispersible granule, dry flowable, aqueous suspension, granule, technical powder, dust, wettable powder, emulsifable suspension, aqueous flowable, bait, and flowable oil (Tamez-Guerra et al., 1996; Navon, 2000), mainly as an insecticide, but with little commercial attention as a bionematicide. Several commercial products have been formulated with Bt such as Dipel 2X, Dipel

DF, Agerin, Agry, Delfin and Protecto (Haggag, 2013). The application of Bt crude suspension and its commercial formulations as Dipel 2X annd Agerin has suppressive action against *Meloidogyne* spp., *T. semipenetrans*, *Helicotylenchus* spp., *P. scribneri*, *Aphelenchoides* spp. and *D. destructor* (Abd-EIgawad, 1995; Mohammed et al. 2008; Yu et al., 2008; EI-Nagdi et al., 2010). Abamectin B1 (Avid 0.15 EC, Novartis, Greensboro, North Carolina) is an acaricide, insecticide and nematicide based on *B. thuringiensis*. LaMondia (1999) found that two weekly sprays of Avid 0.15 EC on the foliage until runoff, controls *A. fragariae* and *D. dipsaci* in *Lamium maculatum* and *Phlox subulat*. Currently, Avid is registered in the USA, in the states of Alabama, Florida, Georgia, Loisiana, Missouri, North Carolina, New Jersey, New Mexico, Ohio, Oklahoma, Pennsylvania, Tennessee, Texas and Virginia for application by drenching or in mixture with fungicides such as Heritage Action® to control sting and ring nematodes on golf courses. Yu et al. (2008) found that the parasporal crystal proteins of ten strains of Bt had nematicidal activity against *D. destructor* and *Aphelenchoides* spp. *Brevibacillus laterosporus* G4 and *Bacillus* protease have nematicidal effects against *B. xylophilus* (Huang et al., 2005; Niu et al., 2006; Tian et al., 2006), and can be used as potential bionematicides in the near future to control leaf nematodes and other PPNs.

Considering that most rhizobacterial-based bionematicides are applied in soil treatment, their application in the aerial part makes their survival more difficult because they are in a different place from their habitat and the nematodes are protected inside the plant tissue. *Serratia plymuthica* M24T3 isolated from nematode pinewood was found highly effective against *B. xylophilus* (Proença et al., 2010; Paiva et al., 2013). In another study, Proença et al. (2019) observed that *S. plymuthica* M24T3 has a high capacity for colonization, stimulation of lateral rooting and growth in *Arabidopsis thaliana* and *Vigna unguiculata*. On the other hand, some PGPBs such as strains 1 to 4 of *B. megaterium*, *B. brevis*, *P. corrugata*, *P. savastanoi*, *Stenotrophomonas maltophilia* and *Serratia plymuthica* can stimulate root growth and suppress *X. index* in *Vitis vinifera* (Aballay et al., 2011).

Some commercial bionematicides have more than one application method. Their success in controling PPNs and their efficiency depends on the antagonist bacteria, their mode of action and concentration on the formulated product, type of formulation and treatment, mode and number of applications to reach the target nematode, interval for root colonization and field conditions (soil type, temperature, soil moisture/irrigation depth, organic matter).

Conclusion

PGPRs have been known for many years as bacteria inhabiting of rhizospheric soil with antagonistic action against pathogenic fungi and PPNs (Burkett-Cadena et al., 2008; Huang et al., 2010; Subedi et al., 2020). The withdrawal of some chemical nematicides from the global market associated with socioeconomic and environmental issues strengthened the interest of multinational companies in the development of new formulations of bionematicides. The global panorama of bionematicides is promising and with expectations in the coming years for new products released by the leading companies operating in the nematicide market or new companies which boost the business of the bionematicide sector worldwide. It generates new opportunities, competition and greater availability of bionematicides to choose from by farmers. As the application method is fundamental to the success in biocontrol, companies have partnered with agricultural companies, e.g., those that manufacture spray nozzles for better product performance in field conditions.

In the bionematicides market, the marketing done by the companies of the commercialized products and the field results have often presented contradictory results such as Econem™ (Luc et al., 2010; Crow et al., 2011). A very few information related to the application methods and control efficiency data against PPNs has been made available. Moreover, several commercial products based on nematophagous bacteria such as *Pasteuria nishizawae*, *B. subitilis*, *B. firmus*, *B, amyloliquefacies*, *B. thuringiensis*, *Burholderia rinojenses* and *Serratia marcescens* are restricted to some countries or locations.

There are few companies promoting the combined use of bionematicides with chemical pesticides, e.g., the VOTiVO product and Poncho®/Votivo® mixture, combined with a synthetic insecticide, Poncho®, recommended for seed treatment. In recent years, fertilizer companies have developed new products in combination with PGPR, antagonistic fungi, secondary metabolite products, enzymes, stimulants and root enhancers with promising results, which still need to be elucidated with respect to nematicidal action for effective nematode management in ecologically sustainable agriculture.

Bionematicides can perform synergistically or additively with other agricultural inputs, being an available tool among other management practices such as exclusion, immunization and crop rotation used in integrated PPN management programs (Askary, 2012; Askary and Martinelli, 2015). The biological control of PPNs requires permanent treatment, harvest after harvest over the years, but the application technology is essential in order to obtain satisfactory results and advances are still needed in field conditions. The successive cultivation of peas in sandy soils, without crop rotation, has been one of the challenges for nematode management with bionematicides in northern Minas Gerais-Brazil.

The information included in this chapter regarding application methods of commercial bionematicides are not recommendations and it is up to a responsible professional for their correct prescription. Trade names are used with the understanding that no endorsement is intentional and no implicit criticism of similar products that are not mentioned in this work, without any conflict of interest from the authors. It is worth mentioning that the application of chemical nematicides or bionematicides must be used according to the instructions on the manufacturer's label and with appropriate personal protective equipment.

Acknowledgments

The authors gratefully acknowledge the financial support from CNPq, CAPES and FAPEMIG and the support from Federal University of Minas Gerais-UFMG, Federal University of Alagoas-UFAL and Federal University of Lavras-UFLA to carry out the research work. Thanks are also due to Marcelo Gonçalves Macho and João Batista de Campos Menezes for the relevant technical information.

References

Aballay, E., Martensson, A. and Persson, P. (2011). Screening of rhizosphere bacteria from grapevine for their suppressive effect on *Xiphinema index* Thorne & Allen on in vitro grape plants. *Plant and Soil*, 347: 313-325.

Abati, J., Zucareli, C., Foloni, J. S. S., Henning, F. A., Brzezinski, C. R. and Henning, A. A. (2014). Treatment with fungicides and insecticides on the physiological quality and health of wheat seeds. *Journal of Seed Science*, 36: 392-398.

Abd-El-Fattah, A. I., Shalaby, N. M. E. L. S., and El-Nagdi, W. M. A. (2013). Impact of two bio-products on *Meloidogyne javanica* and their influence on yield components of some sugar beet varieties. *Journal of Applied Sciences Research*, 9: 4627-4637.

Abd-Elgawad, M. M. M. (1995). Evaluation of *Bacillus thuringiensis* and plant extracts for control of phytonematodes. *Egyptian Journal of Applied Science*, 10: 1-5.

Abd-Elgawad, M. M. M. (2020). Biological control agents in the integrated nematode management of pepper in Egypt. *Egyptian Journal of Biological Pest Control*, 30: 70. https://doi.org/10.1186/s41938-020-00273-6.

Abd-Elgawad, M. M. M. and Askary, T. H. (2018). Fungal and bacterial nematicides in integrated nematode management strategies. *Egyptian Journal of Biological Pest Control*, 28: 74. https://doi.org/10.1186/s41938-018-0080-x.

Abd-Elgawad, M. M. M. and Vagelas, L. K. (2015). Nematophagous bacteria: field application and commercialization. In: *Biocontrol Agents of Phytonematodes* (Eds. Askary, T. H. and Martinelli, P. R. P.). CAB International, Wallingford, UK, pp. 276-309.

Abd-Elgawad, M. M. M., Mohamed, M. M. M. and El-Gamal, N. G. S. (2008). Development of safe chemical and biological formulations for control of nematodes in cucumber. *Egyptian Pharmaceutical Journal*, 7: 41-50.

Abd-El-Khair, H. and El-Nagdi, W. M. A. (2014). Field application of bio-control agents for controlling fungal root rot and root-knot nematode in potato. *Archives of Phytopathology and Plant Protection*, 47: 1218-1230.

Abdel-Sattar, M., Haikal, A. M., and Hammad, S. E. (2020). *Meloidogyne incognita* population control and nutritional status and productivity of Thompson seedless grapevines managed with different treatments. *PLoS ONE*, 15: e0239993. https://doi.org/10.1371/journal.pone.0239993.

Adam, M., Heuer, H. and Hallmann, J. (2014). Bacterial antagonists of fungal pathogens also control root-knot nematodes by induced systemic resistance of tomato plants. *PLoS ONE*, 9: e90402. https://doi.org/10.1371/journal.pone.0090402.

Al-Ani, R. A., Adhab, M. A., Mahadi, M. H. and Abood, H. M. (2012). *Rhizobium japonicum* as a biocontrol agent of soybean root rot disease caused by *Fusarium solani* and *Macrophomina phaseolina*. *Plant Protection Science*, 48: 149-155.

Ameen, H. H., Osman, H. A., Lashein, A. M. S., Hasabo, S. A., and Koura, F. H. (2015). Management of potato production cv. Diamond infected with root knot nematode, *Meloidogyne arenaria* using biological process under filed conditions. *Middle East Journal of Agricultural Research*, 4: 37-41.

Araújo, J. J. S., Muniz, M. F. S., Moura Filho, G., Rocha, F. S. and Castro, J. M. C. (2018). *Bacillus subtilis* in the treatment of banana plantlets infected by nematodes. *Ceres Journal*, 65: 99-103.

Askary, T. H. (2012). Management of root-knot nematode *M. javanica* in pigeon pea through seed treatment. *Indian Journal of Ecology*, 39: 151-152.

Askary, T. H. (2017). Diversity of plant parasitic nematodes in pulses. In: *Plant Diversity: Present Situation and Future Scenario* (Eds. Ansari, A. A., Gill, S. S., Abbas, Z. K. and Naeem, M.). CAB International, Wallingford, UK, pp. 239-274.

Askary, T. H. and Ali, S. S. (2012). Effect of *Meloidogyne javanica* and *Fusarium udum* singly and concomitantly on wilt resistant accessions of pigeon pea. *Indian Journal of Plant Protection*, 40: 167-170.

Askary, T. H. and Martinelli, P. R. P. (2015). *Biocontrol Agents of Phytonematodes*. CAB International, Wallingford, UK, 470 pp.

Backer, R., Rokem, J. S., Ilangumaran, G., Lamont, J., Praslickova, D., Ricci, E., Subramanian, S. and Smith, D. L. (2018). Plant growth-promoting rhizobacteria:

context, mechanisms of action, and roadmap to commercialization of biostimulants for sustainable agriculture. *Frontiers in Plant Science*, 9: 1473.

Burkett-Cadena, M., Kokalis-Burelle, N., Lawrence, K., Santen, E. V. and Kloepper, J. W. (2008). Suppressiveness of root-knot nematodes mediated by rhizobacteria. *Biological Control*, 47: 55-59.

Campos, V. P., Souza, J. T. and Souza, R. M. (1998). Control of plant parasitic nematodes by bacteria. *Annual Revision of Plant Pathology*, 6: 285-327.

Castaneda-Alvarez, C. and Aballay, E. (2016). Rhizobacteria with nematicide aptitude: enzymes and compounds associated. *World Journal of Microbiology and Biotechnology*, 32: 203.

Castillo, J. D. (2012). Biocontrol studies of *Rotylenchulus reniformis* in cotton crops in Alabama. Ph.D. Thesis, Auburn University, Auburn, Alabama, USA.

Castillo, J. D., Lawrence, K. S. and Kloepper, J. W. (2013). Biocontrol of the reniform nematode by *Bacillus firmus* GB-126 and *Paecilomyces lilacinus* 251 on cotton. *Plant Disease*, 97: 967-976.

Cawoy, H., Bettiol, W., Fickers, P. and Ongena, M. (2011). Bacillus-based biological control of plant diseases. In: *Pesticides in the Modern World - Pesticides Use and Management* (Ed. Stoytcheva, M.). Rijeka: IntechOpen, pp. 273-302.

Cayrol, J. C., Djian, C. and Frankowski, J. P. (1993). Efficacy of Abamectin B1 for the control of *Meloidogyne arenaria*. *Fundamental and Applied Nematology*, 16: 239-246.

Channer, G. R. and Gowen, S. R. (1992). Selection for increased host resistance and increased pathogen specificity in the *Meloidogyne-Pasteuria penetrans* interaction. *Fundamental and Applied Nematology*, 15: 331-339.

Chen, Z. X. and Dickson, D. W. (1998). Review of *Pasteuria penetrans*: biology, ecology, and biological control potential. *Journal of Nematology*, 30: 313-340.

Cheng, W., Yang, J., Nie, Q., Huang, D., Yu, C., Zheng, L., Cai, M., Thomashow, L. S., Weller, D. M., Yu, Z. and Zhang, J. (2017). Volatile organic compounds from *Paenibacillus polymyxa* KM2501-1 control *Meloidogyne incognita* by multiple strategies. *Scientific Reports*, 7: 16213. https://doi.org/10.1038/s41598-017-16631-8.

Crow, W. T., Luc, J. E. and Giblin-Davis, R. M. (2011). Evaluation of Econem™, a formulated *Pasteuria* sp. bionematicide, for management of *Belonolaimus longicaudatus* on golf course turf. *Journal of Nematology*, 43: 101-109.

Davies K. (2009). Understanding the interaction between an obligate hyperparasitic bacterium, *Pasteuria penetrans* and its obligate plant-parasitic nematode host, *Meloidogyne* spp. *Advances in Parasitology*, 68: 211–245.

Davies, K. G., Laird, V. E. and Kerry, B. R. (1991). The mobility, development and infection of *Meloidogyne incognita* encumbered with spores of the obligate hyperparasite *Pasteuria penetrans*. *Revue de Nématologie*, 14: 611-618.

Dawar, S., Sattar, A. and Zaki, M. J. (2008). Seed dressing with biocontrol agents and nematicides for the control of root knot nematode on sunflower and okra. *Pakistan Journal of Botany*, 40: 2683-2691.

Duponnois, R., Bâ, A. M. and Mateille, T. (1999). Beneficial effects of *Enterobacter cloacae* and *Pseudomonas mendocina* for biocontrol of *Meloidogyne incognita* with the endospore-forming bacterium *Pasteuria penetrans*. *Nematology*, 1: 95-101.

EI-Nagdi, W. M. A., EI-Khair, H. A. and EI-Ghonaimy, A. M. (2015). Field application of biological control on root-knot nematode and *Fusarium* root rot fungus in banana cv. Grand Naine. *Middle East Journal of Agriculture Research*, 4: 545-554.

EI-Nagdi, W. M. A. and Youssef, M. M. A. (2004). Soaking faba bean seed in some bio-agents as prophylactic treatment for controlling *Meloidogyne incognita* root-knot nematode infection. *Journal of Pest Science*, 77: 75-78.

El-Nagdi, W. M. A., Youssef, M. M. A. and Hafez, O. M. (2010). Effects of commercial formulations of *Bacillus thuringiensis* and *Streptomyces avermitilis* on *Tylenchulus semipenetrans* and on nutrition status, yield and fruit quality of mandarin. *Mediterranean Nematology*, 38: 147-157.

Fabry, C. F. S., Freitas, L. G., Neves, W. S., Coutinho, M. M., Tótola, M. R., Oliveira, J. R., Dallemole-Giaretta, R. and Ferraz, S. (2007). Obtaining bacteria for the biocontrol of Meloidogyne javanica by heating soil and treating it with root filtrates from plants antagonistic to nematodes. *Brazilian Phytopathology*, 32: 79-82.

Favoreto, L. and Meyer, M. C. (2019). *The Nematode of Green Stem.* Londrina, Parana, Brazil: Embrapa Soybean/Epamig, 11p. (Technical circular, 147).

Ferraz, L. C. C. B. and Brown, D. J. F. (Orgs.). (2016). *Plant Nematology: Fundamentals and Importance.* Manaus: Norma Editora, 251 p.

Forghani, F. and Hajihassani, A. (2020). Recent advances in the development of environmentally benign treatments to control rook-knot nematodes. *Frontiers in Plant Science*, 11: 1125. http://doi.org/10.3389/fpls.2020.01125.

Franklin, M. T. (1965). *Aphelenchoides*. In: *Plant Nematology,* 2nd edition (Ed. Southey, J. F.). Ministry of Agriculture, London, Technical Bulletin No. 7, pp. 131–141.

Frederic, J. B., Costa, C. A., Rocha, F. S. and Fernandes, M. F. G. (2020). Application forms and mode of action of biocontroller in the management of *Meloidogyne incognita* in tomato. *Brazilian Horticulture*, 38: 254-260.

Freitas, L. G., Dickson, D. W., Mitchell, D. J. and Hewlett, T. E. (2000). Infectivity and suppression of *Pasteuria penetrans* to *Meloidogyne arenaria* race 1 in tomato following soil fumigation. *Brazilian Nematology*, 24: 157-166.

Garebedlan, S. and Van Gundy, S. D. (1983). Use of avermectins for the control of *Meloidogyne incognita* on tomatoes. *Journal of Nematology*, 15: 503-519.

Gerber, J. F. and White, J. H. (2005). *Materials and methods for the efficient production of Pasteuria.* U.S. Patent 6, 919, 197 B2.

Giannakou, I. O., Anastasiadis, I. A., Gowen, S. R. and Prophetou-Athanasiadou, D. A. (2007). Effects of a non-chemical nematicide combined with soil solarization for the control of root-knot nematodes. *Crop Protection*, 26: 1644-1654.

Giannakou, I. O., Karpouzas, D. G. and Prophetou-Athanasiadou, D. (2004). A novel non-chemical nematicide for the control of root-knot nematodes. *Applied Soil Ecology*, 26: 69-79.

Gowen, S., Davies K. G., and Pembroke, B. (2008). Potential use of *Pasteuria* spp. in the management of plant parasitic nematodes. In: *Integrated Management and Biocontrol of Vegetable and Grain Crops Nematodes* (Eds. Ciancio, A. and Mukerji, K. G.). Springer, Dordrecht, pp. 205-219.

Gowen, S. R., Tzortzakakis, E. A. and Channer, A. G. R. (1998). Control of the root-knot nematode *Meloidogyne javanica* by the parasite *Pasteuria penetrans* as influenced by the initial nematode population densities. *Nematologica*, 44: 369-379.

Gu, Y. Q., Mo, M., Zhou, J., Zou, C. and Zhang, K. (2007). Evaluation and identification of potential organic nematicidal volatiles from soil bacteria. *Soil Biology and Biochemistry*, 39: 2567–2575.

Haggag, K. H. E. (2013). Changes in protein profile of cotton leaf worm, *Spodoptera littoralis*, induced by Bt-formulations stored at cold and hot storage conditions. *Nature and Science*, 11: 77-85.

Hallmann, J., Quadt-Hallmann, A., Miller, W. G., Sikora, R. A., and Lindow, S. E. (2001). Endophytic colonization of plant by the biocontrol agent *Rhizobium etli* G12 in relations to *Meloidogyne incognita* infection. *Phytopathology*, 9: 415-422.

Hooper, D. J. (1972). *Ditylenchus dipsaci*. In: C. I. H. Descriptions of plant-parasitic nematodes, Set 1, No. 14. Commonwealth Institute of Helminthology, St. Albans, England.

Hu, H., Chen, Y., Wang, Y., Tang, Y., Chen, S. and Yan, S. (2017). Endophytic *Bacillus cereus* effectively controls *Meloidogyne incognita* on tomato plants through rapid rhizosphere occupation and repellent action. *Plant Disease*, 101: 448-455.

Hu, H., Gao, Y., Li, X., Chen, S., Yan, S. and Tian, X. (2020). Identification and nematicidal characterization of proteases secreted by endophytic bacteria *Bacillus cereus* BCM2. *Phytopathology*, 110: 336-344.

Hu, H., Wang, C., Li, X., Tang, Y., Wang, Y., Chen, S. and Yan, S. (2018). RNA-Seq identification of candidate defense genes targeted by endophytic *Bacillus cereus*-mediated induced systemic resistance against *Meloidogyne incognita* in tomato. *Pest Management Science*, 74: 2793–2805.

Huang, X. W., Tian, B. Y., Niu, Q. H., Yang, J. K., Zhang, L. M. and Zhang, K. Q. (2005). An extracellular protease from *Brevibacillus laterosporus* G4 without parasporal crystal can serve as a pathogenic factor in infection of nematodes. *Research in Microbiology*, 156: 719-727.

Huang, Y., Xu, C., Ma, L., Zhang, K., Duan, C. and Mo, M. (2010). Characterisation of volatiles produced from *Bacillus megaterium* YFM3.25 and their nematicidal activity against *Meloidogyne incognita*. *European Journal of Plant Pathology*, 126: 417-422.

Islam, Z., Pamplona, R., Atkinson, A. D. and Ezucena, E. J. (2003). *Biological Control of Rice Diseases.* Available at: http://www.vegetableipmasia.org/docs/Rice/Control_of_Rice_Diseases.pdf. (accessed 20 April 2021).

Jensen, J. P. (2018). Avicta and Clariva affect the biology of the soybean cyst nematode, *Heterodera glycines*. *Plant Disease*, 102: 2480-2486.

Johnson, A. W. and Grill, D. L. (1975). Chemical control of foliar nematodes (*Aphelenchoides fragariae*) on 'Fluffy Ruffles' fern. *Plant Disease Reporter*, 59: 772-774.

Johnson, A. W., Young, J. R., Threadgill, E. D., Doeler, C. C. and Sumner, D. R. (1986). Chemigation for crop production management. *Plant Disease* 70: 998-1004.

Keng, J. C. W. and Guilk, T. V. (1985). Application of chemical through a trickle system for soil borne pest control. I-Derivation of Basic physical theory for practical use. *Canadian Agricultural Engineering*, 27: 31-33.

Keng, J. C. W., Vrain, T. C. and Freeman, J. A. (1985). Applications of chemicals through a trickle system for soi-borne pest control. II-The design of a prototype system and test results. *Canadian Agricultural Engineering*, 27: 35-37.

Keren-Zur, M., Antonov, J., Bercovitz, A., Feldman, K., Husid, A., Kenan, G., Markov, N. and Rebhun, M. (2000). *Bacillus firmus* formulations for the safe control of root-knot nematodes. In: *Proceedings of the Brighton Crop Protection Conference on Pests and Diseases.* 2A: 47-52.

Kohl, L. M. (2011). Foliar nematodes: a summary of biology and control with a compilation of host range. *Plant Health Progress*, 12: 23.

Kokalis-Burelle, N. (2015). *Pasteuria penetrans* for control of *Meloidogyne incognita* on tomato and cucumber, and *M. arenaria* on snapdragon. *Journal of Nematology*, 47: 207-213.

Lamichhane, J. R., You, M. P., Laudinot, V., Barbetti, M. J. and Aubertot, J. (2020). Revisiting sustainability of fungicide seed treatments for field crops. *Plant Disease*, 104: 610-623.

LaMondia, J. A. (1999). Efficacy of insecticides for control of *Aphelenchoides fragariae* and *Ditylenchus dipsaci* in flowering perennial ornamentals. *Journal of Nematology*, 31: 644-649.

Li, X., Hu, H., Li, J., Wang, C., Chen, S. and Yan, S. (2019). Effects of the endophytic bacteria *Bacillus cereus* BCM2 on tomato root exudates and *Meloidogyne incognita* infection. *Plant Disease*, 103: 1551-1558.

Lian, L. H., Tian, B. Y., Xiong, R., Zhu, M. Z., Xu, J. and Zhang, K. Q. (2007). Proteases from *Bacillus*: a new insight into the mechanism of action for rhizobacterial suppression of nematode populations. *Letters in Applied Microbiology*, 45: 262–269.

Liu, C. and Timper, P. (2020). Effectiveness of *Pasteuria penetrans* applied to seed or furrow. *Nematropica*, 50: 229-232.

Liu, C., Timper, P., Ji, P. I., Mekete, T. and Joseph, S. (2017). Influence of root exudates and soil on attachment of *Pasteuria penetrans* to *Meloidogyne arenaria*. *Journal of Nematology*, 49: 304-310.

Luc, J. E., Pang, W., Crow, W. T. and Giblin-Davis, R. M. (2010). Effects of formulation and host density on the ability of *in vitro*-produced *Pasteuria* endospores to control its host *Belonolaimus longicaudatus*. *Journal of Nematology*, 42: 87-90.

Mahmood, A., Turgay, O. C., Farooq, M. and Hayat, R. (2016). Seed biopriming with plant growth promoting rhizobacteria: a review. *FEMS Microbiology Ecology*, 92: 8. https://doi.org/10.1093/femsec/fiw112.

MAPA - Ministry of Agriculture, Livestock and Food Supply, Brazil (2021). Coordination of Pesticides and Similar Chemical/DFIA/DAS. Fitossanitary Pesticides System - AGROFIT - List of formulated products. Available at: http://agrofit.agricultura.gov.br/agrofit_cons/principal_agrofit_cons. Accessed on April 13, 2021.

Mateille, T., Duponnois, R., Dabiré, K., N'Diayes, S. and Diop, M. T. (1996). Influence of the soil on the transport of the spores of *Pasteuria penetrans*, parasite of nematodes of the genus *Meloidogyne*. *European Journal of Soil Biology*, 32: 81-88.

Maximiniano, C., Campos, V. P., Souza, R. M. and Almeida, A. R. (2001). Effect of pH and rhizobacterium filtrates on endospores attachement of *Pasteuria penetrans* on *Meloidogyne javanica* juveniles. *Brazilian Nematology*, 25: 21-26.

Metwally, W. E., Khalil, A. E. and Mostafa, F. A. M. (2019). Biopesticides as eco-friendly alternatives for the management of root-knot nematode, *Meloidogyne incognita* on cowpea (*Vigna unguiculata* L.). *Egyptian Journal of Agronematology*, 18: 129-145.

Meyer, S. L. F. (2003). United States Department of Agriculture – Agricultural Research Service research programs on microbes for management of plant-parasitic nematodes. *Pest Management Science*, 59: 665–670.

Meyer S. L. F., Halbrendt, J. M., Carta, L. K., Skantar, A. M., Liu, T., Abdelnabby, H. M. E. and Vinyard, B. T. (2009). Toxicity of 2,4-diacetylphloroglucinol (DAPG) to plant-parasitic and bacterial-feeding nematodes. *Journal of Nematology* 41: 274-280.

Meyer, S. L. F. and Roberts, D. P. (2002). Combinations of biocontrol agents for management of plant-parasitic nematodes and soilborne plant-pathogenic fungi. *Journal of Nematology*, 34: 1-8.

Mhatre, P. H., Karthik, C., Kadirvelu, K., Divya, K., Venkatasalam, E., Srinivasan, S., Ramkumar, G., Saranya, C., and Shanmuganathan, R. (2019). Plant growth promoting rhizobacteria (PGPR): a potencial alternative tool for nematodes bio-control. *Biocatalysis and Agricultural Biotechnology,* 17: 119-128.

Migunova, V. D., and Sasanelli, N. (2021). Bacteria as biocontrol tool against phytoparasitic nematodes. *Plants*, 10: 389.

Mohammed, S. H., El Saedy, M. A., Enan, M. R., Ibrahim, N. E., Ghareeb, A. and Moustafa, S. A. (2008). Biocontrol effciency of *Bacillus thuringiensis* toxins against root-knot nematode, *Meloidogyne incognita*. *Journal of Cell and Molecular Biology*, 7: 57-66.

Mohan, S., Kumar, K. K., Sutar, V., Saha, S., Rowe, J. and Davies, K. G. (2020). Plant root-exudates recruit hyperparasitic bacteria of phytonematodes by altered cuticle aging: implications for biological control strategies. *Frontiers in Plant Science*, 11: 763. doi: 10.3389/fpls.2020.00763.

Moosavi, M. R. and Zare, R. (2015). Factors affecting commercial success of biocontrol agents of phytonemtodes. In: *Biocontrol Agents of Phytonematodes* (Eds. Askary, T. H. and Martinelli, P. R. P.). CAB International, Wallingford, UK, pp. 423-445.

Mputu K. J. and Thonart, P. (2013). Optimisation of production, freeze-drying and storage of *Pseudomonas fluorescens* BTP1. *International Journal of Microbiology*, 5: 370-373.

Munif, A., Hallmann, J. and Sikora, R. A. (2013). The influence of endophytic bacteria on *Meloidogyne incognita* infection and tomato plant growth. *Journal of International Society for Southeast Asian Agricultural Sciences*, 19: 68-74.

Navon, A. (2000). *Bacillus thuringiensis* insecticides in crop protection-reality and prospects. *Crop Protection*, 19: 669-676.

Nimnoi, P. and Ruanpanun, P. (2020). Suppression of root-knot nematode and plant growth promotion of chili (*Capsicum flutescens* L.) using co-inoculation of *Streptomyces* spp. *Biological Control*, 145: 104244. http://doi.org/10.1016/j.biocontrol.2020.104244.

Niu, Q. H., Huang, X. W., Zhang, L., Yang, J. K. and Zhang, K. Q. (2006). A neutral protease from *Bacillus nematocida*, another potential virulence factor in the infection against nematodes. *Archives of Microbiology*, 185: 439-448.

Noronha, M. A., Assunção, M. C., Costa, M. G. S., Muniz, M. F. S., Favoreto, L., Sercero, B. C. and Machado, A. C. Z. (2020). First Report of *Aphelenchoides besseyi* causing leaf spot on yam (*Dioscorea cayenensis*) in Brazil. *Plant Disease*, 104: 3083. https://doi.org/10.1094/PDIS-03-20-0511-PDN.

Orr, J. N., Mauchline, T. H., Cock, P. J., Blok, V. C. and Davies, K. G. (2018). *De novo* assembly of the *Pasteuria penetrans* genome reveals high plasticity, host dependency, and BclA-like collagens. *BioRxiv*: 485748. doi: https://doi.org/10.1101/485748.

Page, A. P., Stepek, G., Winter, A. D. and Pertab, D. (2014). Enzymology of the nematode cuticle: a potential drug target? *International Journal for Parasitology: Drugs and Drug Resistance*, 4: 133–141.

Paiva, G., Proença, D. N., Francisco, R., Verissimo, P., Santos, S. S., Fonseca, L., Abrantes, I. M. O. and Morais, P. V. (2013). Nematicidal bacteria associated to pinewood nematode produce extracellular proteases. *PLoS ONE*, 8: e79705. http://doi.org/10.1371/journal.pone.0079705.

Parveen, G., Noreen, R., Sharfique, H. A., Sultana, V., Ehteshamul-Haque, S. and Athar, M. (2019). Role of rhizobia in suppressing the root diseases of soybean under soil amendment. *Weed Plant, 37: e019172336.*

Perry, R. N., and Curtis, R. H. C. (2013). Behaviour and sensory perception. In: *Plant Nematology,* 2nd Edition (Eds. Perry, R. N. and Moens, M.). CAB International, Wallingford, UK, pp. 246-273.

Philbrick, A. N., Adhikari, T. B., Louws, F. J. and Gorny, A. M. (2020). *Meloidogyne enterolobii*, a major threat to tomato production: current status and future prospects for its management. *Frontiers in Plant Science*, 11: 606395. https://doi.org/10.3389/fpls.2020.606395.

Porto, A. C. F., Pedrosa, E. M. R., Guimarães, L. M. P. and Oliveira, W. J. (2018). Management of plant-parasitic nematodes in sugarcane. In: *Brazilian Meeting of Nematology,* 35, Bento Gonçalves, RS, Brazil. Brasília, DF, Brazil: Embrapa, pp.28-32.

Proença, D. N., Francisco R., Santos, C. V., Lopes, A., Fonseca L., Abrantes, I. M. O., and Morais, P. V. (2010). Diversity of bacteria associated with *Bursaphelenchus xylophilus* and other nematodes isolated from *Pinus pinaster* trees with pine wilt disease. *PLoS One*, 5: e15191.

Proença, D. N., Schwab, S., Vidal, M. S., Baldani, J. I., Xavier, G. R. and Morais, P. V. (2019). The nematicide *Serratia plymuthica* M24T3 colonizes *Arabidopsis thaliana*, stimulates plant growth, and presents plant beneficial potential. *Brazilian Journal of Microbiology*, 50: 777-789.

Raddy, H. M., Fouad, A. F. A., Montasser, S. A., Abdel-Lateef, M. F. and El-Samdisy, A. M. (2013). Efficacy of six nematicides and six commercial bioproducts against root-knot nematodes, *Meloidogyne incognita* on tomato. *Journal of Applied Sciences Research*, 9: 4410-4417.

Ramirez, C. H., Araújo, A. S., Moura Filho, G., Rocha, F. S., Costa, M. G. S. and Muniz, M. F. S. (2021). Biocontrollers in the management of yam dry rot nematodes. *Diversitas Journal*, 6: 24-35.

Rocha, F. S. and Campos, V. P. (2000). Effect of fungal filtrates on adhesion of *Pasteuria penetrans* and on infectivity and parasitism of second-stage juveniles of *Meloidogyne incognita*. *Brazilian Nematology*, 24: 239-244.

Rocha, F. S., Campos, V. P. and Souza, R. M. (2004). Effect of root exudates on endospores of *Pasteuria penetrans* and on second-stage juvenile of *Meloidogyne incognita*. *Brazilian Phytopathology*, 29: 644-650.

Schaefer, P. E., Martin, T. N., Pizzani, R. and Schaefer, E. L. (2019). Inoculation with *Azospirillum brasilense* on corn yield and yield components in an integrated crop-livestock system. *Acta Scientiarum. Agronomy*, 41: e39481.

Sela, S., Schickler, H., Chet, I. and Spiegel, Y. (1998). Purification and characterization of a *Bacillus cereus* collagenolytic/proteolytic enzyme and its effect on *Meloidogyne javanica* cuticular proteins. *European Journal of Plant Pathololog*y, 104: 59–67.

Sharma, K. K., Singh, U. S., Sharma, P., Kumar, A. and Sharma, L. (2015). Seed treatments for sustainable agriculture-A review. *Journal of Applied and Natural Science*, 7: 521-539.

Sharma, R. D. and Stirling, G. R. (2002). Methods of extraction and quantification of Pasteuria penetrans endospores in powdered tomato root material. Planaltina, DF, Brazil: Embrapa Cerrados, 3p (*Technical Communication,* 85). http://ainfo.cnptia.embrapa.br/digital/bitstream/CPAC-009/24251/1/comtec_85.pdf.

Shetlar, D. J. (1999). Application methods in different cropping systems. In: *Proceedings of the workshop on optimal use of insecticidal nematodes in pest management* (Polararapu, S.). New Brunswick, New Jersey: Ohio State University, pp. 31-36.

Siddiqui, I. A., Haas, D. and Heeb, S. (2005). Extracellular protease of *Pseudomonas fluorescens* CHA0, a biocontrol factor with activity against the root-knot nematode *Meloidogyne incognita*. *Applied and Environmental Microbiology*, 71: 5646–5649.

Siddiqui, I. A. and Shaukat, S. S. (2002). Rhizobacteria-mediated induction of systemic resistance (isr) in tomato against *Meloidogyne javanica*. *Journal of Phytopathology*, 150: 469–473.

Siddiqui, I. A., Shaukat, S. S., Sheikh, I. H. and Khan, A. (2006). Role of cyanide production by *Pseudomonas fluorescens* CHA0 in the suppression of root-knot nematode, *Meloidogyne javanica* in tomato. *World Journal of Microbiology and Biotechnology*, 22: 641–650.

Siddiqui, Z. A., Baghel, G. and Akhtar, M. S. (2007). Biocontrol of *Meloidogyne javanica* by *Rhizobium* and plant growth-promoting rhizobacteria on lentil. *World Journal of Microbiology and Biotechnology*, 23: 435–441.

Sikora, R. A. and Roberts, P. A. (2018). Management practices: an overview of integrated nematode management technologies. In: *Plant Parasitic Nematodes in Subtropical and Tropical Agriculture*. 3rd edition (Eds. Sikora, R. A., Coyne, D., Hallmann, J. and Timper, P.). CAB International, Wallingford, UK, pp. 795-838.

Singh, B., Ali, S. S., Naimuddin and Askary, T. H. (2004). Combined effect of *Fusarium udum* and *Meloidogyne javanica* on wilt resistant accessions of pigeon pea. *Annals of Plant Protection Sciences*, 12: 130-133.

Singh, S., Singh, B. and Singh, A. P. (2015). Nematodes: a threat to sustainability of agriculture. *Procedia Environmental Sciences*, 29: 215-216.

Soares, M. R. C., Lopes, A. P. M., Hernandes, I., Miamoto, A., Mioranza, T. M. and Dias-Arieira, C. R. (2018). Alicina-based product in *Meloidogyne javanica* control in soybean. In: *Brazilian Meeting of Nematology,* 35, Bento Gonçalves, RS, Brazil: Embrapa-Brasília, DF, Brazil, p.189.

Stirling, G. R., and Wachtel, M. F. (1980). Mass production of *Bacillus penetrans* for the biological control of root-knot nematodes. *Nematologica*, 26: 308-312.

Stretton, A. O. W., Campbell, W. C. and Babu, J. R. (1987). Biological activity and mode of action of avermectins. In: *Vistas on Nematology* (Eds. Veech, J. A. and Dickson, D. W.). Society of Nematologists, Hyattsville, Maryland, pp. 136-146.

Subedi, P., Gattoni, K., Liu, W., Lawrence, K. S. and Park, S. (2020). Current utility of plant growth-promoting rhizobacteria as biological control agents towards plant-parasitic nematodes. *Plants*, 9: 1167. https://doi.org/10.3390/plants9091167.

Susic, N., Janezic, S., Rupnik, M. and Stare, B. G. (2020). Whole genome sequencing and comparative genomics of two nematicidal *Bacillus* strains reveals a wide range of possible virulence factors. *G3: Genes|Genomes|Genetics*, 10: 881–890.

Tamez-Guerra, P., McGuire, M. R., Medrano-Roldan, H., Galan-Wong, L. J., Shasha, B. S. and Vega, F. E. (1996). Sprayable granule formulations for *Bacillus thuringiensis. Journal of Economic Entomology*, 89: 1424-1430.

Taylor, A. G. and Harman, G. E. (1990). Concepts and technologies of selected seed treatments. *Annual Review of Phytopathology*, 28: 312-339.

Terefe, M., Terefa, T. and Sakhuja, P. K. (2009). Effect of a formulation of *Bacillus firmus* on root-knot nematode *Meloidogyne incognita* infestation and the growth of tomato plants in the greenhouse and nursery. *Journal of Invertebrate Pathology*, 100: 94-99.

Tian, B. Y., Li, N., Lian, L. H., Liu, J. W., Yang, J. K. and Zhang, K. Q. (2006). Cloning, expression and deletion of the cuticle-degrading protease BLG4 from nematophagous bacterium *Brevibacillus laterosporus* G4. *Archives of Microbiology*, 186: 297-305.

Tian, B., Yang, J. and Zhang, K. (2007). Bacteria used in the biological control of plant-parasitic nematodes: populations, mechanisms of action, and future prospects. *FEMS Microbology Ecology*, 61:197-213.

Tzortzakakis, E. A., Channer, A., Gowen, S. R. and Ahmed, R. (1997). Studies on the potential use of *Pasteuria penetrans* as a biocontrol agent of root-knot nematodes (*Meloidogyne* spp.). *Plant Pathology*, 46: 44-55.

Valicente, F. H., Tuelher, E. S., Leite, M. I. S., Freire, F. L. and Vieira, C. M. (2010). Production of *Bacillus thuringiensis* biopesticide using commercial lab medium and agricultural by-products as nutrient sources. *Brazilian Journal of Maize and Sorghum*, 9 (I):1-11.

Wang, B. and Liu, X. Z. (2004). Mass production and formulation of nematode-antagonistic microbes. In: *Biological Control of Plant-Parasitic Nematodes*. (Eds. Liu, X. Z., Zhang, K. Q., and Li, T. F.). China Science and Technology Press, Beijing, China: pp. 285-297.

Williams, A. B., Stirling, G. R., Hayward, A. C. and Perry, J. (1989). Properties and attempted culture of *Pasteuria penetrans*, a bacterial parasite of root-knot nematode (*Meloidogyne javanica*). *Journal of Applied Bacteriology*, 67: 145-156.

Wilson, M. J. and Jackson, T. A. (2013). Progress in the commercialization of bionematicides. *BioControl*, 58: 715-722.

Woods, S. R. and Haydock, P. P. J. (2000). The effect of granular nematicide incorporation depth and potato planting depth on potatoes grown in land infested with potato cyst nematodes *Globodera rostochiensis* and *G. pallida. Annals of Applied Biology*, 136: 27-33.

Xiang, N., Lawrence, K., and Donald, P. A. (2018a). Biological control potential of plant growth-promoting rhizobacteria suppression of *Meloidogyne incognita* on cotton and *Heterodera glycines* on soybean: a review. *Journal of Phytopathology*, 166: 449-458.

Xiang, N., Lawrence, K. S., Kloepper, J. W., Donald, P. A. and Mclnroy, J. A. (2017a). Biological control of *Heterodera glycines* by spore-forming plant growth-promoting rhizobacteria (PGPR) on soybean. *PLoS ONE*, 12: e0181201. https://doi.org/10.1371/journal.pone.0181201.

Xiang, N., Lawrence, K. S., Kloepper, J. W., Donald, P. A., McInroy, J. A. and Lawrence, G. W. (2017b). Biological control *Meloidogyne incognita* by spore-forming plant growth-promoting rhizobacteria on cotton. *Plant Disease*, 101: 774-784.

Xiang, N., Lawrence, K. S., Kloepper, J. W. and Donald, P. A. (2018b). Biological control of *Rotylenchulus reniformis* on soybean by plant-growth-promotion rhizobacteria. *Nematropica*, 48: 116-125.

Youssef, M. M. A., Abd-EI-Khair, H. and EI-Nagdi, W. M. A. (2017). Management of root knot nematode, *Meloidogyne incognita* infecting sugar beet as affected by certain bacterial and fungal suspensions. *Agricultural Engineering International: CIGR Journal*, Special issue: 293-301.

Yu, Z. Q., Wang, Q. L., Liu, B., Zou, X., Yu, Z. N. and Sun, M. (2008). *Bacillus thuringiensis* crystal protein toxicity against plant-parasitic nematodes. *Chinese Journal of Agricultural Biotechnology*, 5: 13-17.

Chapter 11

Success and Failure in Integrated Application of Entomopathogenic Nematodes and Insecticides in Crop Pest Management

Aashaq Hussain Bhat* **and Jasper Ewany**

Experimental Biology Research Group,
Institute of Biology, Faculty of Science,
University of Neuchâtel, Neuchâtel, Switzerland

Abstract

Entomopathogenic nematodes (EPNs) are environmentally friendly, safe and effective biocontrol agents for a variety of insect pests that harm agricultural and horticultural crops. They have been reported from all over the world and mass-produced for utilization against insect pests. When used in conjunction with pesticides, these EPNs have shown better outcomes. The growing interest in organic agriculture as an alternative to synthetic pesticides creates prospects for the integrated use of EPNs and insecticides in agricultural pest management. However, EPNs first need to be enhanced in terms of efficacy, cost savings, and convenience of use. Furthermore, their ability to recycle in the host and their tolerance to various chemical pesticides used in agriculture needs to be explored for long-term pest control. This chapter summarizes significant progress made in the research and application of EPNs in combination with chemical pesticides for insect pest management programs.

Keywords: pesticides, insect, entomopathogenic nematode, biocontrol agent, management

---

* Corresponding Author's Email: aashiqhussainbhat10@mail.com.

In: Pest Management: Methods, Applications and Challenges
Editor: Tarique Hassan Askary
ISBN: 979-8-88697-268-9

1. Introduction

The economy of most countries is agriculture-based and throughout history has been beset by several insect pest outbreaks (Canto et al., 2009; Salih et al., 2020). Numerous insecticides belonging to conventional (organopho-sphates, carbamates, and pyrethroids) and novel classes (insect growth regulators, diamides, spinosyns, and avermectins) are extensively sprayed by farmers to control insect pests (Arias-Este´vez et al., 2008; Heeb et al., 2019). Imprudent over-spraying of pesticides may elevate risks to the environment, impair non-target biodiversity and threat public health. Moreover, intensive spraying and repeated use of insecticides from the same chemical group often results in the development of resistance in insect pests (Kim et al., 2021; Muraro et al., 2021; Niu et al., 2021; Saeed et al., 2021; Silva et al., 2021). Insecticide resistance of insects has been reported throughout the globe and the problem is accelerating with the advancement of time. This requires a better management program in an eco-friendly manner in the coming years.

EPNs are one of the most important biocontrol agents that are utilized against a variety of economically important insect pests. They have been used as classical, conservational, and augmentative biological control agents (Antwi and Reddy, 2016; Yadav et al., 2017; Khan et al., 2018, 2020). Many surveys have been undertaken around the world to identify efficacious EPNs that could be useful in the control of economically important insect pests. In terms of geographical distribution, EPNs have been found in soils all over the world, with at least 100 *Steinernema*, 1 *Neosteinernema*, and 19 *Heterorhabditis* species reported to date (Bhat et al., 2020a; Machado et al., 2021). They are typically categorized in the Heterorhabditidae and Steinernematidae families, which are not closely related, yet share several traits due to convergent evolution. They are arthropod parasites that are only transmitted horizontally in nature and have an infective juvenile (IJ) stage that actively invades the insect host (Lewis and Clarke, 2012). EPNs have symbiotic bacteria in their gut (*Photorhabdus* in case of *Heterorhabditis* spp. and *Xenorhabdus* in case of *Steinernema* spp.) that play a key role in host infection, and as a result of this association, they are able to kill hosts in a relatively short period of time (Boemare et al., 1993; Thomas and Poinar, 1979). The nematode-bacterium relationship is extremely specialized. In nature, the bacteria lack infective abilities and are unable to survive outside of the EPNs or insect hosts, relying instead on the nematode to transport them from one host to another. The bacteria, on the other hand, play a crucial part in the death of the insect host by suppressing the insect immune system and inducing toxemia and septicemia. They also release antibiotics that prevent further host invasions, act as a deterrent to scavengers that would otherwise feast on the host carcass and provide food for the nematodes. The bacteria play a key role in killing the host, whereas the nematodes play a more active role in contributing to the virulence of the nematode-bacterium complex (Lu et al., 2017).

EPNs from genera *Steinernema* and *Heterorhabditis* have been investigated the most for their ability to manage pests such as *Spodoptera* sp. (Negrisoli et al., 2010; Yadav et al., 2017; Khan et al., 2018; Bhat et al., 2017, 2018, 2019), *Plutella xylostella* (Somvanshi et al., 2006; Sunanda et al., 2014), *Tuta absoluta* (Garcia-del-Pino et al., 2013; Amizadeh et al., 2019; Sabino et al., 2019), *Galleria mellonella*, *Pectinophora gossypiella*, *Leucinodes arbonalis* (Khan et al., 2020), *Anomala orientalis*, *Popillia japonica*, *Cyclocephala borealis* (Liesch et al., 2010), *Holotrichia parallela* (Guo et al., 2017), *Helicoverpa armigera* (Kajol et al., 2020; Rana et al., 2020; Bhat et al., 2020b, 2021), *Zophobas morio* (Navarez et al., 2020) and other destructive insects. EPNs were first discovered in the 1920s and began to gain popularity in the 1950s, with commercialization beginning in the 1980s. A rush of scientific and commercial interest in these nematodes has resulted from advances in their mass-production technique, the finding of multiple effective isolates, and the need to reduce pesticide usage. Because of their high level of safety for humans, non-target creatures, and the environment, they have been largely omitted from pesticide registration regulations in many countries (Ehlers, 2005; Radová, 2010; Piedra-Buena et al., 2015; Yan et al., 2020). This has facilitated EPN commercialization, resulting in the production of at least five *Heterorhabditis* species (*Heterorhabditis bacteriophora, H. indica, H. marelata, H. megidis and H. zealandica*) and eight *Steinernema* species (*Steinernema carpocapsae, S. feltiae, S. kraussei, S. kushidai, S. longicaudum, S. riobrave*, *S. scapterisci* and *S. scarabaei*) for commercial use (Susurluk, 2008; Khan et al., 2016; Daughtrey and Buitenhuis, 2020; Shehata et al., 2020). Starting in the 1990s, research on the application, biology, and ecology of EPNs grew at an exponential rate, eventually expanding into substantial basic research in the 2000s.

Previous studies have shown that EPNs may offer an environmentally safe and compatible alternative to synthetic insecticides for the management of different insect pests (Grewal et al., 2005). The use of EPNs to inhibit hazardous insects was formerly limited to laboratory research, but it is now considered an applied science in agriculture (Laznik and Trdan, 2011). Nowadays, combining EPNs with insecticides appears to be a beneficial strategy, as it gives a low-cost alternative for plant protection while also allowing for effective nematode use in IPM (Georgis and Gaugler, 1991). Insecticides provide the acute toxicity needed for plant protection, while nematodes provide environmental safety as well as parasite-like hunting capabilities. IPM, which mixes chemical and biological agents, can thus be utilized to mitigate the potentially harmful environmental effects of some chemicals while still achieving effective and long-term pest control (Clough et al., 2010; Schreiber et al., 2010). IPM method is effective even when chemical insecticides fail, especially in insect pests with a high reproductive capability, a wide temperature tolerance, and larvae that feed latently (Kary et al., 2021). Field studies have shown that control efficacy increases when pesticides and nematodes are administered together. The compatibility of *Steinernema* and *Heterorhabditis* with a variety of insecticides has been extensively investigated by some research workers (Mohan, 2015; Kary et al., 2018; Hassan and Ibrahim, 2019). EPNs in combination with many conventional and non-conventional insecticides have been found promising in controlling a wide range of economically important insect pests of agricultural and horticultural crops (Sunanda et al., 2014; Garcia-del-Pino et al., 2013; Khan et al., 2018; Viteri et al., 2018; Amizadeh et al., 2019; Sabino et al., 2019). Most acaricides, fungicides, herbicides, insecticides, and growth regulators are tolerable to EPNs; therefore, they can be used in conjunction with a variety of pesticides (Rovestiet al., 1988; Laznik and Trdan, 2014). This tolerance to chemical pesticides in IJs could be attributed to the existence of two exterior epidermal layers, as well as the fact that during dormancy, oral and anal apertures of IJs are closed, preventing hazardous microbial organisms and chemicals from entering their bodies (Campbell and Gaugler, 1991). They don't eat during this time; hence their survival is reliant on internal energy sources (reserves). IJs' ability to survive during dormancy is thus determined by their metabolism and internal energy reserves which support physiological and behavioral processes associated with environmental stress adjustment (Glazer, 2002). However, previous research has found that some pesticides can diminish the vitality and infectivity of EPNs (Hara and Kaya, 1983; Zimmerman and Cranshaw, 1990; Zhang et al., 1994; Grewal et al., 1998; Head et al., 2000; Krishnayya and Grewal, 2002). Furthermore, some EPN species have varied susceptibility and sensitivity to different chemical pesticide formulations (Grewal, 2002). This could be related to changes in epidermal structure in different EPN species.

The present chapter focuses on the integrated application of EPN and insecticides in crop pest management programs. Besides, the limitations coming in the way and ideas to overcome those limitations have also been discussed.

2. Entomopathogenic Nematode - Insecticide Compatibility

Globally, the estimated annual crop yield losses that are caused by pests and pathogens range between 20 and 40 percent of the total production (FAO, 2017). These are further exacerbated by the effects of climate change contributing to the emergence of many crop diseases (Zayan, 2019) and insect pest outbreaks (Canto et al., 2009; Salih et al., 2020). Chemical pesticides are an important method that has been used in pest control for a long time and their significant contribution to crop productivity has resulted in an increase in their application (Arias-Estévez 2008; Heeb et al., 2019). However, despite their contribution to food production, a major problem with the usage of chemical pesticides is the development of resistance among insects to many groups of pesticides viz., carbamates, organophosphates and pyrethroids (Li et al., 2016; Grant et al., 2019; Wang et al., 2019; Pu et al., 2020; Sunitha et al., 2020; Xue et al., 2020; Garlet et al., 2021; Gong et al., 2021; Kim et al., 2021; Muraro et al., 2021; Niu et al., 2021; Saeed et al., 2021; Silva et al., 2021).

In an attempt to forestall the usage of chemical pesticides, biocontrol agents have emerged as one of the best options. Among these biocontrol agents, EPNs belonging to the genera *Steinernema* and *Heterorhabditis* have been employed very successfully to manage several insect pests infecting agricultural and horticultural crops (Klein, 1990; Wright et al., 1993). However, their susceptibility varies depending on the nematode species and strain (Jackson and Brooks, 1989). Studies have confirmed that some species, such as *S. carpocapsae*, *S. feltiae*, and *H. bacteriophora*, have suppressed rootworm larvae damage, reduced rootworm adult emergence, and/or safeguarded yield (Thurston and Yule, 1990; Wright et al., 1993). When nematode performances were compared with chemical pesticides, they varied but proved as much effective as the insecticides (Wright et al., 1993). A synergy of insect suppression has been achieved by using EPNs in conjunction with chemical insecticides, soil amendments, and fertilizers (Hatsukade, 1990; Ishibashi, 1993). EPNs are blended with other pesticides, with the goal of increasing the efficiency of suppressing insect pests on plants (Laznik et al., 2012; Laznik and Trdan, 2014). The EPN–insecticide combination has attracted research attention because of the need to develop more efficacious control against pests and reduce insecticide resistance against the target pests, and also to provide an understanding of how the naturally occurring EPN in soils can be better preserved in agroecosystems (Barbosa-Negrisoli et al., 2009).

Research into EPNs and how they can be mixed with different insecticides started many decades ago and many EPNs belonging to the *Steinernematidae* and *Heterorhabditidae* families have been found to survive exposure to a wide range of chemical pesticides (Hara and Kaya, 1983; Rovesti et al., 1988; Rovesti and Deseö, 1990; Head et al., 2000; Glazer, 2002; Koppenhöfer et al., 2002; Krishnayya and Grewal, 2002; De Nardo and Grewal, 2003; Schroer et al., 2005; Laznik et al., 2012; Laznik and Trdan, 2014; Ulu et al., 2016). Such synthetic chemicals that have minimal or no toxicity to EPNs include those in the group of carbamates and organophosphates whose effects ranged from nontoxic to minimal effects on nematode survival, infectivity and reproduction (Zang et al., 994; Gordon et al., 1996). EPN-chemical pesticide combinations can have synergic, additive effects on pest mortality or antagonist effect when applied simultaneously (De Nardo and Grewal, 2003; Laznik et al., 2012). Chemical insecticides, soil additives, and fertilizers are frequently used in conjunction with EPNs. Most acaricides, fungicides, herbicides, and insecticides have been found tolerable to brief (2–6 h) exposures of IJs, allowing them to be used concurrently with a variety of pesticides (Viteri et al., 2018). Combinations of pesticides and EPNs in tank mixes are cost-effective alternatives to turf IPM systems. Many experiments have been undertaken on whether pesticides can be blended with other ingredients such as fertilizers, microbial control agents, and other chemical pesticides (Grewal et al., 1998; Gutiérrez et al., 2008; Ishibashi and Takii, 1993; Koppenhöfer and Fuzy, 2008; Koppenhöfer et al., 2000; Laznik and Trdan, 2014; Laznik et al., 2012; Rovesti and Deseö, 1990; Şahin and Susurluk, 2018; Thurston et al., 1994; Ulu et al., 2016). Interestingly, some researchers have observed that multiple pesticides have detrimental impacts on nematode infectivity and survival, while others found that they work together efficiently (De Nardo and Grewal, 2003; Laznik et al., 2012). The susceptibility of IJs is influenced by a variety of factors which include nematode species, nematode strain, nematode concentrations, application method and time of application.

Many EPN-insecticide studies have revealed the increased efficacy of EPN against pests. Oxamyl synergistically increased the efficacy of *S. carpocapsae* against *Agrotis segatum* in fumigated soil (Ishibashi, 1993) and the chemical was also found compatible with *H. bacteriophora* (El-Ashry and El-Marzoky, 2018). *S. feltiae* and *H. bacteriophora* were compatible with imidacloprid (Koppenhöfer et al., 2002; Koppenhöfer and Fuzy 2008; Yan et al., 2012; Laznik and Trdan 2014; Özdemir et al., 2020) and cyflumetofen (Özdemir et al., 2020). It has been reported that two insecticides, neonicotinoid and imidacloprid were compatible with different EPN species *viz., S. glaseri, H. bacteriophora, H. marelata* and *H. megidis* and their synergies were effective against the third instar larvae of white grubs (Koppenhöfer et al., 2002). EPN in combination with chlorantrani-liprole and imidacloprid has been found efficacious even at lower rates against the first, second and third instar larvae of the white grubs (Koppenhöfer et al., 2020) and also against the second instar larvae of *Holotrichia oblita* (Guo et al., 2016). Imidacloprid is one of the best studied EPN-pesticides combinations (Koppenhöfer et al., 2020) and is one of the pesticides that has been registered for drip irrigation.

S. carpocapsae and *H. bacteriophora* have also been found compatible with fenamiphos, abamectin and oxamyl (El-Ashry and El-Marzoky, 2018). Under laboratory conditions, *H. bacteriophora* HB and *S. brazilense* IBCB-n06 were subjected to maximal exposures to some insecticides *viz*., malathion, spinetoram, phosmet, acetamiprid, or novaluron and the combinations were effective in the management of *C. capitate* (Jean-Baptiste et al., 2021). In Brazil, several registered systemic insecticides with active ingredients neonicotinoid, thiamethoxam, neonicotinoid, imidacloprid, anthranilic diamide and chlorantraniliprole were found to successfully combine with the native EPN, *H. amazonensis* and the combination pesticides were recommended for the control of tomato leaf miner, *Tuta absoluta* (Sabino et al., 2019). In Egypt, the insecticides lambda-cyhalothrin, emamectin benzoate, and indoxacarb in combination with Egyptian local and imported *H. bacteriophora* strains efficiently controlled cabbage butterfly, *Pieris rapae* on cabbage plants (Aioub et al., 2021).

Several insecticides have been found to work in tandem with EPNs for the management of white grubs. For example, chlorantraniliprole could be blended with *H. bacteriophora* against *Anomala orientalis* and *Popillia japonica* (Koppenhöfer and Fuzy, 2008a), while imidacloprid could be mixed with several nematode species against five scarab species (Koppenhöfer et al., 2002, 2003; Koppenhöfer and Grewal, 2005; Koppenhöfer and Fuzy, 2008b). Studies have shown that combining *S. longicaudum* and *H. bacteriophora* with insecticides like chlorantraniliprole, diflubenzuron, or imidacloprid increased the curative control of the white grub, *Holotricha oblita* in field conditions. Some combinations of EPNs and insecticides resulted in much higher grub mortality or percentage reduction than a single dose of EPNs or insecticides.

Furthermore, the EPN–insecticide combinations reduced grub larvae mortality or percentage reduction faster than a single control agent, resulting in a lower percentage of wounded legumes. The compatibility of EPNs with chemical pesticides generally provides a more cost-effective strategy for managing insect pests (Laznik and Trdan, 2014; Khan et al., 2018), and it discourages the overuse of chemical pesticides by farmers (Aioub et al., 2021). Additionally, the pesticides with successful compatibility with EPNs have the potential for use in managing agricultural field pests whilst enhancing the activity of the naturally occurring soil EPNs, as the pesticides have no or minimal negative effects on the survival, infectivity, and reproduction of EPNs.

3. Compatibility with Botanical Insecticides

Compatibility studies involving EPN and biorational insecticides have also been explored with most studies reporting positive results. For instance, Azadirachtin and Spinosad biopesticides combined with the Turkish EPN isolates (*H. bacteriophora* FLH-4H, *S. carpocapsae* KCS-4S and *S. feltiae* KMP-9S) resulted in 70 - 95% mortality against *G. mellonella* under laboratory conditions (Yüksel and Canhilal 2020). Similar compatibility was observed with *S. carpocapsae and H. bacteriophora* whose efficacy increased when combined with *Azadirachtin*, achieving over 90 percent mortality against *Agrotis ipsilon* (Mahmoud et al., 2016). Moreover, *S. carpocapse* proved more virulent strain than *H. bacteriophora* against *A. ipsilon* when combined with *Azadirachtin* in a greenhouse environment (Mahmoud et al., 2016). Successful combinations of *S. carpocapsae* and *H. indica* with spinosad (Tracer® 240SC) were possible both at lower and higher concentrations of up to 800 ppm, providing effective control against two larval instars of *Spodoptera litura* (Khan et al., 2018). *S. carpocapsae* and *H. bacteriophora* were compatible with biopesticides, *Trichoderma album*, *Bacillus megaterium* and *Serratia marcescens*, with *T. album* and *B. megaterium* exhibiting an additive effect against the 6th instar larvae of *G. mellonella* (El-Ashry and El-Marzoky, 2018). Most of the EPN and biorational insecticides combinations have no or minimal effects on the survival of EPNs (Radova, 2011; Laznik and Trdan, 2013; Raheel et al., 2017; El-Ashry and El-Marzoky, 2018), even at higher insecticide rates (Khan et al., 2018) further making the combination control agents suitable for use in sustainable agriculture.

4. Limitations of EPN – Insecticide Combination

One of the most important tactics for the control of above and below-ground insect pests is the use of EPNs in conjunction with insecticides (Hatsukade, 1990; Ishibashi, 1993; Koppenhöfer and Kaya, 1998; Mannion et al., 2000; Kary et al., 2018). However, various factors influence the success or failure of nematode-insecticide combinations, including EPN species, combined agent concentrations, targeted pest developmental stage, exposure mode, and environmental complexity in which the interface takes place.

4.1. Entomopathogenic Nematode Species

Numerous investigations have revealed that the compatibility of EPNs with different chemical pesticides is highly dependent on both the species and strain of EPN (Rovesti et al., 1988; Krishnayya and Grewal, 2002; De Nardo and Grewal, 2003; Laznik et al., 2012; Laznik and Trdan, 2014). The active components, azadirachtin and pirimicarb have no effect on the mortality of *S. feltiae* and *H. bacteriophora* (Laznik and Trdan, 2014), but on the other hand, they have an effect on the mortality of *S. carpocapsae* and *S. kraussei*. Grewal et al. (1998) reported that different EPN species react differently to a particular chemical. *S. feltiae* and *H. bacteriophora* were found tolerant to the majority of 21 pesticides tested and therefore can be tank-mixed for their application (Radová, 2010). Moreover, for both EPNs, the pesticides with the most detrimental effect on them varied as *H. bacteriphora* was highly negatively affected by only methomyl, whereas *S. feltiae* was affected by all the pesticides in the pyrazole chemical groups, which included tebufenpyrad and fenpyroximate (Radová, 2010). The combination of oxamyl pesticide and *H. bacteriophora* showed an antagonistic effect while the same nematicide exhibited an additive effect when combined with *S. carpocapsae* (El-Ashry and El-Marzoky 2018). Studies by Laznik and Trdan (2013) revealed that *H. bacteriophora* was most tolerant to abamectin but *S. carpocapsae* was much more sensitive to the same pesticide. Carbofuran and fenoxycarb were toxic to both *S. feltiae* Umea strain and *S. carpocapsae* (Gordon et al., 1996). The success of EPN-pesticides combinations depends heavily on the bioactivity of the pesticide used, the susceptibility of the nematode species to the pesticide and the EPN's acetylcholinesterase concentration (El-Ashry and El-Marzoky, 2018). The survival rates of EPN species are linked to the differences in acetylcholinesterase levels (Shamseldean et al., 2005). *Steinernema* sp. survived in many pesticides but a similar trend was not observed with *H. bacteriophora* and this may be due to differences in the acetylcholinesterase levels.

4.2. Entomopathogenic Nematode Strains

The compatibility between EPNs and pesticides is strain specific. Laznik et al. (2012) on the basis of an experiment concluded that compatibility between EPNs and pesticides like fungicides is not only a species-specific but also a strain-specific trait. Laznik and Trdan (2014) found that the pesticide compatibility of the numerous strains of *S. carpocapsae* (strain C101, commercial strains) has no strain-specific characteristics. When mixed with Azadirachtin, *B. thuringiensis* var. *kurstaki* toxin, and imidacloprid, the number of IJs of *S. carpocapsae* lowered. However, the findings were contrary to the results obtained by other workers who reported that active ingredients had no effect on the species' survival or infectivity (Barbarossa et al., 1996; Grewal *et al.*, 1998; Koppenhöfer et al., 2003; Alumai and Grewal, 2004). Some other factors which influence the survival of IJs are mentioned under the following subheadings.

4.2.1. Temperature

Temperature has a significant impact on EPN mortality when used in conjunction with chemical insecticides (Laznik et al., 2012; Laznik and Trdan, 2014). Infective juveniles become sluggish at temperatures below 10–15°C, but beyond 30–40°C, they become inactive. They efficiently parasitize insects at temperatures between 20–30°C. *Steinernema feltiae*, which is most effective at temperatures between 12–25°C, and *S. riobrave*, which functions in the same way at temperatures between 25–35°C are exceptions (Koppenhöfer, 2000). The lifespan of IJs is shortened as a result of increased metabolic activity and the use of energy reserves at higher temperatures (Koppenhöfer and Kaya, 2002). Temperatures exceeding 32°C have a deleterious impact on the reproduction, growth, and survival of nematodes. Laznik et al. (2012) observed that temperature played a significant role in the mortality of IJs when IJs were mixed with fungicides. The mortality of IJs of *S. feltiae* and *S. carpocapsae* increased with the increase in temperature. Only *H. downesi* strain 3173 showed the opposite pattern among the EPNs tested. This strain was discovered in Hungary (Tóth, 2006) and found active at temperatures as low as 12°C (Lola-Luz et al., 2005). The ability of EPNs to endure osmotic stressors has been linked to their fungicide sensitivity, resistance, or tolerance; nevertheless, there is scant information on this fascinating biological phenomenon (Thurston et al., 1994; Finnegan et al., 1999). Laznik and Trdan (2014) came to similar conclusions after testing multiple EPN species (*S. feltiae*, *S. carpocapsae*, *S. kraussei* and *H. bacteriophora*) with various insecticides.

4.2.2. Combined Agent Concentrations

Another factor that limits the success of EPN-pesticide combinations is the dosage or concentration of the pesticide applied as different EPN species react differently to pesticide concentrations and dosage. For example, it was found that carbofuran and fenamiphos were toxic to both *H. bacteriophora* and *S. carpocapsae* at field application rates of the pesticides (Rovesti et al., 1988; Rovesti and Deseo, 1990), as the detrimental effects exhibited were attributed to the high pesticide concentration used. Studies by Gaugler and Campbell (1991) demonstrated that carbamate oxamyl induced partial paralysis in the IJs of *S. carpocapsae* when applied at concentration levels of above 50 μg/ ml. The sensitivity of EPNs to different concentrations of pesticides was also demonstrated in EPN-biopesticide combinations. Turkish strains of EPNs *viz.*, *H. bacteriophora*, *S. carpocapsae* and *S. feltiae* were only compatible with Azadirachtin at 40 g/ l and Spinosad at 480 g/ l as these concentrations had the least effects on the EPN virulence and caused higher mortality to *G. mellonella* in the laboratory experiments (Yüksel and Canhilal, 2018). In another experiment, *S. carpocapse* and *H. indica* were combined with Spinosad at higher concentrations of up to 800 ppm. The combination was highly effective against two larval instars of *S. litura* (Khan et al., 2018). The results showed that different pesticides affect EPNs differently at various concentration levels.

4.2.3. Exposure Mode

Time of exposure is another determinant of EPN-pesticide compatibility. Some EPN species can withstand incompatible pesticides at a short exposure time of fewer than 24 hours (Alumai and Grewal, 2004; Atwa et al., 2013). Many studies revealed the compatibility of EPN-pesticides combinations, but with caveats that the mortality of IJs usually increased with an increase in concentration and time of exposure (Mahmoud et al., 2016). These factors may act singly or combine to limit or even enhance the compatibility of EPN-Pesticides combinations and their efficacy in pest control. Carbaryl and diazinon were highly toxic to *H. bacteriophora* at 24 and 48 hours, respectively (Zimmerman and Cranshaw, 1990). Saleh and Sammour (1995) reported a negative effect on the survival of EPN when exposed to a higher concentration of Selecron insecticide for a longer exposure time. On the contrary, even at higher concentrations of Spinosad insecticide, a stronger synergistic interaction was achieved when the insecticide was combined with *S. carpocapse* and *H. indica* and the synergistic effect increased with an increase in the exposure time of up to 96 hours of treatment against the larvae of *S. litura* (Khan et al., 2018).

5. Future Thrusts

Based on the findings of several studies, the usage of EPNs in combination with chemical pesticides can be a significant component of an IPM system. EPNs can be applied using the same equipment as used for chemical pesticides, making them a cost-effective option for controlling hazardous insects of agricultural and horticultural crops. The synchronous application saves time and money (Georgis, 1990) and allows for a reduction in chemical pesticide and EPN rates. As a result, minimizing the number of nematodes and the amount of pesticide necessary to control a pest makes their combined usage in the IPM program an incredibly appealing proposition from an economic perspective. It has been reported that EPNs can effectively inhibit specific cucumber pests, such as western flower thrips (Ebssa et al., 2004; Trdan et al., 2007) and greenhouse whiteflies (Laznik et al., 2011).

In circumstances when mixed suspensions are used to suppress specific hazardous pests, the application of fungicides and EPNs at the same time is justified. Examples are Colorado potato beetles (*Leptinotarsa decemlineata*) (Laznik et al., 2010) and cereal leaf beetles (*Oulema melanopus*) (Laznik et al., 2011) where the simultaneous application of fungicide and EPNs is justified for the control of the aforementioned pests. In the case of active components such as metiram, copper hydroxide, and mancozeb, similar findings were obtained. These compounds are effective against potato blight (*Phytophthora infestans*), lettuce downy mildew (*Bremia lactucae*), wheat leaf blotch (*Septoria tritici*), and tomato early blight (*Alternaria solani*) (Milus, 1994; Stepanović et al., 2009; Stevenson, 2009), as well as insect pests like the Colorado potato beetle (*L. decemlineata*), cabbage armyworm (*Mamestra brassicae*), barley wireworm (*Agriotes fuscicollis*) and cereal leaf (*Tuta absoluta*). However, field experiments must be used to support future adoption of the synchronous application of EPNs and fungicides because laboratory results cannot be extrapolated uncritically to environmental settings (Laznik et al., 2010).

Among other significant factors that will encourage the success of EPN-pesticide combinations is the rise in consumer demand for foods with no pesticide residues (Willer and Kilcher, 2009; Ribeiro et al., 2012) and the strict regulations on food safety and quality in the international market (Damalas and Koutroubas, 2018). This means that farmers will have to adopt environmentally safe methods of pest control to capture the growing market for pesticide-free foods. EPN is compatible with many newer pesticides including neonicotinoids and anthranilic diamides (Koppenhöfer et al., 2020) which are considered to be low-risk insecticides. Additionally, many studies have shown evidence that lower concentrations of pesticides are required for compatibility, which means that EPN-pesticide combinations will in the future provide a more cost-effective strategy for managing insect pests in farmer fields (Laznik and Trdan, 2014; Khan et al., 2018) and encourage less use of chemical pesticides (Aioub et al., 2021). This will allow farmers to conform the regulations on pesticide use and avoid the accumulation of pesticide residues in the final product and also enhance environmental health.

Conclusion

EPNs are used to combat a variety of foliar and soil-dwelling insect pests of agricultural importance. When used in the right direction, these nematodes can be just as effective as chemical insecticides. The combined application of EPN and insecticides is a key component of IPM programs for the management of a variety of agricultural pests, as the combination would improve the level of pest management. This strategy may not only make it easier to use EPNs, improve the pest management efficacy of programs that include EPNs, and lower the cost of EPN applications, but it may also reduce reliance on chemical insecticides, slowing the development of insecticide resistance and preventing negative consequences for humans and the environment. This will also result in the production of high agricultural yields. The accumulation of hazardous compounds in the soil, water bodies, and agricultural produce can be greatly decreased when EPNs and insecticides are used together. EPNs when used in conjunction with lower concentrations of insecticides, can multiply by infecting insect hosts, increasing their density in the soil and reducing the need for reapplication. However, before using nematode-agrochemical combinations in tank mixes as part of an ecologically oriented IPM strategy, the compatibility of these nematodes with new and commonly used agrochemicals must be established.

Acknowledgments

The authors are thankful to all those who have generated the information required for this chapter. Aashaq Hussain Bhat (Grant Nr. 2021.0463 to AHB) and Jaspher Ewany (Grant Nr. 2021.0663 to JE) are thankful for the financial support through the Swiss Government Excellence Scholarship.

References

Aioub, A. A. A., El-Ashry, R. M., Hashem, A. S. Elesawy, A. and Elsobki, A. E. A. (2021). Compatibility of entomopathogenic nematodes with insecticides against the cabbage white butterfly, *Pieris rapae* L. (Lepidoptera: Pieridae). *Egyptian Journal of Biological Pest Control*, 31: 153 https://doi.org/10.1186/s41938-021-00498-z.

Alumai A. and Grewal P. S. (2004). Tank-mix compatibility of the entomopathogenic nematodes, *Heterorhabditis bacteriophora* and *Steinernema carpocapsae*, with selected chemical pesticides used in turf grass. *Biocontrol Science and Technology*, 14: 725–730.

Amizadeh, M., Hejazi, M. J., Niknam, G. and Askari-Saryazdi, G. (2019). Interaction between the entomopathogenic nematode, *Steinernema feltiae* and selected chemical insecticides for management of the tomato leaf miner, *Tuta absoluta*. *Biocontrol*, 64: 709–721.

Antwi, F. B. and Reddy, G. V. P. (2016). Efficacy of entomopathogenic nematodes and sprayable polymer gel against crucifer flea beetle (Coleoptera: Chrysomelidae) on canola. *Journal Economic Entomology*, 109: 1706–1712.

Arias-Estévez, M., López-Periago, E., Martínez-Carballo, E., Simal-Gándara, J., Mejuto, J. C. and García-Río, L. (2008). The mobility and degradation of pesticides in soils and the pollution of groundwater resources. *Agriculture, Ecosystems and Environment*, 123: 247–260. https://doi.org/10.1016/j.agee.2007.07.011.

Atwa, A. A., Shamseldean, M. M. and Yonis, F. A. (2013). The effect of different pesticides on reproduction of entomopathogenic nematodes. *Turkish Journal of Entomology*, 37: 493–502.

Barbosa-Negrisoli, C. R. C., Garcia, M. S., Dolinski, C., Negrisoli Jr, A. S., Bernardi, D. and Nava, D. E. (2009). Efficacy of indigenous entomopathogenic nematodes (Rhabditida: Heterorhabditidae, Steinernematidae), from Rio Grande do Sul, Brazil, against *Anastrepha fraterculus* (Wied.) (Diptera: Tephritidae) in Peach Orchards. *Journal Invertebrate Pathology*, 102: 6–13.

Barbarossa, B., Frulla, R., Grazzi, G. and Rovesti, L. (1996). Compatibility of *Steinernema carpocapsae* with pesticides. *International Organization for Biological and Integrated Control Bulletin*, 19: 136–139.

Bhat, A. H., Chaubey, A. K. and Askary, T. H. (2020a). Global distribution of entomopathogenic nematodes, *Steinernema* and *Heterorhabditis*. *Egyptian Journal of Biological Pest Control*, 30: 31. https://doi.org/10.1186/s41938-020-0212-y.

Bhat, A. H., Chaubey, A. K., Hartmann, J., Nermuť J.and and Püža, V. (2020b). Notes on morphology, bionomy, distribution and efficacy of *Steinernema siamkayai* (Rhabditida: Steinernematidae). *Journal of Nematology,* 23: 817–836.

Bhat, A. H., Chaubey, A. K. and Půža, V. (2018). The first report of *Xenorhabdus indica* from *Steinernema pakistanense*: co-phylogenetic study suggests co-speciation between *X. indica* and its steinernematid nematodes. *Journal of Helminthology*, 93: 81–90.

Bhat, A. H., Chaubey, A. K., Shokoohi, E. and Mashela, P. W. (2019). Study of *Steinernema hermaphroditum* (Nematoda, Rhabditida), from the West Uttar Pradesh, India. *Acta Parasitologica*, 64: 720–737.

Bhat, A. H., Istkhar, R., Chaubey, A. K., Půža, V. and San-Blas, E. (2017). First report and comparative study of *Steinernema surkhetense* (Rhabditida: Steinernematidae) and its symbiont bacteria from subcontinental India. *Journal of Nematology*, 49: 92–102.

Bhat, A. H., Rana, A., Chaubey, A. K., Shokoohi, E. and Machado, R. A. R. (2021). Characterisation of *Steinernema abbasi* (Rhabditida: Steinernematidae) isolated from Indian agricultural soils and their efficacy against insect pests. *Biocontrol Science and Technology,* 31: 1–26.

Boemare, N. E., Akhurst, R. J. and Mourant, R. G. (1993). DNA relatedness between *Xenorhabdus* spp. (Enterobacteriaceae), symbiotic bacteria of entomopathogenic nematodes, and a proposal to transfer *Xenorhabdus luminescens* to a new genus, *Photorhabdus* gen. nov. *International Journal of Systematic Bacteriology*, 43: 249–255.

Campbell, L. R. and Gaugler, R. (1991). Role of the sheath in desiccation tolerance of two entomopathogenic nematodes. *Nematologica*, 37: 324–332.

Canto, T., Aranda, M. A. and Fereres, A. (2009). Climate change effects on physiology and population processes of hosts and vectors that influence the spread of hemipteran-borne plant viruses. *Global Change Biology*, 15: 1884–1894.

Clough, G. H., Rondon, S., DeBano, S. J., David, N. and Hamm P. B. (2010). Reducing tuber damage by potato tuberworm (Lepidoptera: Gelechiidae) with cultural practices and insecticides. *Journal of Economic Entomology*, 103: 1306–1311.

Damalas, C. A. and Koutroubas, S. D. (2018). Current status and recent developments in biopesticide use. *Agriculture*, 8: 13. https://doi.org/10.3390/agriculture8010013.

Daughtrey, M. and Buitenhuis, R. (2020). Integrated pest and disease management in greenhouse ornamentals. In: *Integrated Pest and Disease Management in Greenhouse Crops* (Eds. Gullino, M. L. Albajes, R. and Nicot, P.C.). Springer International Publishing, Switzerland, pp. 625–679.

De Nardo, E. A. and Grewal P. S. (2003). Compatibility of *Steinernema feltiae* (Nematoda: Steinernematidae) with pesticides and plant growth regulators used in glasshouse plant production. *Biocontrol Science and Technology*, 13: 441–448.

Ehlers, R.-U. (2005). Forum on safety and regulation. In: *Nematodes as Biocontrol Agents* (Eds. Grewal, P. S., Ehlers, R.-U. and Shapiro-Ilan, D. I.). CAB International, Wallingford, UK, pp. 107–114.

El-Ashry, R. M. and El-Marzoky, A. M. (2018). Compatibility of entomopathogenic nematodes, *Heterorhabditis bacteriophora* Poinar and *Steinernema carpocapsae* Weiser with some chemical and biopesticides. *Journal of Agricultural Research, 45:* 905–916.

FAO (2017). The state of food and agriculture-leveraging food systems for inclusive rural transformation https://reliefweb.int/sites/reliefweb.int/files/resources/a-I7658e_0.pdf.

Finnegan, M. M., Downes, M. J., O'Regan, M. and Griffin, C. T. (1999). Effect of salt and temperature stresses on survival and infectivity of *Heterorhabditis* spp. infective juveniles. *Nematology*, 1: 69–78.

Garcia-del-Pino, F., Alabern, X. and Morton, A. (2013). Efficacy of soil treatments of entomopathogenic nematodes against the larvae, pupae and adults of *Tuta absoluta* and their interaction with the insecticides used against this insect. *Biological control*, 58: 723–731.

Garlet, C. G., Gubiani, P. D. S., Palharini, R. B., Moreira, R. P., Godoy, D. N., Farias, J. R. and Bernardi, O. (2021). Field-evolved resistance to chlorpyrifos by *Spodoptera frugiperda* (Lepidoptera: Noctuidae): Inheritance mode, cross-resistance patterns, and synergism. *Pest Management Science*, 77: 5367–5376.

Gaugler, R. and Campbell, J. F. (1991). Behavioral response of the entomopathogenic nematodes *Steinernema carpocapsae* and *Heterorhabditis bacteriophora* to oxamyl. *Annals of Applied Biology*, 119–131.

Georgis, R. (1990). Formulation and application technology. In: *Entomopathogenic Nematodes in Biological Control* (Eds. Gaugler, R. and Kaya, H. K.). CRC Press, Boca Raton, Florida, pp. 173–194.

Georgis, R. and Gaugler, R. (1991). Predictability in biological control using entomopathogenic nematodes. *Journal of Economic Entomology*, 84: 713–720.

Glazer, I. (2002). Survival biology. In: *Entomopathogenic Nematology* (Ed. Gaugler, R.). CAB International, Wallingford, UK, pp. 169–187.

Gong, P., Li, X., Gao, H., Wang, C., Li, M., Zhang, Y., Li, X., Liu, E. and Zhu, X. (2021). Field evolved resistance to pyrethroids, neonicotinoids, organophosphates and macrolides in *Rhopalo siphumpadi* (Linnaeus) and *Sitobion avenae* (Fabricius) from China. *Chemosphere*, 269.

Gordon, R., Chippett, J. and Tilley, J. (1996). Effects of two carbamates on infective juveniles of *Steinernema carpocapsae* All strain and *Steinernema feltiae* Umea strain. *Journal of Nematology*, 28: 310–317.

Grant, C., Jacobson, R., Ilias, A., Berger, M., Vasakis, E., Bielza, P., Zimmer, C. T., Williamson, M. S., Ffrench-Constant, R. H., Vontas, J., Roditakise, E. and Bass, C. (2019). The evolution of multiple-insecticide resistance in UK populations of tomato leafminer, *Tuta absoluta. Pest Management Science*, 75: 2079-2085.

Grewal, P. S. (2002). Formulation and application technology. In: *Entomopathogenic Nematology* (Ed. Gaugler, R.). CAB International, Wallingford, UK, pp. 265–288.

Grewal, P. S., Ehlers, R. U. and Shapiro-Ilan, D. I. (2005). *Nematodes as Biological Control Agents.* CAB International, Wallingford, UK.

Grewal, P. S., Webber, T. and Batterley, D. A. (1998). Compatibility of *Steinernema feltiae* with chemicals used in mushroom production. *Mushroom News*, 46: 6–10.

Guo, W., Yan, X., Zhao, G. and Han, R. (2016). Increased efficacy of entomopathogenic nematode-insecticide combinations against *Holotrichia oblita* (Coleoptera: Scarabaeidae). *Journal Economic Entomology*, 110–41-51.

Guo, W., Yan, X., Zhao, G. and Han, R. (2017). Increased efficacy of entomopathogenic nematode–insecticide combinations against *Holotrichia oblita* (Coleoptera: Scarabaeidae). *Journal Economic Entomology*, 110: 41–51.

Gutiérrez, N., Avila, C. M., Moreno, M. T. and Torres, A. M. (2008). Development of SCAR markers linked to zt-2, one of the genes controlling absence of tannins in faba bean. *Australian Journal of Agricultural Research*, 59: 62–68.

Hara, A. H. and Kaya, H. K. (1983). Development of the entomogenous nematode, *Neoaplectana carpocapsae* (Rhabditida: Steinernematidae), in insecticide-killed beet armyworm (Lepidoptera: Noctuidae). *Journal of Economic Entomology*, 76: 423–426.

Hassan, H. M. and Ibrahim, S. A. (2019). Combined efficacy of insecticides and entomopathogenic nematodes in the management of cotton leaf worm *Spodoptera littoralis*. *Pakistan Journal of Nematology*, 37: 45–52.

Hatsukade, M. (1990). Application of entomogenous nematode for the important turfgrass insects. In: *Development of Biological Integrated control of Agricultural pests by beneficial nematodes* (Ed. Ishibashi, N.). pp 89-98. Ministry of Education, Japan, Grant no. 62860006. Saga University, Japan.

Head, J., Walters, K. F. A. and Langton, S. (2000). The compatibility of the entomopathogenic nematode, *Steinernema feltiae*, and chemical insecticides for the control of the South American leafminer, *Liriomyza huidobrensis. Biological Control*, 45: 345–353.

Heeb, L., Jenner, E., and Cock, M. J. (2019). Climate-smart pest management: building resilience of farms and landscapes to changing pest threats. *Journal of Pest Science, 92:* 951–969.

Ishibashi, N. (1993). Integrated control of insect pests by *Steinernema carpocapsae.* In: *Nematodes and the Biological Control of Insects* (Eds. Bedding, R., Akhurst, R. and Kaya, H. K.). CSIRO, Australia, pp. 105–113.

Ishibashi, N. and Takii, S. (1993). Effects of insecticides on movement, nictation and infectivity of *Steinernema carpocapsae*. *Journal of Nematology*, 25: 204–13.

Jackson, J. J. and Brooks, M. A. (1989). Susceptibility and immune response of western corn rootworm larvae (Coleoptera: Chrysomelidae) to the entomogenous nematode, *Steinernema feltiae* (Rhabditida: Steinernematidae). *Journal of Economic Entomology*, 82: 1073–1077.

Jean-Baptiste, M. C., de Brida, A. L., Bernardi, D., Dias, S. D. C, Pazini, J. D. B., Leite, L. G. Wilcken, S. R. S. and Garcia, F. R. M. (2021). Effectiveness of entomopathogenic nematodes against *Ceratitis capitata* (Diptera: Tephritidae) pupae and nematode compatibility with chemical insecticides. *Journal of Economic Entomology*, 114: 248–256.

Kajol, Y., Bhat, A. H., Aasha, R. and Chaubey, A. K. (2020). Biochemical and molecular characterization of associated *Photorhabdus* symbiont of Indian strain of *Heterorhabditis indica* and its efficacy. *Pakistan Journal of Nematology,* 38: 15–24.

Kary, E. N., Sanatipour, Z., Mohammadi, D. and Dillon, A. B. (2021). Combination effects of entomopathogenic nematodes, *Heterorhabditis bacteriophora* and *Steinernema feltiae*, with abamectin on developmental stages of *Phthorimaea operculella* (Lepidoptera, Gelechiidae). *Crop Protection*, 143:105543. https://doi.org/10.1016/j.cropro.2021.105543.

Kary, N. E., Chahardoli, S., Mohammadi, D. and Dillon, A. B. (2018). Effects of abiotic factors on the osmotic response of alginate-formulated entomopathogenic nematode, *Heterorhabditis bacteriophora* (Nematoda: Rhabditida). *Biocontrol Science and Technology*, 28: 688–701.

Khan, B., Javed, N., Khan, S. A., Rajput, N. A., Atiq, M., Jabbar, A., Rehman, A., Moosa, A. and Ali, M. A. (2020). Potential of entomopathogenic nematode (*Steinernema kraussei)* against last instar larvae of different lepidopteran insect pests. *Pakistan Journal of Zoology*, 52: 1275–1281.

Khan, M. A., Ahmad, W., Paul, B., Paul, S., Khan, Z. and Aggarwal, C. (2016). Entomopathogenic nematodes for the management of subterranean termites. In: *Plant, Soil and Microbes* (Eds. Hakeem, K., Akhtar, M. and Abdullah, S.). Springer, Switzerland, pp. 317–352.

Khan, R. R., Ali, R. A., Ali, A., Arshad, M., Majeed, S., Ahmed, S., Khan, S. A. and Arshad, M. (2018). Compatibility of entomopathogenic nematodes (Nematoda: Rhabditida) and the biocide, spinosad for mitigation of the armyworm, *Spodoptera litura* (F.) (Lepidoptera: Noctuidae). *Egyptian Journal of Biological Pest Control*, 28: 58. https://doi.org/10.1186/s41938-018-0063-y.

Kim, J., Nam, H. Y., and Kwon, M. (2021). Novel diamide resistance-linked mutation in Korean *Spodoptera exigua* and a LAMP assay based on a mutation-associated intronic InDel. *Journal of Pest Science*, 94: 1017–1029.

Klein, M. G. (1990). Efficacy against soil-inhabiting insect pests. In: *Entomopathogenic Nematodes in Biological Control* (Eds. Gaugler, R., Kaya, H. K.). CRC Press, Boca Raton, Florida, pp. 193–214.

Koppenhöfer, A. M. (2000). Nematodes. In: *Field Manual of Techniques in Invertebrate Pathology* (Ed. Kaya, H. K.). Kluwer Academic Publishers, The Netherlands, pp. 283–301.

Koppenhöfer, A. M., Cowles, R. S., Cowles, E. A., Fuzy, E. M. and Kaya, H. K. (2003). Effect of neonicotinoid insecticide synergists on entomopathogenic nematode fitness. *Entomologia Experimentalis et Applicata*, 106: 7-18.

Koppenhöfer, A. M., Cowles, R. S., Cowles, E. A., Fuzy, E. M., and Baumgartner, L. (2002). Comparison of neonicotinoid insecticides as synergists for entomopatho-genic nematodes. *Biological Control*, 24: 90–97.

Koppenhöfer, A. M. and Fuzy, E. M. (2008a). Effects of chlorantraniliprole on *Heterorhabditis bacteriophora* (Rhabditida: Heterorhabditidae) efficacy against and reproduction in white grubs (Coleoptera: Scarabaeidae). *Biological Control*, 45: 93-102.

Koppenhöfer, A. M., and Fuzy, E. M. (2008b). Early timing and new combinations to increase the efficacy of neonicotinoid–entomopathogenic nematodes (Rhabditida: Heterorhabditidae) combinations against white grubs (Coleoptera: Scarabaeidae). *Pest Management Science*, 64: 725–735.

Koppenhöfer, A. M. and Grewal, P. S. (2005). Compatibility and interactions with agrochemicals and other biocontrol agents. In: *Nematodes as Biocontrol Agents* (Eds. Grewal, P. S., Ehlers, R.-U. and Shapiro-Ilan, D. I.). CAB International, Wallingford, UK, pp. 363–381.

Koppenhöfer, A. M., Grewal, P. S. and Kaya, H. K. (2000). Synergism of imidacloprid and entomopathogenic nematodes against white grubs: the mechanism. *Entomologia Experimentalis et Applicata*, 94: 283–293.

Koppenhöfer, A. M. and Kaya, H. K. (1998). Synergism of imidacloprid and an entomopathogenic nematode: a novel approach to white grub (Coleoptera: Scarabaeidae) control in turf grass. *Journal of Economic Entomology*, 91: 618–623.

Koppenhöfer, A. M. and Kaya, H. K. (2002). Entomopathogenic nematodes and insect pest management. *Microbial Biopesticides*, 15: 277–305.

Koppenhöfer, A. M., Shapiro-Ilan, D. I., and Hiltpold, I. (2020). Entomopathogenic nematodes in sustainable food production. *Frontiers in Sustainable Food Systems* 4:125. https://doi.org/10.3389/fsufs.2020.00125.

Krishnayya, P. V. and Grewal, P. S. (2002). Effect of neem and selected fungicides on viability and virulence of the entomopathogenic nematode *Steinernema feltiae*. *Biocontrol Science and Technology*, 12: 259–266.

Laznik, Ž., Tóth, T., Lakatos, T., Vidrih, M. and Trdan, S. (2010). Control of the Colorado potato beetle (*Leptinotarsa decemlineata* Say) on potato under field conditions: a comparison of the efficacy of foliar application of two strains of *Steinernema feltiae* (Filipjev) and spraying with thiametoxam. *Journal of Plant Diseases and Protection*, 117: 129–135.

Laznik, Ž. and Trdan, S. (2011). Entomopathogenic nematodes (Nematoda: Rhabditida) in Slovenia: from tabula rasa to implementation into crop production systems. In: *Insecticides - Advances in Integrated Pest Management* (Ed. Perveen, F.) In Tech, Rijeka, pp. 627–656.

Laznik, Ž. and Trdan, S. (2013). An investigation on the chemotactic responses of different entomopathogenic nematode strains to mechanically damaged maize root volatile compounds. *Experimental Parasitology*, 134: 349–355.

Laznik, Ž. and Trdan, S. (2014). The influence of insecticides on the viability of entomopathogenic nematodes (Rhabditida: Steinernematidae and Heterorhabditidae) under laboratory conditions. *Pest Management Science*, 70: 784–789.

Laznik, Ž., Vidrih, M. and Trdan, S. (2012). The effects of different fungicides on the viability of entomopathogenic nematodes *Steinernema feltiae* (Filipjev), *S. carpocapsae* Weiser, and *Heterorhabditis downesi* Stock, Griffin & Burnell (Nematoda: Rhabditida) under laboratory conditions. *Chilean Journal of Agricultural Research*, 72: 62–67.

Laznik, Ž., Žnidarčič, D. and Trdan, S. (2011). Control of *Trialeurodes vaporariorum* (Westwood) adults on glasshouse-grown cucumbers in four different growth substrates: an efficacy comparison of foliar application of *Steinernema feltiae* (Filipjev) and spraying with thiamethoxam. *Turkish Journal of Agriculture and Forestry*, 35: 631–640.

Lewis, E. E. and Clarke, D. J. (2012). Nematode parasites and entomopathogens. In: *Insect Pathology* (Eds. Vega, F. E. and Kaya, H. K.). Academic Press, USA, pp. 395–443.

Li, Y., Xu, Z. F., Shi, L., Shen, G. M. and He, L. (2016). Insecticide resistance monitoring and metabolic mechanism study of the green peach aphid, *Myzuspersicae* (Sulzer) (Hemiptera: Aphididae), in Chongqing, China. *Pesticide Biochemistry and Physiology*, 132: 21–28.

Liesch, P. J. and Williamson, R. C. (2010). Evaluation of chemical controls and entomopathogenic nematodes for control of *Phyllophaga* white grubs in a Fraser fir production field. *Journal Economic Entomology*, 103: 1979–1987.

Lola-Luz, T., Downes, M. and Dunne, R. (2005). Control of black vine weevil *Otiorhynchus sulcatus* (Fabricus) (Coleoptera: Curculionidae) in grow bags outdoors with nematodes. *Agricultural Forest Entomology*, 7: 121–126.

Lu, D., Macchietto, M., Chang, D., Barros, M. M., Baldwin, J., Mortazavi, A. and Dillman, A. R. (2017). Activated entomopathogenic nematode infective juveniles release lethal venom proteins. *PLOS Pathogens* 13(4): e1006302. https://doi.org/10.1371/journal.ppat.1006302.

Machado, R. A. R., Bhat, A. H., Abolafia, J., Muller, A., Bruno, P., Fallet, P., Carla, C. M. A., Turlings, T. C. J., Bernal, J. S., Kajuga, J., Waweru, B. and Toepfe, S. (2021). Multi-locus phylogenetic analyses uncover species boundaries and reveal the occurrence of two new entomopathogenic nematode species, *Heterorhabditis ruandica* n. sp. and *Heterorhabditis zacatecana* n. sp. *Journal of Nematology*, 53: e2021-89. https://doi.org/10.21307/jofnem-2021-089.

Mahmoud, M. F., Mahfouz, H. M. and Mohamed, K. M. (2016). Compatibility of entomopathogenic nematodes with neonicotinoids and azadirachtin insecticides for controlling the Black Cutworm, *Agrotis ipsilon* (Hufnagel) in canola plants. *International Journal of Research, 2:* 11–18.

Mannion, C. M., Winkler, H. E. Shapiro, D. I. and Gibb, T. (2000). Interactions between halofenozide and *Heterorhabditis marelatus* for control of the Japanese beetle, *Popillia japonica*. *Journal Economic Entomology*, 93: 48–53.

Milus, E. A. (1994). Effect of foliar fungicides on disease control, yield and test weight of soft red winter wheat. *Crop Protection*, 13: 291–295.

Mohan, S. (2015). Entomopathogenic nematodes and their bacterial symbionts as lethal bioagents of lepidopteran pests. In: *Biocontrol of Lepidopteran Pests* (Eds. Sree, K. and Varma, A.). Springer, New York, pp. 273–288.

Muraro, D. S., de Oliveira Abbade Neto, D., Kanno, R. H., Kaiser, I. S., Bernardi, O. and Omoto, C. (2021). Inheritance patterns, cross-resistance and synergism in *Spodoptera frugiperda* (Lepidoptera: Noctuidae) resistant to emamectin benzoate. *Pest Management Science*, 77: 5049–5057.

Navarez, M. L., Sangcopan, R., Arya, S., Sumaya, N. P. D., Bhat, A. H. and Sumaya, N. H. (2020). Native Philippine *Heterorhabditis indica* isolates from banana and rice fields and preliminary results of their virulence against the larvae of super worm (*Zophobas morio* Fabricius Coleoptera: Tenebrionidae). *Egyptian Journal of Biological Pest Control* 31: 96. https://doi.org/10.1186/s41938-021-00388-4.

Negrisoli, A. S., Garcia, M. S., Barbosa Negrisoli, C. R. C., Bernardi, D. and da Silva, A. (2010). Efficacy of entomopathogenic nematodes (Nematoda: Rhabditida) and insecticide mixtures to control *Spodoptera frugiperda* (Smith, 1797) (Lepidoptera: Noctuidae) in corn crops. *Crop Protection*, 29: 677–683.

Niu, Y., Oyediran, I., Yu, W., Lin, S., Dimase, M., Brown, S., Reay-Jones, F. P. F., Cook, D., Reisig, D., Thrash, B., Ni, X., Paula-Moraes, S. V., Zhang, Y., Chen, J. S., Wen, Z. and Huang, F. (2021). Populations of *Helicoverpa zea* (Boddie) in the Southeastern United States are commonly resistant to Cry1Ab, but still susceptible to Vip3Aa20 expressed in MIR 162 Corn. *Toxins*, 13: 2–14.

Özdemir, E., İnak, E., Evlice, E., and Laznik, Ž. (2020). Compatibility of entomopathogenic nematodes with pesticides registered in vegetable crops under laboratory conditions. *Journal of Plant Diseases and Protection*, 127: 529–535.

Piedra-Buena, A., López-Cepero, J. and Campos-Herrera, R. (2015). Entomopathogenic nematode production and application: regulation, ecological impact and non–target effects. In: *Nematode Pathogenesis of Insects and Other Pests* (Ed. Campos-Herrera, R.). Springer, Switzerland, pp. 255–282.

Pu, J., Wang, Z. and Chung, H. (2020). Climate change and the genetics of insecticide resistance. *Pest Management Science*, 76: 846–852.

Radová, Š. (2010). Effect of selected pesticides on the vitality and virulence of the entomopathogenic nematode *Steinernema feltiae* (Nematoda: Steinernematidae). *Plant Protection Science*, 46: 83–88.

Radová, Š. (2011). Effect of selected pesticides on survival and virulence of two nematodes species. *Polish Journal of Environmental Studies*, 20: 181–185.

Raheel, M., Javed, N., Khan, S. A. and Ahmed, S. (2017). Exploiting the biocontrol potential of entomopathogenic nematodes in combination with chemicals against greater wax moth (*Galleria mellonella* L.). *Journal of Animal and Plant Science*, 27: 877–881.

Rana, A., Bhat, A. H., Chaubey, A. K., Shokoohi, E. and Machado, R. A. R. (2020). Morphological and molecular characterization of *Heterorhabditis bacteriophora* nematodes isolated from Indian soils and their biocontrol potential. *Zootaxa*, 4878: 077–102. https://doi.org/10.11646/zootaxa.4878.1.3.

Ribeiro, A. R., Gonçalves, V. M. M., Maia, A. S., Carvalho, M. F., Castro, P. M. and Tiritan, M. E. (2012). Microbial degradation of pharmaceuticals followed by a simple hplc-dad method. *Journal of Environmental Science* and *Health*-Part A, Toxic/Hazardous Substances and Environmental Engineering, 47: 2151–2158.

Rovesti, L. and Deseo, K. V. (1990). Compatibility of chemical pesticides with entomopathogenic nematodes, *Steinernema carpocapsae* Weiser and *S. feltiae* Filipjev (Nematoda: Steinernematidae). *Nematologica*, 36: 237–245.

Rovesti, L., Heinzpeter, E. W., Tagliente, F. and Deseő, K. V. (1988). Compatibility of pesticides with the entomopathogenic nematode *Heterorhabditis bacteriophora* Poinar (Nematoda: Heterorhabditidae). *Nematologica*, 34: 462–476.

Sabino, P., Negrisoli, A. S., Andaló, V., Filgueiras, C. C., Moino, A., Jr. and Sales, F. S. (2019). Combined application of entomopathogenic nematodes and insecticides in the control of leaf-miner *Tuta absoluta* (Meyrick) (Lepidoptera: Gelechiidae) on tomato. *Neotropical Entomology* 48: 314–322.

Saeed, R., Abbas, N. and Hafez, A. M. (2021). Fitness cost of imidacloprid resistance in the cotton-staining bug, *Dysdercus koenigii*. *Chemospherei*, 265: 129118. https://doi.org/10.1016/j.chemosphere.2020.129118.

Şahin, Y. S. and Susurluk, I. A. (2018). Effects of some inorganic fertilizers on the entomopathogenic nematodes *Steinernema feltiae* (Tur-S3) and *Heterorhabditis bacteriophora* (HBH). *Turkish Journal of Biological Control*, 9: 102–109.

Saleh, M. M. E. and Sammour, E. A. (1995). Interactions of three insecitices and two entomopathogenic nematodes against *Spodoptera littoralis* (Boisd.) larvae. *Egyptian Journal of Biological Pest Control*, 5: 119–122.

Salih, A. A. M., Baraibar, M., Mwangi, K. K. and Artan, G. (2020). Climate change and locust outbreak in East Africa. *Natural Climate Change*, 10: 584–585. https://doi.org/10.1038/s41558-020-0835-8.

Schreiber, A., Jensen, A., Pike, K., Alvarez, J. and Rondon, S. I. (2010). *Integrated Pest Management guidelines for insects and mites in Idaho, Oregon, and Washington Potatoes*. http://oregonstate.edu/potatoes/ipm/publications.htm.

Schroer, S., Ziermann, D. and Ehlers, R. U. (2005). Mode of action of a surfactant-polymer formulation to support performance of the entomopathogenic nematode *Steinernema carpocapsae* for control of diamond back moth larvae (*Plutella xylostella*). *Biocontrol Science and Technology*, 15: 601–613.

Shamseldean, M. M., Atwa, A. A. and Yonis, F. A. (2005). Effect of different pesticides on the survival of entomopathogenic nematodes. In: *Proceedings of The Third Conference of Applied Entomology* held at Faculty of science, Cairo University, Egypt from March 23-24, pp. 111–123.

Shehata, I. E., Hammam, M. M. A., El-Borai, F. E., Duncan, L. W. and Abd-Elgawad, M. M. M. (2020). Traits of the entomopathogenic nematode, *Heterorhabditis bacteriophora* (Hb-EG strain), for potential biocontrol in strawberry fields. *Egyptian Journal of Biological Pest Control* 30: 40. https://doi.org/10.1186/s41938-020-00233-0.

Silva, J. E., Silva, W. M., Silva and T. B. M. (2021). High resistance to insect growth disruptors and control failure likelihood in Brazilian populations of the tomato pinworm *Tuta absoluta*. *Phytoparasitica*, 49: 1–13.

Somvanshi, V. S., Ganguly, S. and Paul, A. V. N. (2006). Field efficacy of the entomopathogenic nematode *Steinernema thermophilum* Ganguly and Singh (Rhabditida: Steinernematidae) against diamond back moth (*Plutella xylostella* L.) infesting cabbage. *Biological Control*, 37: 9–15.

Stepanovic, M., Vuksa, P., Potocnik, I., Milijasevic, S., Rekanovic, E. and Todorovic, B. (2009). (2009). Integrated management of viral sensitivity of *Alternaria solani* isolates to some fungicides. *Acta Horticulturae*, 808: 377–380.

Stevenson, W. R. (2009). Late blight control strategies in the United States. *Acta Horticulturae*, 834: 83–86.

Sunanda, B. S., Jeyakumar, P. and Jacob, V. V. (2014). Bioefficacy of different formulations of entomopathogenic nematode *Steinernema carpocapsae* against diamond back moth (*Plutella xylostella*) infesting cabbage (*Brassica oleracea* var. *capitata*). *Journal of Biopesticides*, 7: 210–215.

Sunitha, V., Singh, T. V. K., Supriya, G. B. and Satyanarayana, J. (2020). Insecticide resistance monitoring of diamond back moth, *Plutella xylostella* (Linn.) in Delhi population. *Journal of Entomology and Zoology Studies*, 8: 1706–1712.

Susurluk, A. (2008). Potential of the entomopathogenic nematodes *Steinernema feltiae*, *S. weiseri* and *Heterorhabditis bacteriophora* for the biological control of the sugar beet weevil *Bothynoderes punctiventris* (Coleoptera: Curculionidae). *Journal of Pest Science*, 81: 221–225.

Thomas, G. M. and Poinar, G. O. Jr. (1979). *Xenorhabdus* gen. nov., a genus of entomopathogenic, nematophilic bacteria of the family enterobacteriaceae. *International Journal of Systemetic Bacteriology*, 29: 352–360.

Thurston, G. S. and Yule, W. N. (1990). Control of larval northern corn rootworm (*Diabrotica barberi*) with two steinernematid nematode species. *Journal of Nematology*, 22: 127–131.

Thurston, G. S., Ni, Y. and Kaya, H. K. (1994). Influence of salinity on survival and infectivity of entomopathogenic nematodes. *Journal of Nematology*, 26: 345–351.

Tóth, T. (2006). Collection of entomopathogenic nematodes for biological control of insect pests. *Journal of Fruit Ornamental and Plant Research*, 14: 225–230.

Trdan, S., Žnidarcˇicˇ, D. and Vidrih, M. (2007). Control of *Frankliniella occidentalis* on glasshouse-grown cucumbers: an efficacy comparison of foliar application of *Steinernema feltiae* and spraying with abamectin. *Russian Journal of Nematology*, 15: 25–34.

Ulu, T. C., Sadic, B. and Susurluk, I. A. (2016). Effects of different pesticides on virulence and mortality of some entomopathogenic nematodes. *Invertebrate Survival Journal*, 13: 111–115.

Viteri, D. M., Linares, A. M. and Flores, L. (2018). Use of the entomopathogenic nematode *Steinernema carpocapsae* in combination with low-toxicity insecticides to control fall armyworm (Lepidoptera: Noctuidae) larvae. *Florida Entomologist*, 101: 327–329.

Wang, X., Lou, L. and Su, J. (2019). Prevalence and stability of insecticide resistances in field population of *Spodoptera litura* (Lepidoptera: Noctuidae) from Huizhou, Guangdong Province, China. *Journal of Asia-Pacific Entomology*, 22: 728–732.

Willer, H. and Kilcher, L. (2009). The world of organic agriculture—Statistics and emerging trends 2009. *FIBLIFOAM Report*. IFOAM, Bonn, FiBL, Frick, Switzerland, ITC, Geneva.

Wright, R. J., Witowski, J. F., Echtenkamp, G. and Georgis, R., (1993). Efficacy and persistence of *Steinernema carpocapsae* (Rhabditida: Steinemematidae) applied through a center-pivot irrigation system against larval corn rootworms (Coleoptera: Chrysomelidae). *Journal Economic Entomology*, 86: 1348–1354.

Xue, W., Snoeck, S., Njiru, C., Inark, E., Dermauw, W. and Leeuwen, T. V. (2020). Geographical distribution and molecular insights into abamectin and milbemectin cross-resistance in European field populations of *Tetranychus urticae*. *Pest Management Science*, 76: 2569-2581.

Yadav, S., Patil, J. and Sharma, H. K. (2017). Bio-efficacy of *Steinernema carpocapsae* against *Spodoptera litura* under laboratory condition. *Journal of Pure Applied Biosciences*, 5: 165–172.

Yan, J., Sarkar, S. C., Meng, R., Reitz, S. and Gao, Y. (2020). Potential of *Steinernema carpocapsae* (Weiser) as a biological control agent against potato tuber moth, *Phthorimaea operculella* (Zeller) (Lepidoptera: Gelechiidae). *Journal of Integrative Agriculture*, 19: 389–393.

Yan, X., Moens, M., Han, R., Chen, S. and De Clercq, P. (2012). Effects of selected insecticides on osmotically treated entomopathogenic nematodes. *Journal of Plant Diseases and Protection*, 119: 152–158.

Yuksel, E. and Canhilal, R. (2018). Evaluation of local isolates of entomopathogenic nematodes for the management of black cutworm, *Agrotis ipsilon* Hufnagel (Lepidoptera: Noctuidae). *Egyptian Journal Biological Pest Control*, 28: 82. https://doi.org/10.1186/s41938-018-0087-3.

Yüksel, E. and Canhilal, R. (2020). The effects of some commonly used biopesticides on the survival and virulence of native Turkish entomopathogenic nematode isolates. *Turkish Journal of Biological Control*, 11: 35–41.

Zayan, S. A. (2019). Impact of climate change on plant diseases and IPM strategies. In: *Plant Pathology and Management of Plant Diseases*. IntechOpen, London pp. 1–12. https://doi.org/10.5772/intechopen.87055.

Zhang, L., Shono, T., Yamanaka, S. and Tanabe, H. (1994). Effects of insecticides on the entomopathogenic nematode *Steinernema carpocapsae* Weiser. *Applied Entomology and Zoology*, 29: 539–547.

Zimmerman, R. J. and Cranshaw, W. C. (1990). Compatibility of three entomogenous nematodes (Rhabditida) in aqueous solutions of pesticides used in turf grass maintenance. *Journal of Economic Entomology*, 83: 97–100.

Chapter 12

Baculoviruses for Insect Pest Management: Applications, Challenges and Opportunities

Mudasir Gani[1,*], Insha Shafi[1], Taskeena Hassan[2], Mohammad Ayoub Mantoo[1], Mushtaq Ahmad Guroo[3], Rakesh Kumar Gupta[3] and Kamlesh Bali[3]

[1]Division of Entomology, Faculty of Agriculture,
Sher-e-Kashmir University of Agricultural Sciences and Technology of Kashmir, Jammu and Kashmir, India
[2]Department of Zoology,
Aligarh Muslim University, Uttar Pradesh, India
[3]Division of Entomology, Faculty of Agriculture,
Sher-e-Kashmir University of Agricultural Sciences and Technology of Jammu, Jammu and Kashmir, India

Abstract

Baculoviruses are natural pathogens of insects that have been reported from insects belonging to different insect orders such as Lepidoptera, Diptera and Hymenoptera. They are eco-friendly, species specific, safe to non-target organisms and have been found effective against a variety of crop insect pests on a large scale. A number of baculovirus formulations are available in the market as biological insecticides for eco-friendly management of insect pests of agricultural, horticultural and forest crops. The commercial production and application of baculovirus based biopesticides in the Americas, Europe, Asia, Australia and Africa have increased in the last two decades. However, slow action, narrow

[*] Corresponding Author's Email: mudasir32@gmail.com.

In: Pest Management: Methods, Applications and Challenges
Editor: Tarique Hassan Askary
ISBN: 979-8-88697-268-9

host range, technical and economical issues during *in vitro* commercial production and inactivation on exposure to sunlight or high temperatures are few limitations in the use of baculoviruses. Moreover, the dose dependent mortality, inconsistent performance under field conditions and farmers' attitude towards the use of baculoviruses for pest management are other such problems. Hence, further research is needed to develop novel formulations of baculoviruses with improved insecticidal efficacy, search for more virulent isolates and application of genetic engineering and nanotechnology in order to expand their use in insect pest management programs.

Keywords: baculoviruses, formulations, genetic engineering, insects, management.

1. Introduction

Baculoviruses are a large group of double-stranded DNA viruses that have been reported from Lepidoptera, Diptera and Hymenoptera orders of the class Insecta (de Souza et al., 2007). They are occluded viruses, pathogenic to invertebrates and their DNA genome is packaged into rod shaped nucleocapsids with the genome size of 80 to 200 kilo-base-pair (kbp) (Shang et al., 2017). Natural isolates of baculoviruses generally consist of homogenous or heterogenous populations of genotypes (Wennmann et al., 2020; Gani et al., 2021b). Baculoviruses are typically 40-50 nano metre (nm) in diameter and 200-400 nm in length (Pidre et al., 2013). These viruses belong to the family baculoviridae, which on the basis of phylogenetic evidence, molecular characteristics and comparison of genomic sequences is classified into four genera i. Alpha baculoviruses (lepidopteran specific NPVs), ii. Beta baculoviruses (lepidopteran specific GVs), iii. Gamma baculoviruses (hymenopteran specific NPVs) and iv. Delta baculoviruses (dipteran-specific NPVs) (Jehle et al., 2006). The lepidopteran NPVs are further divided into group I and group II NPVs on the basis of polyhedrin gene sequences and utilization of two different budded virus (BV) envelope fusion proteins. Group I NPVs contain the major envelope glycoprotein *GP64,* which mediates membrane fusion whereas Group II NPVs lack *GP64* protein, but contain a functional homolog designated as F protein.

The baculovirus virions are enveloped and their two phenotypes are i. occlusion derived virus (ODV) and ii. budded virus (BV). These two types of virions contain the same genome but differ in the morphogenesis and composition of their envelopes and functions in the virus life cycle. Baculovirus infection cycle can be split into three phases: early (0–6 hours after infection), late (6–24 hours after infection), and very late (18–24 to 72 hours after infection) (Figure 1). BVs are produced in the late phase of infection, obtain their envelope from the cell membrane, and they require fusion protein *GP64* (Monsma et al., 1996) or another unrelated protein termed as F protein (Westenberg et al., 2004) that facilitates systemic infection. The fusion protein forms structures called peplomers at one end of the budded virus particle, but they are not present in ODVs (Monsma et al., 1996). Several ODV envelope proteins have been identified as essential for primary infection of midgut cells of insect larvae and others as ODV components whose specific location and function have not yet been determined (Pijlman et al., 2003; Slavicek and Popham, 2005; Fang et al., 2009). ODVs are produced in the very late phase of the infection when nucleocapsids become enveloped within the nucleus and are subsequently occluded in a protein crystal structure (polyhedrin in NPV or granulin in GV) forming the occlusion bodies (OBs). This structure is quite resistant to diverse environmental conditions and therefore facilitates persistence and horizontal transmission of the virus in nature. Evidence suggests that they can survive passage through the gut of insect predators and gastrointestinal tract of birds, which can facilitate their dispersal (Gupta et al., 2013). Depending on the number of nucleocapsids per ODV, virions of Alphabaculoviruses are designated single (S) or multiple (M) occlusion-derived viruses, whereas delta- and gamma- baculoviruses normally contain a single nucleocapsid per ODV (Volkman, 1995; Theilmann et al., 2005). The commonly used detection methods for baculoviruses rely on laboratory-based tests which include microscopy and nucleic-acid amplification (Gani et al., 2021a).

According to FAO report, pests are responsible for up to 40 per cent loss of global agricultural production each year. Plant diseases cost the world economy more than $220 billion per year, while invasive insects cost at least $70 billion (Secretariat, 2021). In most of the developing countries including India, synthetic pesticides have been the primary method of managing the insect pests on various crops. Although chemical control allows farmers to achieve remarkable results in a short period of time, however, an over-reliance on pesticides and their indiscriminate use for a long time has resulted in a myriad of issues, such as the risk of environmental contamination, biodiversity loss, the development of insecticide-resistant pest populations, resurgence, secondary pest outbreaks, increased chemical inputs, and toxicological hazards due to the accumulation of pesticide residues in the food chain (Armes et al., 1992, Kranthi et al., 2002a). As a result, it is critical to rationalise the use of chemical pesticides for the control of insect pests. Moreover, growing public concern about the potential health risks of synthetic pesticides, as well as a sharp rise in the expense of cultivation, has prompted researchers to look for alternate and eco-friendly pest management strategies. Under these circumstances, biopesticides are one of the best alternatives. They are derived from natural plants or animals that can interfere in the life cycle of insect pests to reduce crop damage. The microbial pathogens such as viruses, bacteria, fungi, protozoa, and nematodes have gained significance for use as biopesticides primarily due to ease in production and application. Among the viruses, baculoviruses are the intriguing alternatives as biological control agent under integrated pest management (IPM) program and have been utilised with remarkable success for more than two decades (Zhang, 1989).

The present chapter is an attempt to highlight the significance of baculoviruses in crop insect pest management, limitations in their application and latest ideas to overcome the problems.

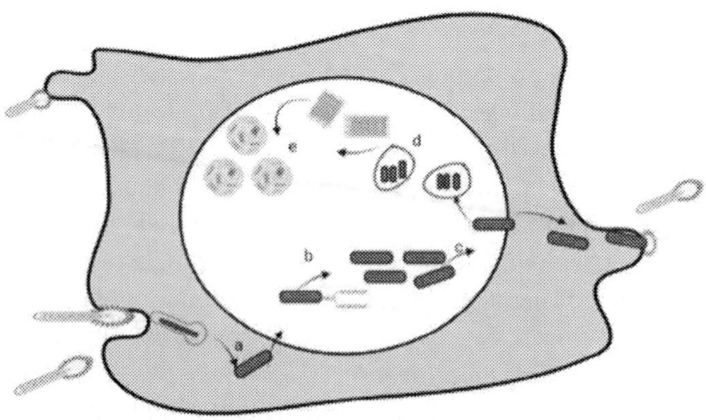

Source: Modified from Sosa-Gomez et al. (2020).

Figure 1. *Baculovirus Infection phases:* Baculovirus infection is categorized into early, late and very late phases. In the early phase, the budded virus (BV) form of the baculovirus enters the cell via endocytosis and (a) after fusion of the virus membrane with the endosomal membrane, the virus particles are released into the cytoplasm and then directed towards the cell nucleus. Within the nucleus, the virus releases its DNA that (b) starts transcription by the cell machinery producing early virus genes that prepare the cell for the late phase of infection. In the late phase, new virus particles are produced and released to the cell's cytoplasm and then to the outside of the cell via budding (c) In very late stages of the infection, virus particles inside the cell nuclei begin to envelop, forming the occlusion derived virus (ODV) form, and (d) occluded in a (e) proteinaceous matrix, composed mainly of a highly conserved protein called polyhedron, forming the occlusion body (OB).

2. Life Cycle of Baculoviruses

The viral life cycle begins when a susceptible host ingests OBs that have been deposited on foliage by a previously infected host, resulting in the release of hundreds of ODVs in the gut. In the host midgut, alkaline environment of the midgut triggers the dissolution of polyhedra (OB) and the release of ODV into the midgut. The released ODVs then pass through the peritrophic membrane, attach to the microvilli, and subsequently initiate primary infection of mature columnar epithelial cells within the midgut. BV produced in these cells initiates secondary infections, spreading throughout the host. The nucleocapsids are released from the endosomes and are transported to the nucleus, where viral transcription, DNA replication, and assembly of progeny nucleocapsids occur, resulting in the production of BV and ODV.

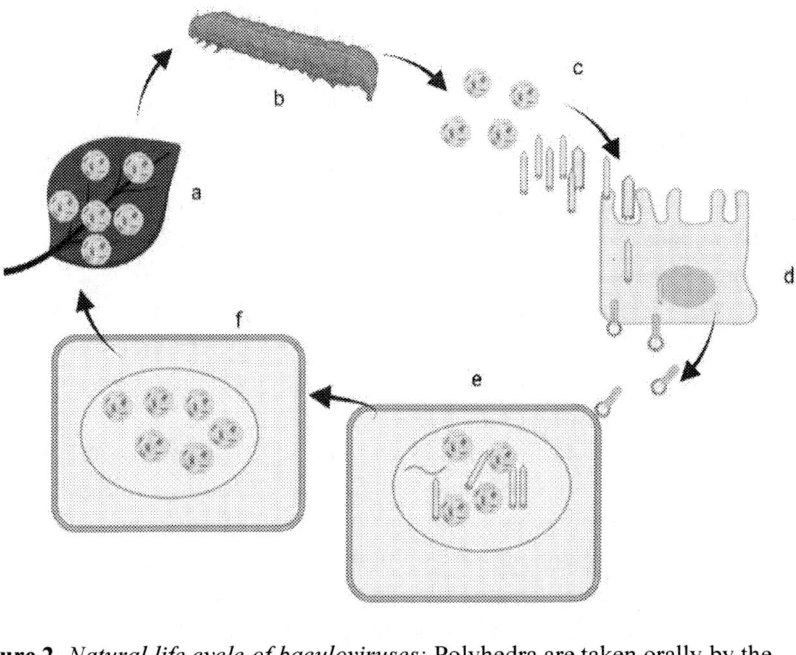

Figure 2. *Natural life cycle of baculoviruses:* Polyhedra are taken orally by the larvae while feeding on virus contaminated plant material and (a) are dissolved in the alkaline pH of the midgut (b) ODVs are liberated and (c) infect midgut epithelial cells (d) virus replicates, and budded viruses (BVs) are produced, and they infect other tissues (e) After secondary infection, polyhedra are accumulated and finally (f) the larval body liquifies and disintegrates, and millions of new polyhedra are released to the environment.

In the final stage of infection, most of the nucleocapsids remain in the nucleus and become occluded in a protein matrix to form OBs. Progeny OBs are released upon death and disintegration or liquefaction of the infected insect. Subsequently, a new round of infection to other hosts begins. The terminally infected insect can migrate to a higher elevation on the branch of a plant, facilitating dispersal of the occlusion bodies (Kamita et al., 2005; Rohrmann, 2019). The consecutive steps of this complex process of infection are shown in Figure 1 and the life cycle of baculovirus is shown in Figure 2.

3. Baculoviruses in Insect Pest Management

Baculoviruses play a major role in the suppression of a variety of insect pests in agro-ecosystems. These naturally occurring viruses are known to infect over 600 insect species worldwide (Slack and Arif, 2007) and can initiate virus outbreaks in the field populations of insects without additional inputs. As such the baculoviruses are an important component of the biological control programs of many insect pests and have been used for more than 20 years in different parts of the world with great success (Zhang, 1989). The virus particles of baculoviruses are present in the environment and usually found on the surface of plants or in the soil. Insects become infected by consuming plant material contaminated with viral particles on the surface or by contact with the soil. They act as natural regulators of pest population dynamics of economically important insect pests primarily in the order Lepidoptera such as *Helicoverpa armigera*, *Spodoptera litura*, *Autographa californica* and *Cydia pomonella* (Williams et al., 2017).

Among the insect viruses found in nature, those belonging to the baculoviruses (Family: Baculoviridae) are considered for the development of commercial viral biopesticides. The advantages of using baculoviruses as microbial insecticides include their specificity, persistence, cheap availability and safety to non-target organisms such as beneficial insects and vertebrates (Monobrullah and Nagata 1999; Nakai et al., 2003; Ashour et al., 2007). The majority of viruses have an extremely narrow host range, coupled with high virulence and low environmental impact, making the most natural baculoviruses (both NPVs and GVs) highly selective pest control agents (Copping and Menn, 2000; Moscardi et al., 2011). Besides, their compatibility with other pest control methods makes them an attractive alternative to chemical insecticides. Despite the considerable number of programs worldwide utilizing baculoviruses as biopesticides, actual usage of baculoviruses in pest management has been disappointing, particularly with the microbial insecticide approach, primarily for three reasons: economics, slow speed of kill and adverse effects of environment on the viruses. However, the recombinant-DNA revolution has greatly increased the prospects for baculoviruses in insect pest management.

4. Application of Baculoviruses for Management of Insect Pests

Baculoviruses are naturally occurring pathogens of insects that have been isolated and formulated as biopesticides for the management of insect pests in different countries of the world (Fauquet et al., 2005). They have been successfully used in insect pest management applications owing to their highly pathogenic nature and more stability in the environment (Hunter-Fujita et al., 1998). From an environmental risk perspective, the baculoviruses are species specific and show no toxicity against vertebrates, plants and beneficial insects such as silkworm and honeybees (Laird et al., 1990). They have the capacity to persist in the environment and can be subjected to mass production, formulation, packaging, storing and marketing in a similar way to chemical insecticides. Many of these viruses have been characterized and applied successfully in the field as one of the components of IPM programs.

Steinhaus and his collaborators (1950–1970) applied nucleopoly-hedrovirus (NPV), a baculovirus, to control the alfalfa caterpillar, *Colias eurytheme*. However, the first introduction of baculovirus resulting in successful regulation of an insect pest took place in 1930 when European spruce sawfly, *Gilpinia hercyniae* occurred accidentally in Canada (Bird and Elgee, 1957). Gani et al. (2016, 2017) isolated, characterized and evaluated the NPV of *Lymantria obfuscata* and *Myrnoteras indicum* under field conditions. 4

First aerial applications of NPV against *C. eurytheme*, were attempted in California in the late 1940s. Elcar was the first commercial product developed and marketed by Sandoz for use in agriculture against *Helicoverpa* species on cotton (Ignoffo et al., 1973; Ignoffo and Couch, 1981). In the last two decades, many countries of the world have prepared baculovirus based biopesticides that have been successfully commercialized (Table 1).

Table 1. List of commercial baculovirus-based products manufactured by different countries

Virus	Host	Crop	Product Name	Manufacturer	Country
Anticarsia gemmatalis MNPV	*Anticarsia gemmatalis*	Soybean	Baculo-soja	Nova Era Biotecnologia	Brazil
			Baculovirus Nitral	Nitral Urbana	
			Coopervirus SC	COODETEC	
			Protege	Milenia	
			Multigen	EMBRAPA	
Autographa californica NPV	*Autographa californica*, *Trichoplusia ni*	Broccoli, cabbage	Loopex	Sylvar Technologies Incorporation	Canada
Autographa californica MNPV + Spodoptera albula NPV	*Autographa californica*, *Trichoplusia ni*, *Pseudoplusia includens*, *Heliothis virescens*, *Spodoptera exigua*, *Estigmene acrea*, *Plutella xylostella*	Alfalfa, vegetable Crops	VPN-ULTRA	Agricola El Sol	Guatemala
Spodoptera sunia NPV	*Spodoptera* spp.	Vegetables	VPN 82	Agricola El Sol	Guatemala
Cydia pomonella GV	*Cydia pomonella*, *Grapholita molesta*	Apple, pear, Walnut, peach	Carpovirus Plus	NPP-Arysta Life Science	Argentina
			Madex	Andermatt Biocontrol	Argentina
			Carpovirusine	NPP-Arysta Life Science	Chile
			Madex Twin	Andermatt Biocontrol	Uruguay
Cydia pomonella GV	*Cydia pomonella*	Apple, crab apple, pear, quince	Madex Top	Andermatt Biocontrol	UK
Cydia pomonella GV (Mexican strain)	*Cydia pomonella*	Apple, crab apple, pear, quince	Madex Plus	Andermatt Biocontrol	Switzerland

Virus	Host	Crop	Product Name	Manufacturer	Country
Adoxophyes orana GV	*Adoxophyes orana*	Apple, crab apple, pear, cherry, plum, apricot, quince	Capex	Andermatt Biocontrol	UK
Erinnyis ello GV	*Erinnyis ello*	Cassava, Rubber trees	Baculovirus erinnyis	Empresa de Pesquisa Agropecuaria e Extensao Rural de Santa Catarina S.A.	Brazil
				BioCaribe SA and CORPOICA	Columbia
Helicoverpa zea SNPV	*Helicoverpa* spp.	Maize, tomato, cotton and tobacco	Gemstar	Certis USA	Mexico
			HzNPV CCAB	AgBiTech Australia	Brazil
Helicoverpa armigera NPV	*Helicoverpa* spp.	Tomato, sweet pepper, maize, soybean, tobacco, vegetable crops	Diplomata	Koppert	Brazil
			Helicovex	Andermatt Biocontrol	Brazil
Lymantria dispar MNPV	*Lymantria dispar*	Forest and fruit trees (Apple, pear)	Gypchek	USDA Forest Service	USA
Phthorimaea operculella GV	*Phthorimaea Operculella, Tecia solanivora*	Potato	Baculovirus Corpoica	CORPOICA	Colombia
			PTM baculovirus	SENASA Peru	Peru
			PTM baculovirus	INTA Costa Rica	Costa Rica
Phthorimaea operculella GV + *Bacillus thuringiensis*	*Phthorimaea operculella, Tecia solanivora, Symmetrischema tangolias*	Potato	Matapol Plus	PROINPA Foundation	Bolivia
			Bacu-Turin	INIAP, Ecuador	Ecuador
Spodoptera exigua NPV	*Spodoptera exigua*	Tomato, chilli, Egg plant	SPOD-X LC	Certis USA-SUMMIT AGRO, Mexico	Mexico

Table 1. (Continued)

Virus	Host	Crop	Product Name	Manufacturer	Country
Spodoptera exigua NPV	*Spodoptera exigua*	Tomato, cabbage	Spexit	Andermatt Biocontrol	Switzerland
Spodoptera frugiperda MNPV	*Spodoptera frugiperda*	Maize, sorghum, soybean	Spodovir	Embrapa and Andermatt Biocontrol	Brazil
Helicoverpa armigera NPV	*Helicoverpa armigera, Helicoverpa spp.*	Vegetable crops, sweet corn, cotton	Helicovex	Andermatt Biocontrol, Andermatt France	Switzerland, France, Brazil
Cryptophlebia leucotreta GV	*Thaumatotibia leucotreta*	Citrus, stone fruits	Cryptex	Andermatt Biocontrol	South Africa
Spodoptera littoralis NPV	*Spodoptera littoralis, Spodoptera frugiperda*	Cotton, paprica, strawberry, maize	Littovir	Andermatt Biocontrol	Switzerland
Neodiprion abietis NPV	*Neodiprion abietis*	Balsam fir	Abietiv	Andermatt Biocontrol	Canada
Phthorimaea operculella GV	*Tuta absoluta*	Tomato, eggplant, pepper, tobacco, potato	Tutavir	Andermatt Biocontrol	Switzerland

4.1. *Anticarsia gemmatalis* (Ag) MNPV

The velvetbean caterpillar, *Anticarsia gemmatalis* Hubner is a soybean defoliation pest, distributed in the United States and Argentina (Sosa Gomez, 2004). The control of this pest is usually achieved by applying synthetic insecticides such as pyrethroids or insect growth regulators, and by growing soybeans that express the insecticidal proteins of *Bacillus thuringiensis* in areas where the use of transgenic crops is permitted (Panizzi, 2013). *A. gemmatalis* is naturally infected with a nucleopolyhedrovirus (AgMNPV; Baculoviridae: Alphabaculovirus) (de Castro Oliveira et al., 2006). The infection process begins when the larvae consume the foliage contaminated with OBs and the virus breaks down in the midgut of the insect and release virions (ODVs) which infect the epithelial cells of the insect midgut. The infection thereby spreads throughout the body of the larva and death of the host occurs within few days. Viral OBs are released from infected cadavers for horizontal transmission. Anticarsia gemmatalis (Ag) MNPV applied against *A. gemmatalis* infesting soybean under field condition in Brazil was found very successful in controlling the pest (Moscardi, 1999). Later on, this virus was exploited against the pests in other South American countries and neighbouring countries, such as Paraguay and Uruguay (Haase et al., 2015a). AgMNPV has also been utilized in the United States (Fuxa and Richter, 1999), as well as in Argentina and Colombia (Haase et al., 2015a). However, the use of AgMNPV has been declined, as transgenic soybeans have become increasingly common in some regions (Moscardi et al., 2011). The virus is highly virulent to host larvae at a low dose (1.5×10^{11} OB/ha) and get transmit efficiently in the host population by natural enemies and abiotic factors. Usually, one AgMNPV application is sufficient to control the insect, compared with a mean of 1.8 chemical applications. The implementation of a soybean IPM program in Brazil in the mid-1970s facilitated farmers' acceptance of the AgNPV, which is now being used on approximately 1 million ha annually in Brazil (Monobrullah and Kaul, 2005).

4.2. *Autographa californica* NPV

Autographa californica multiple nucleopolyhedrovirus (AcMNPV) is the type species of the genus *Alpha baculovirus*. The Autographa californica NPV (AcNPV) was discovered from a single larva collected on alfalfa in southern desert of California. It has a wide range that covers 43 species of Lepidoptera belonging to 11 families. The virus has been found extremely pathogenic to *A. californica, Helicoverpa virescens, Spodoptera exigua and Trichoplusia ni,* but it is up to 40-fold less pathogenic in other key pests, including *Helicoverpa zea* (Payne, 1986). In Canada, AcNPV applied @ 1.5 x 10^{12} OB/ litre with 2-6 applications per year in small plots infested with *T. ni* resulted in effective control of the pest (Jaques, 1975).

AcNPV has been found to infect the pink bollworm *Pectinophora gossypielia* (Saunders). However, the application of this virus in aqueous suspension reduced the infection of pink bollworm in field trials (Vail et al., 1972). Pink bollworm larvae enter plant shortly after hatching and eat very little on the surface. Bell and Kanavel (1975) suggested that the incorporation of this virus in the bait formulation can increase its effectiveness by increasing the uptake of this virus by the larvae.

4.3. *Helicoverpa armigera* NPV

The American boll worm, *Helicoverpa armigera* Hubner (Lepidoptera: Noctuidae) is a highly polyphagous insect pest that attacks broad spectrum of crops like chickpea, pigeon pea, soybean, sunflower and cotton (Fitt, 1989). The intensive and indiscriminate use of chemical pesticides against *H. armigera* has led to the development of host resistance towards organic insecticides *viz.*, organophosphates and carbamates (Harries et al., 1971). The first report on the development of resistance in *H. armigera* to pyrethroids was published in the late 1980s (Dhingra et al., 1988; McCaffery et al., 1989). Kranthi et al. (2002b) reported high level of resistance in *H. armigera* to many commonly used insecticides. In China, intensive spraying of chemical insecticide has eliminated natural enemies and reduced natural control agents of cotton borers (Wanxue, et al., 2000; Yang et al., 2000). These problems reaffirmed the need to focus attention on alternate pest management strategies.

Helicoverpa armigera NPV (HearNPV) has been used to control *H. armigera* in a variety of crops, including oil seeds (Rabindra et al., 1985), legumes (Rabindra, 1992), cotton (Jones, 1994) and vegetables (Jones et al., 1998). HearNPV has been found more effective than natural or commercially manufactured insecticides in the treatment of *H. armigera* on chickpea (Cherry et al., 2000). The HearNPV have been reported to cause almost 98 per cent mortality in *H. armigera* larvae under laboratory conditions (Figueiredo et al., 2009; Elamathi et al., 2012). However, better results were obtained by combining HearNPV with endosulfan under field conditions (Mir et al., 2010; Siddique et al., 2010). The local HearNPV isolates have been mass produced and applied to different crops for management of *H. armigera* in India, Australia and China (Moscardi et al., 2011). In China, HearNPV was widely used in an area of around 100,000 hectares, with only little negative effects on the biotic ecosystem (Chen et al., 2000). The unpurified suspension of HearNPV was found to be more effective than the Emulsifiable Concentrate (EC), ULV suspension and microencapsulated preparation in the management of *H. armigera* (Cherry et al., 2000). HearNPV applied @ 1.15 x 10^7 OBs/ ml reduced 100 per cent infestation of *H. armigera* on tomato grown in greenhouse (Moore et al., 2004). Under field conditions, HearNPV was found equally effective as Indoxacarb 14.5 SC (Avant ®) and Endosulfan 35 EC (Endocil®) against *H. armigera*, and also compatible with other biocontrol agents such as *Metarhizium anisopliae* (Metchnikoff) and the parasitoid, *Campoletus chloridae* (Uchida) (Doddabasappa et al., 2013).

A combined application of bio-pesticide, HearNPV and synthetic pesticide, Spinosad 240 SC was undertaken by Nawaz et al. (2020) for the management of *H. armigera* under controlled conditions. Different concentrations of HearNPV and Spinosad were used singly and concomitantly (HearNPV + Spinosad). The insect mortality increased with increase in the concentration of the pesticides. Hence, it was concluded that the HearNPV is an effective biocontrol agent and can be incorporated in the Integrated Pest Management (IPM) programme of *H.* armigera.

Interesting research confirming the efficacy and persistence of a binary mixture of HearSP1B and HearLB6 variants co-occluded into OBs was evaluated by Arrizubieta et al., (2016) as the basis for a biological insecticide for controlling H. armigera on tomato crops grown under greenhouse and open-field conditions in southern Spain and Portugal. It was concluded that the application of HearSP1B: LB6 OB is as effective as commercial insecticides in controlling the damage caused by *H. armigera* on tomato.

4.4. *Spodoptera* spp. NPV

The genus *Spodoptera* (Lepidoptera: Noctuidae) includes several species such as *S. litura*, *S. exigua, S. littoralis* and *S. frugiperda*, which attack a wide range of agricultural, horticultural, fibre, vegetables crops and wild plants as well as weeds (Zhou et al., 2011). *S. litura* larvae can cause yield losses varying between 26-100 per cent in the field (Tuan et al., 2014). It is distributed worldwide, especially in North America, the islands of Oceania, Africa and Asia (Elbendary and ElHelaly, 2013). The species specific NPV of *Spodoptera litura, S. littoralis, S. exigua* and *S. frugiperda* have been successfully isolated and applied for themanagement of insect pests. Spodoptera exigua MNPV was found highly pathogenic both in field and the laboratory (Caballero et al., 2009). The high cash value of many crops attacked by the *S. exigua* led to development of' the S. exigua NPV (SeNPV) as the viral insecticide, Spodex. SeNPV was found highly effective when applied @ 2-4 x 10^{12} polyhedra/ ha, in the field on lettuce and in glasshouses on chrysanthemums (Gelernter et al., 1986). S. exigua NPV is being produced in the laboratory and distributed to farmers for field production of the virus in Thailand (Monobrullah and Kaul, 2005).

SeMNPV is sensitive to ultraviolet light. Yan et al. (2021) on the basis of an experiment reported that coating SeMNPV with polydopamine PDA enhanced the UV resistance. PDA-modified SeMNPV outperformed SeMNPV (without coating), thus providing a novel strategy to develop viral insecticides. The control of *S. litura* in India has been studied on various crops *viz.,* banana, blackgram, castor, cauliflower, cotton and tobacco using the NPV alone or in combination with a lower dose of insecticide. It was found that the application of 9.3 x 10^7 OB/ ml or half this dose with 0.02% carbaryl on tobacco reduced the seedling damage significantly (Ramakrishnan et al., 1981). The field application of S. litura NPV has shown excellent results on strawberries in the polyhouses (Nemoto and Okada., 1990). The application of S. litura NPV in China yielded good results against *S. litura* in vegetables, cotton, rice and peanuts (Yi and Li., 1989).

Combining biorational insecticides with microbes is a pre- mixing strategy to combat pesticide overuse. Sarwar et al., (2021) aimed to devise a novel method for the control of *Spodoptera* spp. using Spodoptera NPV alone as well as in combination with an insecticide, Chlorantraniliprole. The integrated application was found more toxic for adult emergence and egg eclosion as compared to their individual application. Moreover, the sub-lethal concentration of Chlorantraniliprole + high concentration of NPV proved to be a better strategy to manage insecticidal resistance. Another example of the integrated application of bio- and synthetic pesticide (NPV and Emamectin benzoate, Proclaim® 19 EC) against *S. litura* proved more fatal than their sole application in controlling this pest (Yasin et al., 2020).

4.5. *Cydia pomonella* GV

Codling moth, *Cydia pomonella* is found in almost all countries where apples are grown and has an almost worldwide distribution, making it one of the most devastating pests (Thaler et al., 2008). In the past, control of this pest has relied on the intensive use of pesticides, which eventually led to the development of resistance and a decline in the population of beneficial insect species that were once the only natural regulators of pest populations in apple orchards.

CpGV (Betabaculovirus: Baculoviridae) is highly pathogenic to codling moth, *C. pomonella* (Crook et al., 1997; Jehle et al., 2006). It is the only beta baculovirus whose replication has been thoroughly investigated both under *in vivo* and *in vitro* conditions (Winstanley and Crook, 1993; Hilton and Winstanley, 2007). The first isolate of CpGV was collected from infected codling moth larvae in Mexico (Tanada, 1964). Its safety to non-target organisms and specificity for codling moth and a few closely similar species have been thoroughly proven (Falcon et al., 1968; Groner, 1986). It is considered as the most virulent baculoviruses, with an LD_{50} of 1.2 to 17 granules/larva in neonate larvae (Laing and Jaques, 1980). Balaško et al. (2020) stated that for more than 30 years, commercial CpGV products have been successfully applied to control codling moth in organic and integrated fruit production.

To infect and kill susceptible species, CpGV must be consumed. While the virus is usually acquired when larvae consume fruit, it is also possible for larvae to become infected by strolling or browsing on CpGV-sprayed leaf surfaces in as little as 3.5 minutes (Ballard et al., 2000). Few surveys utilizing molecular methods for detection of the virus reported that the CpGV is widespread wherever codling moth is found (Vincent et al., 2007). Many countries utilize the CpGV to manage codling moth. In places where the codling moth is univoltine, such as many European countries, CpGV is extremely effective, however where there are multiple annual generations, such as California, and especially when such generations do not have perfectly synchronised phenology, the situation is more difficult. Several commercial products of CpGV are being marketed in Europe. It includes Carpovirisine, Madex, Granupom and Granusal (Helsen et al., 1992). In the Netherlands, Granusal was found much effective as chemical pesticide. The field comparison of Carpovirisine and Granupom in Italy suggested that both were as effective as standard insecticide (Pasqualini et al., 1994).

Morisset et al., (2021) examined the compatibility and complementarity of *Trichogramma minutum* and CpGV. The combined efficacy of both biocontrol agents in controlling codling moth in the orchard was 77 and 79 per cent in the first and second year, respectively. Ultravoilet radiations can affect the persistence of CpGV. Research are being conducted on materials that can provide protection from UV in the field before a target insect consumes it. Wu et al., (2015) confirmed that ZnO and TiO_2 can act as potential UV protectants for pathogenic CpGV-ZY at low concentrations.

5. Challenges in the Application of Baculoviruses

Baculoviruses are potent regulators of the insect pest population, but their use as biological control agents in agricultural, horticultural and forest ecosystems is limited by biological or technical reasons. They are considered imperfect insecticides from an agro-industrial point of view and have several potential limitations (Possee et al., 1997; Inceoglu et al., 2006). The high costs and difficult production, problems encountered in registration and marketing as well as shorter field stability have inhibited their development (Morales et al., 2001, Claus et al., 2012). The limitations in the application of baculovirus based biopesticides have been described below:

1. Baculoviruses have a slow speed of kill, which may take 5-15 days to kill the host.
2. They are highly specific to the target insect pest.
3. The pest mortality is dose dependent and very high doses are often necessary for adequate management of the pest.
4. These viruses are very effective against early larval instars; the late larval instars are less susceptible to virus infection.
5. The field stability of virus particles is too low and they get rapidly inactivated on exposure to direct sunlight or at higher temperatures.
6. Baculoviruses have a short shelf life and their mass production is a costly affair.
7. The cultural practices can hide the viral particles in the soil thus affecting viral persistence.

Despite these limitations, the adoption of baculovirus based biopesticides is likely to increase in the near future due to health hazard problems caused with the use of chemical pesticides.

6. Opportunities

The difficulties with respect to the poor persistence ability of baculoviruses can be solved through biotechnological approaches (genetically modifying the baculovirus using recombinant DNA technology) and by developing novel baculovirus formulations with suitable adjuvants. Strategies to counteract some of the limitations of baculoviruses have been investigated and have been found promising. Such strategies can enhance viral activity and increase speed of kill as well as reduce larval feeding activity.

Attempts to develop UV-resistant baculoviruses through strain selection or genetic modification have been investigated (Akhanaev et al., 2017) and the research is still going on. The optical brighteners derived from stilbene compounds (Tinopal and Blankophor) absorb UV radiation and are used as UV protectors in baculovirus formulations (Hamm and Shapiro, 1992). The SfMNPV OBs with optical brighteners (Tinopal CBS and Tinopal C1101) and without optical brighteners were found effective against the *S. frugiperda* larvae for more than 240 minutes and up to 15 minutes, respectively when exposed to UV light, (Mondragon et al., 2007). The stability and insecticidal activity of S. exigua NPV was increased with the addition of optical brighteners (Martínez et al., 2003). The optical brightener, Calcofluor M2R was reported to enhance insecticidal activity of SfMNPV formulation against *S. frugiperda* larvae (Martinez et al., 2003). Boric acid was reported to increase the speed of kill of *A. gemmatalis* MNPV (Morales et al., 1997) and SfMNPV (Cisneros et al., 2002) with no detrimental effect on the natural enemies. Diaminostilbene disulfonic acid based optical florescent brighteners (BLANKOPHOR) and metal oxides (titanium dioxide and zinc dioxide) have been used to enhance the UV stability of baculovirus biopesticides (Shapiro and Farrar, 2003; Farrar et al., 2003).

Now-a-days encapsulation is being used as a means of protecting baculoviruses against harmful UV radiations (Pemsel, et al., 2010; Wilson et al., 2020). The C. pomonella GV (CpGV) was microencapsulated with Titanium dioxide and this baculovirus formulation was reported to be biologically viable with enhanced UV protection (Pemsel et al., 2010). The SfNPV was more resistant to UV-inactivation by the encapsulation in Eudragit-S100 microparticles (Villamizar et al., 2010). Wilson et al., (2020) developed a new baculovirus formulation in which the S. littoralis nucleopolyhedrovirus (SpliNPV) was micro-encapsulated with ENTOSTAT wax and combined with titanium dioxide (TiO_2) as UV absorbant. The new formulation technology (ENTOSTAT-TiO_2-NPV formulation) has the potential to reduce the UV degradation and improve the efficacy of baculovirus based biopesticides.

The work on the genetically modified (GM) baculoviruses was started by Summers and Smith (1987) and Miller (1988) who established the baculovirus expression vector systems (BEVS) using AcMNPV. The GM baculoviruses can easily become an integral part of insect pest management strategies and may become effective alternatives in the management of those insect pests that have become resistant to synthetic chemical pesticides (Bonning and Hammock, 1996; Harrison and Bonning, 2000; Inceoglu et al., 2001; Bonning et al., 2002). Genetic engineering has been utilized to improve the speed of kill of baculoviruses via insertion of a foreign gene, deletion of an endogenous gene or replacement of an endogenous gene from the baculovirus genome (Kamita et al., 2005). These approaches have also been used in combination to increase the speed of kill of baculoviruses and enable the larvae to stop feeding on the host immediately after infection (Kamita et al., 2005). RNA interference (RNAi) and other RNA-based strategies have also been conducted for management of lepidopteran insect pests (Terenius et al., 2011; Burand and Hunter, 2013; Nandety et al., 2015).

There are reports of GM baculovirus designed to express cathepsin B-like proteases (Shao et al., 2008) and cathepsin L-like proteases (Sun et al., 2009; Gramkow et al., 2010) in the laboratory and field conditions. A recombinant AcMNPV expressing a growth-blocking peptide from *S. exigua* showed that the median lethal time of this recombinant virus is about 1 day faster as compared to wild type AcMNPV in larval *S. exigua* (Wan et al., 2015). The genetic modification of baculoviruses by inserting toxin genes from scorpions (Jinn et al., 2006; Tang et al., 2011), spiders (Ardisson-Araujo et al., 2013; Ali et al., 2015) and *B. thuringiensis* (Kim et al., 2005; El-Menofy et al., 2014) have also been investigated. The genetically modified viruses with a high speed of kill can be mass-produced in cell culture (Haase et al., 2015b). Although the efficacy of current GM baculoviruses as biopesticides is promising, there is possibility of further improvement in the efficacy of baculoviruses for crop protection through advanced genome engineering approaches and improving formulations with suitable chemical or biological substances.

Conclusion

Baculoviruses are natural pathogens that regulate the population dynamics of insect pests, prevent insects from attaining pest status and also reduce their damage potential. Their safety towards non-target organisms, compatibility with other integrated pest management strategies and increased utilization for the management of insect pests will significantly reduce the risks associated with the use of chemical pesticides. The wider use of registered baculovirus based biopesticides and the development of new formulations will provide impetus to the sustainable agriculture. However, there are few constraints that prevent the large-scale application of baculoviruses. It is possible to overcome these problems with the improvements in the *in vitro* commercial production, development of formulations with adjuvants, searching for the new virulent isolates and through genetic engineering and nanotechnology to expand their use in insect pest management programs. The awareness among the farmers and the collaborative efforts of the government agencies, research institutions, grower associations, private and public sector extension agencies are needed for successful implementation of baculovirus based biological control programs. This will increase the presence of baculovirus based biopesticides in the global market and will meet the need for a safer and ecologically acceptable insect pest management strategy for sustainable agriculture.

Acknowledgments

We are grateful to the Head of Division of Entomology, SKUAST-Kashmir and the Head of Division of Entomology, SKUAST- Jammu for providing the facilities for conducting the research work on baculoviruses. Thanks to Professor Subhash Chander, Director, ICAR-NRCIPM, New Delhi and Professor Johannes A. Jehle, Director, Institute for Biological Control, Dossenheim, Germany, for their valuable suggestions and guidance during the preparation of this manuscript.

References

Akhanaev, Y. B., Belousova, I. A., Ershov, N. I., Nakai, M., Vyacheslav, V. M. and Glupov, V. V. (2017). Comparison of tolerance to sunlight between spatially distant and genetically different strains of *Lymantria dispar* nucleopolyhedrovirus. *Plos one,* 12: e0189992. https://doi.org/10.1371/journal.pone.0189992.

Ali, M. P., Kato, T. and Park, E. Y. (2015). Improved insecticidal activity of a recombinant baculovirus expressing spider venom cyto-insectotoxin. *Applied Microbiology and Biotechnology*, 23: 10261-10269.

Ardisson-Araújo, D. M. P., Morgado, F. D. S., Schwartz, E. F., Corzo, G. and Ribeiro, B. M. (2013). A new theraphosid spider toxin causes early insect cell death by necrosis when expressed *in vitro* during recombinant baculovirus infection. *PloS one*, 8: e84404. https://doi.org/10.1371/journal.pone.0084404.

Armes, N. J., Jadhav, D. R., Bond, G. S. and King, A. B. S. (1992). Insecticide resistance in *Helicoverpa armigera* in South India. *Pesticide Science*, 34: 355-364.

Arrizubieta, M., Simon, O., Torres-Vila, L. M., Figueiredo, E., Mendiola, J., Mexia, A., Caballero, P. and Williams, T. (2016). Insecticidal efficacy and persistence of a co-occluded binary mixture of *Helicoverpa armigera* nucleopolyhedrovirus (HearNPV) variants in protected and field grown tomato crops on the Iberian Peninsula. *Pest Management Science*, 72: 660-670.

Ashour, M. B., Ragheb, D. A., El-Sheikh, E. S. A., Gomaa, E. A. A., Kamita, S. G. and Hammock, B. D. (2007). Biosafety of recombinant and wild type nucleopolyhedroviruses as bioinsecticides. *International Journal of Environmental Research and Public Health*, 4: 111-125.

Balaško, M. K., Bažok, R., Mikac, K. M., Lemic, D. and Živković, I. P. (2020). Pest management challenges and control practices in codling moth: a review. *Insects,* 11: 38. https://doi.org/10.3390/insects11010038.

Ballard, J., Ellis, D. J. and Payne, C. C. (2000). The role of formulation additives in increasing the potency of *Cydia pomonella* granulovirus for codling moth larvae, in laboratory and field experiments. *Biocontrol Science and Technology*, 10: 627-640.

Bell, M. R. and Kanavel, R. F. (1975). Potential of bait formulations to increase effectiveness of nuclear polyhedrosis virus against the pink bollworm. *Journal of Economic Entomology*, 68: 389-391.

Bird, F. T. and D. E. Elgee (1957). A virus disease and introduced parasites as factors controlling the European spruce sawfly, *Diprion hercyniae* (Htg.), in Central New Brunswick1. *The Canadian Entomologist*, 89: 371-378.

Bonning, B. C., Boughton, A. J., Hailing J. and Harrison, R. L. (2002). Genetic enhancement of baculovirus insecticides. In: *Advances in Microbial Control of Insect Pests* (Ed. Upadhyay, R.K.). Springer, USA, pp. 109-125.

Bonning, B. C. and Hammock, B. D. (1996). Development of recombinant baculoviruses for insect control. *Annual Review of Entomology*, 41: 191-210.

Burand, J. P. and Hunter, W. B. (2013). RNAi: future in insect management. *Journal of Invertebrate Pathology*, 112: 68-74.

Caballero, P., Murillo, R., Muñoz, D. and Williams, T. (2009). The nucleopolyhedrovirus of *Spodoptera exigua* (Lepidoptera: Noctuidae) as a biopesticide: analysis of recent advances in Spain. *Revista Colombiana de Entomología*, 35: 105-115.

Chen, X., Sun, X., Hu, Z., Li, M., O'Reilly, D. R., Zuidema, D. and Vlak, J. M. (2000). Genetic engineering of *Helicoverpa armigera* single-nucleocapsid nucleopolyhedrovirus as an improved pesticide. *Journal of Invertebrate Pathology*, 76: 140-146.

Cherry, A. J., Rabindra, R. J., Parnell, M. A., Geetha, N., Kennedy, J. S. and Grzywacz, D. (2009). Field evaluation of *Helicoverpa armigera* nucleopolyhedrovirus formulations for control of the chickpea pod-borer, *H. armigera* (Hubn.), on chickpea (*Cicer arietinum* var. Shoba) in southern India. *Crop Protection*, 19: 51-60.

Cisneros, J., Pérez, J. A., Penagos, D. I., Ruiz, J., Goulson, D., Caballero, P., Cave, R. D. and Williams, T. (2002). Formulation of a nucleopolyhedrovirus with boric acid for control of *Spodoptera frugiperda* (Lepidoptera: Noctuidae) in maize. *Biological Control*, 23: 87-95.

Claus, J. D., Gioria, V. V., Micheloud, G. A. and Visnovsky, G. (2012). Production of insecticidal baculoviruses in insect cell cultures: potential and limitations. In: *Insecticides-Basic and Other Applications* (Ed. Soloneski, S.). IntechOpen Access, pp. 127-152. https://doi.org/10.5772/29551.

Copping, Leonard G., and Menn, J. J. (2000). Biopesticides: a review of their action, applications and efficacy. *Pest Management Science*: *Formerly Pesticide Science*, 56: 651-676.

Crook, N. E., James, J. D., Smith, I. R. and Winstanley, D. (1997). Comprehensive physical map of the *Cydia pomonella* granulovirus genome and sequence analysis of the granulin gene region. *Journal of General Virology*, 78: 965-974.

de Souza, M. L., de Castro, M. E. B., Sihler, W., Krol, E. and Szewczyk, B. (2007). Baculoviruses: a safe alternative in pest control? *Pest Technology*, 1: 53-60.

Dhingra, S., Phokela, A. and Mehrotra, K. N. (1988). Cypermethrin resistance in the populations of *Heliothis armigera* Hubner. *National Academy Science Letters*, 4: 123-125.

Doddabasappa, B., Chakravarthy, A. K., Narabenchi, G. B., Devi, S. S. G. and Rajagopal, D. (2013). Evaluation and validation of HearSNPV for fruit borer, *Helicoverpa armigera* (Noctuidae: Lepidoptera) management on tomato. *Indian Journal of Agricultural Science,* 83: 103-109.

Elamathi, E., Cholan, J. R. R., Vijayakumar, N. and Ramamourti, A. (2012). Formulation and optimisation of various nuclear polyhedrosis virus isolates and assessment of their insecticidal activity against *Helicoverpa armigera* (Hubner) (Lepidoptera: Noctuidae) larvae. *Archives of Phytopathology and Plant Protection*, 45: 750-765.

El-Bendary, H. M. and El-Helaly, A. A. (2013). First record nanotechnology in agricultural: Silica nano-particles a potential new insecticide for pest control. *Applied Sciences*, 4: 241-246.

El-Menofy, W., Osman, G., Assaeedi, A. R. and Salama, M. (2014). A novel recombinant baculovirus overexpressing a *Bacillus thuringiensis* Cry1Ab toxin enhances insecticidal activity. *Biological Procedures Online*, 16: 1-7.

Falcon, L. A., Kane, W. R. and Bethell, R. S. (1968). Preliminary evaluation of a granulosis virus for control of the codling moth. *Journal of Economic Entomology*, 61: 1208-1213.

Fang, M., Nie, Y., Harris, S., Erlandson, M. A. and Theilmann, D. A. (2009). Autographa californica multiple nucleopolyhedrovirus core gene ac96 encodes a per os infectivity factor (pif-4). *Journal of Virology*, 83: 12569-12578.

Farrar, R. R., Shapiro, M. and Javaid, I. (2003). Photostabilized titanium dioxide and a fluorescent brightener as adjuvants for a nucleopolyhedrovirus. *BioControl*, 48: 543-560.

Fauquet, C. M., Mayo, M. A., Maniloff, J., Desselberger, U. and Ball, L. A. (2005). *Virus Taxonomy: VIIIth Report of the International Committee on Taxonomy of Viruses. Academic Press*, London, UK.

Figueiredo, E., Muñoz, D., Murillo, R., Mexia, A. and Caballero, P. (2009). Diversity of Iberian nucleopolyhedrovirus wild-type isolates infecting *Helicoverpa armigera* (Lepidoptera: Noctuidae). *Biological Control*, 50: 43-49.

Fitt, G. P. (1989). The ecology of *Heliothis* species in relation to agroecosystems. *Annual Review of Entomology*, 34: 17-53.

Fuxa, J. R. and Arthur, R. R. (1999). Classical biological control in an ephemeral crop habitat with *Anticarsia gemmatalis* nucleopolyhedrovirus. *BioControl*, 44: 405-421.

Gani, M., Gupta, R. K., Kaul, V., Zargar, S. M., Bali, K. and Khan, G. (2016). Management of the tent caterpillar (*Malacosoma indicum*) using nucleopolyhedrovirus in Jammu and Kashmir, India. *Materials Today: Proceedings*, 3: 3914-3924.

Gani, M., Gupta, R. K., Zargar, S. M., Kour, G., Monobrullah, M., Kandasamy, T. and Mohanasundaram, A. (2017). Molecular identification and phylogenetic analyses of multiple nucleopolyhedrovirus isolated from *Lymantria obfuscata* (Lepidoptera: Lymantriidae) in India. *Applied Entomology and Zoology*, 52: 389-399.

Gani, M., Hassan, T., Saini, P., Bhat, K. H., Gupta, R. K. and Bali, K. (2021a). Biosensor as a potential tool for on-site detection of insect pathogens. *Biosensors in Agriculture: Recent Trends and Future Perspectives* (Eds. Pudake, R. N., Jain, U. and Kole, C.). Springer, Switzerland, pp. 465-493.

Gani, M., Sergei, S., Lokanath, S., Saini, P., Bali, K., Gupta, R., Sivaprasad, V., Jehle, J. A. and Wennmann, J. T. (2021b). Patterns in genotype composition of Indian isolates of the *Bombyx mori* nucleopolyhedrovirus and *Bombyx mori* Bidensovirus. *Viruses*, 13: 901. https://doi.org/10.3390/v13050901.

Gelernter, W. D. and Federici, B. A. (1986). Isolation, identification, and determination of virulence of a nuclear polyhedrosis virus from the beet armyworm, *Spodoptera exigua* (Lepidoptera: Noctuidae). *Environmental Entomology*, 15: 240-245.

Gramkow, A. W., Perecmanis, S., Sousa, R. L. B., Noronha, E. F., Felix, C. R., Nagata, T. and Ribeiro, B. M. (2010). Insecticidal activity of two proteases against *Spodoptera frugiperda* larvae infected with recombinant baculoviruses. *Virology Journal*, 7: 1-10.

Groner, A. (1986). Specificity and safety of baculoviruses. In: *The Biology of Baculoviruses* (Eds. Granados, R. R. and Federici, B. A.). CRC Press, Boca Raton, Florida, USA, pp. 177-202.

Gupta, R. K., Gani, M., Jasrotia, P. and Srivastava, K. (2013). Development of the predator *Eocanthecona furcellata* on different proportions of nucleopolyhedrovirus infected *Spodoptera litura* larvae and potential for predator dissemination of virus in the field. *BioControl*, 58: 543-552.

Haase, S., McCarthy, C. B., Ferrelli, M. L., Pidre, M. L., Sciocco-Cap, A. and Romanowski, V. (2015b). Development of a recombination system for the generation of occlusion positive genetically modified *Anticarsia gemmatalis* Multiple Nucleopolyhedrovirus. *Viruses*, 7: 1599-1612.

Haase, S., Sciocco-Cap, A. and Romanowski, V. (2015a). Baculovirus insecticides in Latin America: historical overview, current status and future perspectives. *Viruses*, 7: 2230-2267.

Hamm, John J. and Shapiro, M. (1992). Infectivity of fall armyworm (Lepidoptera: Noctuidae) nuclear polyhedrosis virus enhanced by a fluorescent brightener. *Journal of Economic Entomology*, 85: 2149-2152.

Harries, F. A., Graves, J. B., Nemes, S. J., Vindson, S. D. and Wolfer Barger, D. A. (1971). Distribution abundance and control of *Heliothis* spp. In cotton and other host plants. *Insect Resistance, Southern Operative Series*, 119: 17-27.

Harrison, R. L. and Bonning, B. C. (2000). Use of scorpion neurotoxins to improve the insecticidal activity of *Rachiplusia ou* multicapsid nucleopolyhedrovirus. *Biological Control*, 17: 191-201.

Helsen, H. H. M., Blommers, L. and Vaal, F. (1992). Efficacy and implementation of granulosis virus against codling moth in orchard IPM. Medical Faculty Landbou. *Rijksuniversiteit Faculteit Landbouwwetenschappen, Gent*. 569-573.

Hilton, S. and Winstanley, D. (2007). Identification and functional analysis of the origins of DNA replication in the *Cydia pomonella* granulovirus genome. *Journal of General Virology*, 88: 1496-1504.

Hunter-Fujita, F. R., Entwistle, P. F., Evans, H. F. and Crook, N. E. (1998). *Insect Viruses and Pest Management.* John Wiley and Sons, New Jersey, USA.

Ignoffo, C. M. and Couch, T. L. (1981). Nucleopolyhedrosis virus of *Heliothis* species as a microbial insecticide. *Microbial Control of Pests and Plant Diseases*, 70-98.

Ignoffo, C. M., Parker, F. D., Boening, O. P., Pinnell, R. E. and Hostetter, D. L. (1973). Field stability of the *Heliothis* nucleopolyhedrosis virus on corn silks. *Environmental Entomology*, 2: 302-303.

Inceoglu, A. B., Kamita, S. G., Hinton, A. C., Huang, Q., Severson, T. F., Kang, K. D. and Hammock, B. D. (2001). Recombinant baculoviruses for insect control. *Pest Management Science*, 57: 981-987.

Jaques, R. P. (1975). Persistence, accumulation, and denaturation of nuclear polyhedrosis and granulosis viruses. In: *Baculoviruses for Insect Pest Control: Safety Considerations* (Eds. Summers, M. D., Engler, R., Falcon, L.A., Vail, P. V.). American Society of Microbiology, Washington, DC, USA, pp. 90-99.

Jehle, J. A., Blissard, G. W., Bonning, B. C., Cory, J. S., Herniou, E. A., Rohrmann, G. F., Theilmann, D. A., Thiem, S. M. and Vlak, J. M. (2006). On the classification and nomenclature of baculoviruses: a proposal for revision. *Archives of Virology*, 151: 1257-1266.

Jinn, T. R., Tu, W. C., Lu, C. I. and Tzen, J. T. C. (2006). Enhancing insecticidal efficacy of baculovirus by early expressing an insect neurotoxin, LqhIT2, in infected *Trichoplusia ni* larvae. *Applied Microbiology and Biotechnology*, 72: 1247-1253.

Jones, K. A. (1994). Use of baculoviruses for cotton pest control. In: *Insect Pest of Cotton* (Eds. Matthews, G. A. and Tunstall, J. P.). CAB International, Wallingford, UK, pp. 477-504.

Jones, K. A., Zelazny, B., Ketunuti, U., Cherry, A. and Grzywacz, D. (1998). A World survey of virus control of insect pests: South East Asia and Western Pacific. In: *Insect Viruses and Pest Management: Theory and Practice* (Eds. Hunter-Fujita, F. R., Entwistle, P. F., Evans, H. F. and Crook, N. E.). John Wiley and Sons, Chichester, UK, pp. 244-257.

Kamita, S. G., Koukichi, N., Chua, J. W., Shimada, T., Mita, K., Kobayashi, M., Maeda, S. and Hammock, B. D. (2005). A baculovirus-encoded protein tyrosine phosphatase gene induces enhanced locomotory activity in a lepidopteran host. *Proceedings of the National Academy of Sciences*, 102: 2584-2589.

Kim, J. S., Cho, J. Y., Chang, J. H., Shim, H. J., Roh, J. Y., Jin, B. and Je, Y. H. (2005). Characterization of an improved recombinant baculovirus producing polyhedra that contain *Bacillus thuringiensis* Cry1Ac crystal protein. *Journal of Microbiology and Biotechnology,* 15: 710-715.

Kranthi, K. R., Jadhav, D. R., Kranthi, S., Wanjari, R. R., Ali, S. S. and Russell, D. A. (2002a). Insecticide resistance in five major insect pests of cotton in India. *Crop Protection,* 21: 449-460.

Kranthi, K. R., Russell, D., Wanjari, R., Kherde, M., Munje, S., Lavhe, N. and Armes, N. (2002b). In-season changes in resistance to insecticides in *Helicoverpa armigera* (Lepidoptera: Noctuidae) in India. *Journal of Economic Entomology*, 95: 134-142.

Laing, D. R. and Jaques, R. P. (1980). Codling moth: techniques for rearing larvae and bioassaying granulosis virus. *Journal of Economic Entomology*, 73: 851-853.

Laird, M., Lacey, L. A. and Davidson, E. W. (1990). *Safety of Microbial Insecticides*. CRC Press, Boca Raton, Florida, USA.

Martínez, A. M., Simón, O., Williams, T. and Caballero, P. (2003). Effect of optical brighteners on the insecticidal activity of a nucleopolyhedrovirus in three instars of *Spodoptera frugiperda*. *Entomologia Experimentalis et Applicata*, 109: 139-146.

McCaffery, A. R., King, A. B. S., Walker, A. J. and El-Nayir, H. (1989). Resistance to synthetic pyrethroids in the bollworm, *Heliothis armigera* from Andhra Pradesh, India. *Pesticide Science,* 27: 65-76.

Miller, L. K. (1988). Baculoviruses as gene expression vectors. *Annual Reviews in Microbiology*, 42: 177-199.

Mir, M., Gaurav, S. S., Prasad, C. S. and Tyagi, Ashish (2010). Field efficacy of HaNPY against *Helicoverpa armigera* on Tomato. *Annals of Plant Protection Sciences,* 18: 301-303.

Mondragón, G., Pineda, S., Martínez, A. and Martínez, A. M. (2007). Optical brightener Tinopal C1101 as an ultraviolet protectant for a nucleopolyhedrovirus. *Communications in Agricultural and Applied Biological Sciences*, 72: 543-547.

Monobrullah, M. and Kaul, V. (2005). Prospects of using Baculoviruses in pest management. *Environment and Ecology*, 23: 187-209.

Monobrullah, M. and Nagata, M. (1999). Immunity of lepidopteran insects against baculoviruses. *Journal of Entomological Research*, 23: 185-194.

Monsma, S. A., Oomens, A. G. and Blissard, G. W. (1996). The GP64 envelope fusion protein is an essential baculovirus protein required for cell-to-cell transmission of infection. *Journal of Virology,* 70: 4607-4616.

Moore, S. D., Pittaway, T., Bouwer, G. and Fourie, J. G. (2004). Evaluation of *Helicoverpa armigera* nucleopolyhedrovirus (HearNPV) for control of *Helicoverpa armigera* (Lepidoptera: Noctuidae) on citrus in South Africa. *Biocontrol Science and Technology*, 14: 239-250.

Morales, L., Moscardi, F., Gómez, D. R. S., Paro, F. E. and Soldorio, I. L. (1997). Enhanced activity of *Anticarsia gemmatalis* Hüb.(Lepidoptera: Noctuidae) nuclear polyhedrosis virus by boric acid in the laboratory. *Anais da Sociedade Entomológica do Brasil*, 26: 115-120.

Morales, L., Moscardi, F., Sosa-Gómez, D. R., Paro, F. E. and Soldorio, I. L. (2001). Fluorescent brighteners improve *Anticarsia gemmatalis* (Lepidoptera: Noctuidae) nucleopolyhedrovirus (AgMNPV) activity on AgMNPV-susceptible and resistant strains of the insect. *Biological Control*, 20: 247-253.

Morisset, O., Cabrera, P., Todorova, S., Cormier, D., Chouinard, G. and Lucas, E. (2021). Compatibility of the egg parasitoid Trichogramma minutum (Hymenoptera: Trichogrammatidae) and the granulovirus (CpGV) used in combination against the codling moth *Cydia pomonella* (Lepidoptera: Tortricidae). *Phytoprotection*, 101: 6-13.

Moscardi, F., de Souza, M. L., de Castro, M. E. B., Moscardi, M. L. and Szewczyk, B. (2011). Baculovirus Pesticides: present state and future perspectives. In: *Microbes and Microbial Technology* (Ahmad, I., Ahmad, F., Pichtel, J.). Springer, USA, pp. 415-445.

Moscardi, F. (1999). Assessment of the application of baculoviruses for control of Lepidoptera. *Annual Review of Entomology*, 44: 257-289.

Nakai, M., Goto, C., Kang, W. K., Shikata, M., Luque, T. and Kunimi, Y. (2003). Genome sequence and organization of a nucleopolyhedrovirus isolated from the smaller tea tortrix, *Adoxophyes honmai*. *Virology*, 316: 171-183.

Nandety, R. S., Kuo, Y. W., Nouri, S. and Falk, B. W. (2015). Emerging strategies for RNA interference (RNAi) applications in insects. *Bioengineered*, 6: 8-19.

Nawaz, A., Ali, H., Sufyan, M., Gogi, M. D., Arif, M. J., Ranjha, M. H., Arshid, M., Waseem, M., Mustafa, T., Qasim, M., Rizwan, M., Zaynab, M., Khan, K. A. and Ghramh, H. A. (2020). Comparative bio-efficacy of nuclear polyhedrosis virus (NPV) and Spinosad against American bollwormm, *Helicoverpa armigera* (Hubner). *Revista Brasileira de Entomologia*, 63: 277-282.

Nemoto, H. and Okada, M. (1990). Pest management for strawberries grown in the greenhouse: microbial control of the tobacco cutworm, *Spodoptera litura*, and cultural, chemical and physical control of aphids. *Bulletin SROP,* 13: 149-152.

Oliveira, D. C., Velasco, J., Wolff, J. L. C., Garcia-Maruniak, A., Ribeiro, B. M., de Castro, M. E. B., de Souza, M. L., Moscardi, F., Maruniak, J. E. and Zanotto, P. M. D. A. (2006). Genome of the most widely used viral biopesticide: Anticarsia gemmatalis multiple nucleopolyhedrovirus. *Journal of General Virology*, 87: 3233-3250.

Panizzi, A. R. (2013). History and contemporary perspectives of the integrated pest management of soybean in Brazil. *Neotropical Entomology*, 42: 119-127.

Pasqualini, E., Antropoli, A. and Faccioli, G. (1994). Performance tests of granulosis virus against *Cydia pomonella* L. (Lepidoptera, Oleuthreutidae). *Bulletin OILB SROP*, 17: 113-119.

Payne, C. C. (1986). Insect pathogenic viruses as pest control agents. *Fortschritte der Zoologie,* 32: 183-200.

Pemsel, M., Schwab, S., Scheurer, A., Freitag, D., Schatz, R. and Schlücker, E. (2010). Advanced PGSS process for the encapsulation of the biopesticide *Cydia pomonella* granulovirus. *The Journal of Supercritical Fluids*, 53: 174-178.

Pidre, M. L., Ferrelli, M. L., Haase, S. and Romanowski, V. (2013). Baculovirus display: a novel tool for vaccination. In: *Current Issues in Molecular Virology-Viral Genetics and Biotechnological Applications* (Romanowski, V.). IntechOpen Access, pp. 137-164. https://doi.org/10.5772/55572.

Pijlman, G. P., Pruijssers, A. J. P. and Vlak, J. M. (2003). Identification of pif-2, a third conserved baculovirus gene required for per os infection of insects. *Journal of General Virology*, 84: 2041-2049.

Possee, R. D. and Rohrmann, G. F. (1997). Baculovirus genome organization and evolution. In: *The Baculoviruses* (Ed. Miller, L. K.). Springer, USA, pp. 109-140.

Rabindra, R. J. (1992). Genetic improvement of baculoviruses. In: *Emerging Trends in Biological Control of Phytophagous Insects* (Ed. Ananthakrishnan, T. N). Oxford and FBH Publishing Cooperative Private Ltd., New Delhi, India, pp. 183-186.

Rabindra, R. J., Jayaraj, S. and Balasubramanian, M. (1985). Efficacy of nuclear polyhedrosis virus for the control of *Heliothis armigera* (Hubner) infesting sunflower. *Journal of Entomological Research*, 9: 18-19.

Ramakrishnan, N., Chaudhari, S., Kumar, S., Rao, R. S. N. and Satyanarayana, S. V. V. (1981). Field efficacy of nuclear polyhedrosis virus against the tobacco caterpillar, *Spodoptera litura* (F). *Tobacco Research*, 7: 129-134.

Rohrmann, G. F. (2019). *Baculovirus Molecular Biology*, 4th edition. National Center for Biotechnology Information, USA.

Sarwar, G., Maan, N. A., Ayub, M. A., Shahid, M. R., Malik, M. A. and Farooq, M. (2021). Evaluation of indigenous the nucleopolyhedrovirus (NPV) of *Spodoptera litura* (Fabricius) (Lepidoptera: Noctuidae) in combination with chlorantraniliprole against Spodoptera species. *Egyptian Journal of Biological Pest Control*, 31: 58. https://doi.org/10.1186/s41938-021-00403-8.

Secretariat, I. P. P. C., Gullino, M. L., Albajes, R., Al-Jboory, I., Angelotti, F., Chakraborty, S., Garrett, K. A., Hurley, B. P., Juroszek, P., Makkouk, K., Pan, X. And Stephenson, T. (2021). *Scientific review of the impact of climate change on plant pests: A global challenge to prevent and mitigate plant-pest risks in agriculture, forestry and ecosystems.* FAO, Rome, Italy. http://www.fao.org/3/cb479en/online/cb4769en.html.

Shang, Y., Wang, M., Xiao, G., Wang, X., Hou, D., Pan, K., Liu, S., Li, J., Wang, J., Arif, B. M., Vlak, J. M., Chen, X., Wang, H., Deng, F. and Hu, Z. (2017). Construction and rescue of a functional synthetic baculovirus. *ACS Synthetic Biology*, 6: 1393-1402.

Shao, H. L., Dong, D. J., Dong, H. J., Wang, J. X. and Zhao, X. F. (2008). Construction of the recombinant baculovirus AcMNPV with cathepsin B-like proteinase and its insecticidal activity against *Helicoverpa armigera. Pesticide Biochemistry and Physiology*, 91: 141-146.

Shapiro, M. and Farrar Jr., R. R. (2003). Fluorescent brighteners affect feeding rates of the corn earworm (Lepidoptera: Noctuidae) and act as enhancers and sunlight protectants for its nucleopolyhedrovirus. *Journal of Entomological Science*, 38: 286-299.

Siddique, S. S., Babu, R. and Arif, M. (2010). Efficacy of *Trichogramma brasiliense*, nuclear polyhedrosis virus and endosulfan for the management of *Helicoverpa armigera* on tomato. *Journal of Experimental Zoology*, 13: 177-180.

Slack, J. and Arif, B. M. (2007). The baculoviruses occlusion-derived virus: virion structure and function. *Advances in Virus Research*, 69: 99-165.

Slavicek, J. M. and Popham, H. J. R. (2005). The *Lymantria dispar* nucleopolyhedrovirus enhancins are components of occlusion-derived virus. *Journal of Virology*, 79: 10578-10588.

Sosa-Gomez, D. R., Morgado, F. S., Correa, R. F. T., Silva, L. A., Ardisson-Araujo, D. M. P., Rodrigues, B. M. P., Oliveira, E. E., Aguiar, R. W. S. and Ribeiro, B. M. (2020). Entomopathogenic viruses in the neotropics: current status and recently discovered species. *Neotropical Entomology*, 49: 315-331.

Sosa-Gomez, D. R. (2004). Intraspecific variation and population structure of the velvetbean caterpillar, *Anticarsia gemmatalis* Hubner, 1818 (Insecta: Lepidoptera: Noctuidae). *Genetics and Molecular Biology*, 27: 378-384.

Summers, M. D. and Smith, G. E. (1987). *A Manual of Methods for Baculovirus Vectors and Insect Cell Culture Procedures.* Texas Agricultural Experiment Station Bulletin, USA, Bulletin 1555.

Sun, X., Wu, D., Sun, X., Jin, L., Ma, Y., Bonning, B. C., Peng, H. and Hu, Z. (2009). Impact of *Helicoverpa armigera* nucleopolyhedroviruses expressing a cathepsin L-like protease on target and non target insect species on cotton. *Biological Control*, 49: 77-83.

Tanada, Y. (1964). A granulosis virus of the codling moth, *Carpocapsa pomonella* (Linnaeus) (olethreutidae, Lepidoptera). *Journal of Insect Pathology*, 6: 378-380.

Tang, X. C., Lu, H. R. and Ross, T. M. (2011). Baculovirus-produced influenza virus-like particles in mammalian cells protect mice from lethal influenza challenge. *Viral immunology*, 24: 311-319.

Terenius, O., Papanicolaou, A., Garbutt, J. S., Eleftherianos, I., Huvenne, H., Kanginakudru, S. and Albrechtsen, M. (2011). RNA interference in Lepidoptera: an overview of successful and unsuccessful studies and implications for experimental design. *Journal of Insect Physiology*, 57: 231-245.

Thaler, R., Brandstätter, A., Meraner, A., Chabicovski, M., Parson, W., Zelger, R., Dalla Via, J. and Dallinger, R. (2008). Molecular phylogeny and population structure of the codling moth (Cydia pomonella) in Central Europe: II. AFLP analysis reflects human-aided local adaptation of a global pest species. *Molecular Phylogenetics and Evolution*, 48: 838-849.

Theilmann, D. A., Blissard, G. W., Bonning, B., Jehle, J. A., O'Reilly, D. R., Rohrmann, G. F., Thiem, S. and Vlak, J. M. (2005). Baculoviridae. In: *Virus Taxonomy—Classification and Nomenclature of Viruses. 8th Report of the International Committee on the Taxonomy of Viruses.* Elsevier, Amsterdam, pp. 177 – 185.

Tuan, S. J., Lee, C. C. and Chi, H. (2014). Population and damage projection of *Spodoptera litura* (F.) on peanuts (*Arachis hypogaea* L.) under different conditions using the age-stage, two-sex life table. *Pest Management Science*, 70: 805-813.

Vail, P. V., Jay, D. L., Hunter, D. K. and Staten, R. T. (1972). A nuclear polyhedrosis virus infective to the pink bollworm, *Pectinophora gossypiella*. *Journal of Invertebrate Pathology*, 20: 124-128.

Villamizar, L., Barrera, G., Cotes, A. M. and Martínez, F. (2010). Eudragit S100 microparticles containing *Spodoptera frugiperda* nucleopolyehedrovirus: physico-chemical characterization, photostability and *in vitro* virus release. *Journal of Microencapsulation*, 27: 314-324.

Vincent, C., Andermatt, M. and Valéro, J. (2007). Madex® and VirosoftCP4®, viral biopesticides for codling moth control. In: *Biological Control: a global perspective*

(Eds. Vincent, C., Goettel, M. and Lazarovits, G.). CAB International, Wallingford, UK, pp: 336-343.

Volkman, L. E. (1995). Baculovirus bounty. *Science*, 269: 1834.

Wan, N. F., Ji, X. Y., Zhang, H., Yang, J. H. and Jiang, J. X. (2015). Nucleopolyhedrovirus infection and/or parasitism by *Microplitis pallidipes* Szepligeti affect hemocyte apoptosis of *Spodoptera exigua* (Hübner) larvae. *Journal of Invertebrate Pathology*, 132: 165-170.

Wanxue, Liu, Wan Fanghao, Zhang Fan, Meng Zhaojun, and Wang Fulian. (2000). Evaluation on role of predators in *Helicoverpa armigera* control. *Chinese Journal of Biological Control*, 16: 97-101.

Wennmann, J. T., Jiangbin F., and Jehle, J. A. (2020). Bacsnp: Using single nucleotide polymorphism (SNP) specificities and frequencies to identify genotype composition in baculoviruses. *Viruses*, 12: 625. https://doi.org/10.3390/v12060625.

Westenberg, M., Veenman, F., Roode, E. C., Goldbach, R. W., Vlak, J. M. and Zuidema, D. (2004). Functional analysis of the putative fusion domain of the baculovirus envelope fusion protein F. *Journal of Virology*, 78: 6946-6954.

Williams, T., Virto, C., Murillo, R. and Caballero, P. (2017). Covert infection of insects by baculoviruses. *Frontiers in Microbiology*, 8: 1337.

Wilson, K., Grzywacz, D., Curcic, I., Scoates, F., Harper, K., Rice, A., Paul, N. and Dillon, A. (2020). A novel formulation technology for baculoviruses protects biopesticide from degradation by ultraviolet radiation. *Scientific Reports*, 10: 1-10.

Winstanley, D. and Crook, N. E. (1993). Replication of *Cydia pomonella* granulosis virus in cell cultures. *Journal of General Virology*, 74: 1599-1609.

Wu, Z. W., Fan, J. B., Yu, H., Wang, D. and Zhang, Y. L. (2015). Ultraviolet protection of the *Cydia pomonella* granulovirus using zinc oxide and titanium dioxide. *Biocontrol Science and Technology*, 25: 97-107.

Yan, G., Wang, H., Lv, J., Li, C. and Zhang, B. (2021). Surface modification of nucleopolyhedrovirus with polydopamine to improve its properties. *Pest Management Science*, 78: 456-466.

Yang, Y., Pang, X. and Liang G. (2000). Lifetable of cotton bollworm under different control condition. *The Journal of Applied Ecology*, 11: 856-860.

Yasin, M., Qazi, M. S., Wakil, W. and Qayyum, M. A. (2020). Evaluation of Nuclear Polyhedrosis Virus (NPV) and Emamectin Benzoate against *Spodoptera litura* (F.) (Lepidoptera: Noctuidae). *Egyptian Journal of Biological Pest Control*, 30: 1-6.

Yi, P. and Li., Z. (1989). Microbial pest control in China: viruses and fungi. In: *Symposium on Sino-American Biocontrol Research.* San Antonio, Texas, USA.

Zhang, G. (1989). Commercial viral insecticide-*Heliothis armigera* viral insecticide in China. *The IPM Practitioner*, 11: 13.

Zhou, C., Liu, Y. Q., Weili, Y., Deng, Z., Gao, M., Liu, F. and Mu, W. (2011). Resistance of *Spodoptera exigua* to ten insecticides in Shandong, China. *Phytoparasitica*, 39: 315-324.

Chapter 13

Challenges in Field Application of Biopesticides for Successful Crop Pest Management

Mahfouz M. M. Abd-Elgawad*
Plant Pathology Department,
National Research Centre, Dokki, Giza, Egypt

Abstract

Accelerated concern over chemical pesticides has created wideset interest in the use and potential of biopesticides in sustainable agriculture across a wide range of ecosystems and economically important plant species. However, many economic, biotic and abiotic factors represent challenges for biopesticidal application and have frequently relegated them to niche markets or made them unexploited by growers. To aid with grasping their potential in sustainable agro-ecosystems, this chapter discusses such factors and offers overviews on the biopesticides- biology and ecology, commercial production and formulation, and application technology. It highlights their interactions with each other and with other management tools to optimize their use. For successful broad pest management, biopesticides should further be developed with respect to their efficacy, familiarity, costs, and ease of use. Awareness of farmers, extensions, and stakeholders of various aspects for biopesticidal use should be raised via technology transfer and acquaintance of related issues. The potential of biocontrol agents to recycle in host populations are good notions to be further employed for long term pest suppression.

Keywords: biopesticides, biocontrol agent, formulation, application methods, pest control

---

* Corresponding Author's Email: mahfouzian2000@yahoo.com.

In: Pest Management: Methods, Applications and Challenges
Editor: Tarique Hassan Askary
ISBN: 979-8-88697-268-9

1. Introduction

It goes without saying that huge losses are annually caused by various pests on many economically important crops. Due to the enormity of these losses, some analyzers and stakeholders prefer to divide them into different orders or levels. These latter may comprise the country/community, consumer, and farm producer. The primary level is the producer who endures the lack of revenue or the elevated production cost due to the quality and/or quantity decrease incited by pests in crop yields. Otherwise, the producer is faced by costs of pesticides and their application. In addition, it is stressed that the factor of uncertainty that the grower must challenge when costs of pesticidal applications cannot afford the expected value or returns, these biopesticides are not profitable (Abd-Elgawad and Askary, 2015). Another level of losses is incurred that can reach to the consumer, household and even industry. It stems from the elevated price or the degradation in quality of the consumable product or both together. Unfortunately, health hazards might sometimes be implied on applying banned pesticides or consuming produce treated with them (Abd-Elgawad, 2019a). Moreover, the widest level is the societal loss in terms of wasted resources that have both direct and indirect adverse effects on the community. As these costs are usually shown on broad levels, the targeted countries may tackle them in various approaches, e.g., providing relevant assistance to the growers, paying subsidies to the farmers and/or consumers. The country resources so diverted could have been used in other required objectives to optimize the use of a society's economic supplies and resources. Such losses could be significant in economically weak countries especially after corona pandemic. Although crop losses due to pests are a main challenge worldwide, the related losses may spoil the efforts to enhance production and efficiency in developing and agriculturally backward countries. Nevertheless, the losses sustained by the world community concerning these countries that are struggling against population growth, food and raw material shortages, and slow development are not restricted to food crops. They also have an impact on the cash crops that form a main source of revenue to the stakeholders and the industry.

As pest losses can be categorized into different levels, there are also challenges related to the order/rank to which the pests causing these losses belong. Insect pests of the phylum Arthropoda may comprise many groups and orders such as aphids, caterpillars, cutworms, grasshoppers and locusts, thrips, weevils, and whiteflies. Arthropod pests may also have various groups of mites. Also, plant-parasitic nematodes (PPNs) are graded as animal pests.

Clearly, the biopesticides intended herein are generally those that can be used against pests, weeds, and plant diseases. These latter are frequently caused by various pathogens, e.g., bacteria, fungi, nematodes, and viruses. Moreover, a serious type of loss that negatively impacts all the above-mentioned levels is that which possesses a persistent character. It usually leaves residual effects in cultivated soil and/or plant residues that will affect farm quality and health for prolonged seasons or years. This often occurs with uncontrolled PPN-infested fields so severely that the growers are enforced to re-allocate their resources. Therefore, when a soil-dwelling pest/pathogen is a prominent issue, selecting profitable management methods are necessary to maintain the pest/pathogen population levels below the damaging or the economic threshold level.

Contrary to chemical pesticides, biopesticides are intended to be ecofriendly alternative materials for pest/pathogen management resulting in clean crop production and protection of the environment. They are planned to be a way forward in sustainable agriculture for safe disease control and crop protection. Thus, various biopesticides are either biological or biologically-derived agents. Biopesticides are defined by the European Union (EU) as a form of pesticide based on micro-organisms or natural products (European Commission, 2008) and realized by the US Environmental Protection Agency (EPA) to include naturally occurring substances that control pests (biochemical pesticides), microorganisms that control pests (microbial pesticides), and pesticidal substances obtained from plants containing joined genetic material as incorporated protectants (EPA, 2012). Originally, biocontrol agents possess many groups such as predators, parasitoids, parasites, and pathogens. Many of their bioactive compounds may also be used as biopesticides. The primary objective of biopesticides is to combat pests and protect crops almost as chemical pesticides do, but they are supposed also to achieve these goals within the frame of sustainable agricultural systems and in an environmentally friendly approach. Therefore, this chapter sheds light on the current directions of research relevant to field application of various biopesticides indicating their advantages and disadvantages of different biopesticidal groups in general. It clarifies the current status of these pesticides, mentioning the challenges they face and the opportunities that can be exploited to meet these challenges. To avoid prolongation and repetition, suffice it to point herein to major examples of such challenges vividly expressing their counterpart and how to successfully face them relying on mentioning all the details in biopesticidal applications on various economically important crops in the other chapters of this book. Hence, definite avenues that need to be further improved with respect to different biopesticidal issues such as efficacy, reduced costs, and ease of use are discussed herein.

2. Resources and Directions Relevant to Effective Biopesticidal Applications

Scientists have been doing much research worldwide resulting in interesting findings to enhance farmers' income and quality of life via replacing unhealthy chemical pesticides by safe biopesticidal alternatives. Therefore, it is important to summarize dependable resources and new directions of research focusing on the data produced to document their useful applications and direct attention towards their vast untapped potentials. These latter should also be targeted to offer lasting, ecologically sound, and socially favourable control measures against most key pests on economic crops, and invasive organisms degrading global biodiversity. Admittedly, the use of natural enemies against pests is considered a basic source, but the practical uses of biopesticides go beyond that to apply other important sources. These include the activation of certain natural anti-pest products such as bacterial, fungal, or plant extracts or materials as well as the development of other agents. These may also comprise pests-trapping plants, antagonistic or incompatible plants, and developing plant varieties/cultivars resistant to pests through different genetic methods. Admittedly, current microbial control agents include many commercial products that may be based on:

- Entomopathogenic viruses (e.g., *cydia pomonella* granulovirus),
- Entomopathogenic fungi (e.g., *metarhizium* spp. And *beauveria bassiana*),
- Beneficial nematodes (e.g., predacious, entomopathogenic, and slug parasitic nematodes), and
- Plant disease control agents (e.g., *trichoderma* spp. And *bacillus* spp.). Other pests as weeds and rodents have also been controlled with microbial agents. Also, natural compounds used as biopesticides may comprise numerous products based on fungal and plant extracts, pheromones and other semiochemicals, fermentation products such as spinosad, or chitosan (Matthews et al., 2014).

There are many directions for biopesticidal applications, or on the way to be broadly used. A common direction is based on using a biopesticide, e.g., biological control agent (BCA), that can properly manage one type of harmful pests, on one hand, and also directly enhance crop yield, on the other (Abd-Elgawad 2020a). Another direction can control more than one or a set of pests. Other orientations comprise products which contain BCA(s) that are also used not only as biopesticides but also as plant growth promoters, plant strengtheners, or soil conditioners in order to enhance plants' ability to tolerate or resist the pest attack. These products are practically used by many market segments with different standpoints regarding their application as biopesticides too. Clearly, numerous products contain BCAs such as fungi or bacteria but are also commercialized as non-biopesticides, examples are Agrinos' proprietary High Yield Technology™ products, and many products containing different species of *Trichoderma*. However, such products can support plants' defence against pest attack (Wilson and Jackson, 2013). Furthermore, there is a conflict of opinions in differentiating between some chemical pesticides and biopesticides especially concerning pesticides which are produced by living BCAs. In this vein, Syngenta market the pesticide abamectin that is generated during fermentation of the actinomycete *Streptomyces avermitilis*. As the active ingredient is thoroughly purified prior to sale, the product needs to be registered as a pesticide, and materials are considered as chemical nematicides rather than biopesticides (Wilson and Jackson, 2013). On the contrary, abamectin is commercialized as biopesticide elsewhere (Anonymous, 2018). Meanwhile, other candidate biopesticides are still under experimentation worldwide as it is necessary to undergo laboratory, semi-field, and field experiments to confirm the multiple hypotheses of testing their validity.

Admittedly, it is impossible to recognize or address all available biopesticides given the many unregistered ones in many developing countries of Asia, Africa, and Latin America. A lot of small industries using cheap labour are there to manufacture microbial products including biopesticides at low cost primarily for domestic markets, although they might be exported as well. A great deal of these biopesticides may be efficacious, but they have not often undergone the firm and costly registration procedures always required by developed countries in North America and Europe. On the other hand, biopesticides are usually not examined for their possible function in plant growth, photosynthesis, or other basic aspects of plant physiology. Therefore, researchers are recently trying to explore their impacts on vital aspects such as plant nutrition which proved to have significant roles in the health and productivity of the biopesticide-treated plant (Haryuni et al., 2020). Nevertheless, most experimentations and applications are utilizing biopesticides as active materials against biological pests. Moreover, many scientists who call for expanded success in biopesticidal applications are urging to fit them into existing or emerging integrated pest management (IPM) strategies and organic farming systems (Dara, 2019; Koppenhöfer et al., 2020a; Abd-Elgawad, 2021a) especially via novel and effective approaches. They may consider other materials like various antifeedants (chemical compounds) that are generated by plants to guard them against pests as biopesticides (Valcárcel et al., 2021). These compounds and similar others (Pal and Kumar, 2013; Cheng et al., 2017; Forghani and Hajihassani, 2020) have several related merits because they are renewable alternatives and biodegradable. Moreover, some of them, e.g., 5-Aminolevulinic acid, were reported as potential herbicide, insecticide, nematicide, or plant-growth promoting agent (Cheng et al., 2017). Hence, they are suggested for economic and practical use against pests in organic farming.

Eventually, growers and stakeholders should check local guidance and directions as well as legal requirements before applying any biopesticidal product or related specific technique. Labels on the biopesticidal products can always offer guidelines and devices to help farmers and growers to prepare, apply, and store these biopesticides successfully. Scientists' ambition calls for optimism in expanding the use of effective and safe biopesticides by increasing growers' awareness based on browsing the relevant websites as well. The stakeholders should increasingly be able to consult the internet for relevant information and challenging techniques. On the other hand, websites with relevant information and simplified techniques on biopesticides should be available via international, governmental and commercial organizations as well as universities.

3. Biopesticidal Categorization

Biopesticides contain many groups but can be categorized into the following main divisions:

- Microbial pesticides which contain various organisms such as bacteria, fungi, nematodes, viruses, and protozoa. Linked subgroup of microbial pesticides may comprise the metabolites or bioactive compounds that fungi or bacteria generate,

- Bio-derived chemicals that are naturally occurring substances. Currently, they contain main commercial sets as neem oil, pyrethrum, essential oils, and rotenone. Many of these bio-derived chemicals are utilized as biopesticides to directly kill pests and/or phyto-pathogens. Some of them may be used as deterrents, repellants or through toxicity post-ingestion as well as used as pheromones to monitor or attract pests too (Isman, 2006; Tembo et al., 2018; Abd-Elgawad, 2021b),
- Genetically modified (GM) crops or called plant-incorporated protectants where plants possess genetic material incorporated from other species into their genetic material. Their use is debatable as some European countries are conservative to use GM crops (Liu et al., 1999; Matthews et al., 2014), and
- RNA interference (RNAi) pesticides, where RNA molecules can participate in sequence-specific suppression of gene, expressed by double-stranded RNA. Its action is done via translation or transcriptional repression and may also be called as post-transcriptional gene silencing, co-suppression, or quelling (Fire et al., 1998; Daneholt, 2006).

Nowadays, RNAi and its regulatory potentials are being further explored as RNAi has tremendous potential in suppressing the targeted genes. It is validated as accurate, effective, stabilized, and better than antisense therapy for gene suppression (Saurabh et al., 2014). Antisense RNA, generated intracellularly through an expression vector could be promoted and included as new types of therapeutic agents (Weiss et al., 1999). Some RNAi pesticides are utilized as topical, whereas others are absorbed by the crop. They are being further studied for possible application as a spray-on insecticide by several companies. Their usage as sprays does not change the genome of the target plant though the RNA could be adjusted to maintain its impact as the target species develop tolerance. As RNA is a relatively fragile molecule, it usually degrades within a few days of application (Geldhof et al., 2007). It has been applied against weeds that tolerate the Roundup® (herbicide, grass and weed killer product). Being mixed with a silicone surfactant, the RNA molecules could enter air-exchange openings and stomata of the plant's surface to interrupt the gene for weed tolerance. Thus, the molecules affected the gene long enough to let Roundup work. This tactic could allow the continued usage of glyphosate-based pesticides but would not per se assist herbicide rotation strategy that relied on alternating this herbicide with others (Regalado, 2015). Its mechanism is capable to manage plant pests and pathogens too; e.g., the root-knot nematodes (Fletcher et al., 2020). The EPA approved the first plant-incorporated protectant (PIP), named SmartStax Pro which is supposed to help US growers to control Western corn rootworm (*Diabrotica virgifera virgifera*) and Northern corn rootworm (*Diabrotica barberi*), destructive corn pests with resistance to some insecticides (Kumar et al., 2021).

4. Challenges in Applying Biopesticides

The current momentum in the advancement of the biopesticide research and industry is due to the problems resulting from livestock and human diseases, and environmental pollution caused by the excessive use of chemical pesticides. In addition, some pests were able to develop resistance against some chemical pesticides, which sometimes led to outbreaks in their population levels and the exacerbation of damages and losses in their host crops. Furthermore, development of pest populations/biotypes that break host plant resistance has frequently been reported (Dolinski et al., 2012; Matthews et al., 2014; Abd-Elgawad and Askary, 2018). These problems and technical advances in biopesticide production and practical use will offer further chances to integrate the utilization of biopesticides in pest management programs. This is especially timely since biological pesticides are viewed as safe alternatives that can achieve their main goal in pest control/management without bringing about these issues and shortcomings related to chemical pesticides.

There are many approaches to addressing these challenges and encouraging effective application of biopesticides. Dolinski et al., (2012) portrayed selected agricultural industries in, Korea, the United States, and Brazil where entomopathogenic nematode (EPN) use has been applied and recommended for pest management. In each case, they focused on the basis on which recommendations were made and the magnitude of adopting these biopesticides by growers and stakeholders. Interestingly, they offered their perceived reasons for the current challenges to upgrade the status of commercial EPN products. Dolinski et al., (2012) stressed that research should focus on developing cultural practices that exploit the potential for enhanced pest management using EPN conservation methods.

A few types of biopesticides may have additional merits (Table 1). These bionematicides are less expensive than chemical nematicides for controlling PPNs that attack tomato (Abd-Elgawad, 2020b), potato (Abd-Elgawad, 2020c), pepper (Abd-Elgawad, 2020d), and eggplant (Abd-Elgawad, 2021c). Biocontrol agents have the capacity not only to kill the targeted pest but they can also grow and recycle on the nutrients found in the pest. Thus, their ability to recycle in host populations bodes well to be further employed for long term pest suppression. Meanwhile, growers can gain healthy and productive crop without detection of harmful biopesticidal residues, biopesticides are usually rapidly biodegradable. Many biopesticides could also demonstrate capacities to induce resistance of plant hosts via physiological processes which can make these crops more tolerant to a variety of biotic and abiotic stresses (Belgacem et al., 2021).

Table 1. Key commercially available bionematicides and chemical nematicides, their applications rates, and prices in Egypt*

Active ingredient	Product name	Application rate (product/Feddan^{-1})$^{+}$	Price per Feddan
Abamectin produced during the fermentation process of *Streptomyces avermitilis* (soluble concentrate at 20 g/l)	Tervigo 2% SC	2.5 L / Feddan	Egyptian pound (L.E.) 2000 ($ 134)
10^9 CFU/ml of *Serratia* sp., *Pseudomonas* sp., *Azotobacter* sp., *Bacillus circulans* and *B. thuringiensis*	Micronema	30 L / Feddan (thrice) / year	L.E. 600 ($ 40)
10^8 units/ml *Purpureocillium lilacinum*	Bio-Nematon	2 L/ Feddan/ year	L.E. 500 ($ 33)
10^9 bacterium cells of *Serratia marcescens* /ml water	Nemaless	10 L/ Feddan (thrice)/ year	L.E. 600 ($ 40)
Cadusafos (O-ethyl S,S-bis (1-methylpropyl) phosphorodithioate)	Rugby 10 G	24 Kg/ Feddan	L.E. 6480 ($ 432)
Oxamyl (methyl 2-(dimethylamino)-*N*-(methylcarbamoyloxy)-2 oxoethanimidothioate)	Vydate 24% SL	4 L/ Feddan (twice) / year	L.E. 2800 ($ 187)

*Figures are given for comparative purposes, Feddan = 4200 m^2 (Adapted from Abd-Elgawad, 2020b).

On the other hand, biopesticides are frequently slower acting and generally less effective than chemical pesticides (Abd-Elgawad, 2020a). In order to overcome these general shortcomings, stakeholders and researchers should adequately understand the complex network of interactions among biotic and abiotic factors in intimate contact with these biopesticides to optimize their efficacy using safe and advanced technology for refined applications. Details of these defects and approaches to face them can be clarified as follows:

4.1. High Specificity

As a biopesticide is usually specific to a single group or species of pests, it does not affect directly natural enemies, e.g., parasitoids and predators and is less likely to harm non-target and beneficial species. However, the unfavorable impact of the pest specificity is the limited markets. Consequently, the evolution, registration and mass-production costs cannot be expanded over a broad range of pest control sales, e.g., insect viruses are not broadly available (Koul, 2011). The case herein requires best biopesticide-host matching for superior efficacy too. Nonetheless, the use of multiple biopesticide products can avert both the specificity of the action and the demerit of low efficacy in case of missing such matching. In this regard, simultaneous manipulation of closely related or compatible species of BCAs could challenge this issue and offer optimum biocontrol efficacy, domination in the soil microbiome, and/or better root protection. Combining *Bacillus subtilis* with *B. pumilus* gave more average increment in growth parameters of *Meloidogyne incognita*-infected cowpea plants than either alone (El-Nagdi et al., 2019). Hence, it is probable that different *Bacillus* spp. may prove unlike anti-root-knot nematode (RKN, *Meloidogyne* spp.) mechanisms that open novel avenues to enhance the biopesticidal efficacy. Another approach is the sequential application of pesticides. Dahlin et al., (2019) reported that fluopyram (a nematicide) reduced *M. incognita* population level on tomato at planting. Then, adding *Purpureocillium lilacinum* strain PL251 during the growing season could reinforce the reduction. The reduction in number of *M. incognita*-second stage juveniles (J_2) was 56 and 68% when PL251 was applied alone and after fluopyram, respectively. In another study, fewer *M. incognita* galls were found as *Bacillus firmus* preceded synthetic nematicides in consecutive tomato cycles too. The galling severity assessed by the galling index (scale of 0-10) was reduced from 7.7 in the control roots to 5.9, 4.3, and 4.0 for *B. firmus*, *B. firmus* + oxamyl, and *B. firmus* + fosthiazate, respectively (d'Errico et al., 2019). These favourable results should be extended and documented as BCAs are compatible with other agricultural inputs (Table 2). In this concern, available fungal and bacterial nematicide with their comprehensive references in terms of their targeted nematode species and crop, types of formulations, product name, active ingredient, producer, and country of origin were presented (Abd-Elgawad and Askary, 2018).

Table 2. Fungal and bacterial biocontrol agents used with other components in integrated nematode management

Fungus/ Bacterium	Integrated with	Nematode Managed	Crop	Results
Pasteuria penetrans	Carbofuran	*Meloidogyne javanica*	Tomato	Combined application of *P. penetrans* and Carbofuran reduced root galling by 50%.
P. penetrans	Neem cake	*M. incognita*	Tomato	*P. penetrans* in combination with neem cake was found most effective in parasitizing the nematode juveniles and adults as compared to individual treatment.
P. penetrans	Neem cake	*M. incognita*	Babchi	*Nematode infection was least when P. penetrans and neem extract was applied in combination. Minimum number of juveniles per root system was observed.*
P. penetrans	Neem cake	*M. javanica*	Tomato	Combined application of *P. penetrans* and neem (*Azadirachta indica*) cake reduced nematode population by 75%.
P. penetrans	Castor cake	*M. incognita*	Chilli	Combination of castor and *P. penetrans* caused greater reduction in galling index (84.75%) and soil population (85.74%) of *M. incognita* as compared to control (*M. incognita* alone).
P. penetrans	Nematicides	*M. javanica*	Tomato	Combined effect of *P. penetrans* with nematicides (carbofuran, aldicarb, miral, scbufos, and phorate) reduced the root galling.
P. penetrans	Carbofuran	*M. incognita*	Tomato	Combined application of *P. penetrans* and carbafuran reduced gall formation on tomato roots by 63.02%.
P. penetrans	Carbofuran	*Heterodera cajani*	Pigeon pea	The penetration to host root was minimum when *P. penetrans* was applied with carbofuran. Number of healthy cysts, eggs/cyst and total nematode population were significantly reduced in this treatment.
P. penetrans	Neem cake	*M. incognita*	Tomato	Increase in plant growth parameters and parasitism of *M. incognita* female was found when *P. penetrans* was applied in combination with neem cake.
P. penetrans	Neem cake	*M. javanica*	Tomato	Combined application of *P. penetrans* and neem cake reduced root galling by 32% as compared to nematode alone treatment. The spore-encumbered juveniles were more susceptible to the effects of the neem.

Table 2. (Continued)

Fungus/ Bacterium	Integrated with	Nematode Managed	Crop	Results
Pseudomonas fluorescens	Organic manure	*M. incognita*	Tomato	*P. fluorescens* GRP3 with organic manure was the best combination for the management of *M. incognita* on tomato.
P. fluorescens	Carbofuran	*M. graminicola*	Rice	Combined application of *P. fluorescens* and Carbofuran 3G increased plant height, root length, grain yield and decreased nematode population by 79.34%.
Trichoderma harzianum	Neem cake	*M. incognita*	Tomato	A significant increase in plant growth and decrease in root galling and final population of *M. incognita* were observed in tomato seedlings transplanted in neem cake amended soil incorporated with *T. harzianum.*
T. viride	Carbofuran	*M. graminicola*	Rice	Combined application of *T. viride* and Carbofuran increased plant height, root length, grain yield and decreased nematode population by 69.17%.

* Partially adapted from Abd-Elgawad and Askary (2018).

4.2. Slow Acting

Initially, it is recommended not to use such slow acting biopesticides against pests infesting crops with a low economic threshold level. Such crop-pest systems may make biopesticides unsuitable especially if a pest outbreak is an immediate threat to a crop. Yet, several approaches may be followed to circumvent and overcome serious pests. An apparent method is to introduce the antagonists in combination with conservation practices or other agricultural inputs to establish and boost the efficacy of the introduced organisms. Applying BCAs as fungal antagonists with organic amendments has resulted in accelerated rates of PPN suppression than applications of amendments only. Also, rotations that support root-knot nematode reproduction should maintain higher population levels of an introduced *Pasteuria penetrans*, as BCA, than rotations which restrict RKN reproduction (Timper, 2014). Also, the introduction of timely acting biopesticides is critical in crop cultivation for effective pest control and high yielding. Jaffuel et al., (2020) reported negligible pest control when rootworm attacking maize started a week before the EPN application. On the contrary, they reported that well-timed application of alginate capsules (beads that encapsulate the EPNs) caused considerable reduction in the banded cumber beetle *Diabrotica balteata* Le Conte-inflicted root damage comparable ($P < 0.05$) to that of EPNs in water suspension. Likewise, the mortality of bait insects in field plots with EPN-infective juvenile (IJ) inoculation soon after peanut seeding was mostly higher than that in plots with later IJ inoculation. This is likely because of the non-contact of IJs with the chemicals for 7 days (Shehata et al., 2021). Moreover, they concluded that relative increase of peanut germination rate in plots early treated with favorable *Steinernema glaseri* implies EPN effectiveness against soil insects that feed on cultivated or germinated peanut seeds. This implication is corroborated by the fact that *S. glaseri* is quite effective against the black cutworm *Agrotis ipsilon* (Radhakrishnan et al., 2017), which was detected in many of the examined samples. Hence, less peanut germination rate was found in plots that were lately treated with *S. glaseri* (Shehata et al., 2021).

4.3. Inconsistent (Variable) Efficacy

This is probably due to the impacts of various biotic and abiotic factors especially on the BCAs and to a less degree their bioactive compounds. As soil fauna and flora may contain a variety of myriad factors representing an unforeseeable environment, the intended effect is occasionally difficult to attain. Abd-Elgawad (2021a) stressed that biological control of PPNs using an introduced BCA may not be as effective nematicide in various settings as that of indigenous BCA due to ecological validity comprising the edaphic and biotic factors. Shapiro-Ilan et al., (2012a) reported that cultivating a clover cover crop boosted the persistence and efficacy of an indigenous population of the entomopathogen *Beauveria bassiana* relative to bare soil, but not an introduced strain of the same fungus. They assumed that the inconsistency between the indigenous and introduced fungus was due to differences in the physiology and/or biology of the *B. bassiana* strains. For bioinsecticides, soil moisture and texture (Duncan et al., 2013), salinity (Nielsen et al., 2011), mulching (Hussaini, 2017), and pH (Campos-Herrera et al., 2013) were found to modulate EPN populations directly or indirectly by influencing their arthropod hosts or enemies (Campos-Herrera et al., 2019).

Thus, various aspects to improve biopesticidal efficacy may include production technology, BCA strain enhancement, formulation and application technology, and improving the environment for boosted biological control (Shapiro-Ilan et al., 2020). Duncan et al., (2013) modified the soil environment by amending texture, citrus trees were cultivated in islands of soil that were more conducive to EPN efficacy than endemic soils. Other methods that reduce exposure to detrimental biotic or abiotic stressors or enhance BCA persistence, multiplication, or virulence will expand or stabilize biocontrol efficacy.

Admittedly, there are a lot of unmeasured, hidden, biotic and physical variables that can differently affect whatever biopesticide, especially BCA, are softly released in an environment. While it is aimed at managing a definite pest species, there is quite different possibility that this BCA will switch to another process or target; even a different but beneficial species. A striking instance would be introducing predatory nematodes (PNs), as BCAs, of course, to control PPN pests into soil having EPNs too. Non-selective PNs may prey on the beneficial EPNs. Another possible risk is to disrupt the existing food chain by introducing a new species or BCA to an ecosystem (Abd-Elgawad 2020a). Therefore, it is likely that the advanced application of biopesticides for successful crop pest management should not only consider the aforementioned factors but also set management plans on a case-by-case basis for attaining the best matching for each intended pest and its settings. Clearly, information is still meager regarding the biotic and abiotic factors impacting the efficacy of many biopesticides. Consequently, some efficacy results may sound unclear, or even controversial (Abd-Elgawad 2017a).

As mentioned above, introducing the biopesticide in combination with conservation agricultural practices to establish and promote the biopesticide may also lead to consistent or appropriate efficacy. In several programs, applying fungal antagonists with organic amendments has led to greater rates of parasitism or PPN suppression than applications of amendments only (Timper, 2014). Rotating crop species that boost proliferation in the plant rhizosphere of an applied fungal parasite as BCA should also enhance biopesticide efficacy (Atkins et al., 2003). In another trend, splitting the used biopesticide in two at half rates as well as its combinations with compatible chemical insecticide could improve suppression of insect pests. Single applications of EPNs have been too low and variable for control of the annual bluegrass weevil, *Listronotus maculicollis* Kirby, larvae (a major golf course turf pest in Eastern North America with widespread insecticide resistance). However, splitting EPN applications in two at half rates approximately one week apart as well as combinations of the nematodes with the neonicotinoid imidacloprid upgraded EPN performance against *L. maculicollis* (Koppenhöfer et al., 2020a).

4.4. Evolving Pest-Resistance

Many pest populations can evolve and elevate their resistance or tolerance to biological, physical, chemical, or any other method for pest control. If the target population is not erased or at least rendered incapable of multiplication, the remaining living population can acquire a resistance/tolerance of whatever pressures are brought to sustain, leading to resistance-breaking pathotypes or the so called an evolutionary arms race. The genetic change in populations of these pathotypes gradually accumulates and explains a constant adaptation to have lower fitness costs to the introduced pressure. Consequently, they can avoid extinction and become resistant (Van Valen, 1973). A prominent example is the recently developed resistance against *Bacillus thuringiensis* (*Bt*) toxin in genetically modified plants which occupies 90 per cent of the bioinsecticide market (Abd-Elgawad 2017a). Numerous crop pests have been evolving resistance against *Bt* toxins.

These comprise the following economically significant insect pests:

- *Heliothis virescens* (tobacco budworm)
- *Ostrinia nubilalis* (European corn borer)
- *Culex quinquefasciatus* (mosquito)
- *Pectonophora gossypiella* (pink bollworm moth)
- *Aedes aegypti* (yellow fever mosquito)
- *Trichloro plusiani* (tiger moth)
- *Leptinotarsa decemlineata* (Colorado potato beetle)
- *Spodoptera exigua* (beet armyworm)
- *Spodoptera littoralis* (Egyptian cotton leaf worm), and
- *Chryosomela scripta* (cottonwood leaf beetle) (Liu et al., 1999; Maithri et al., 2013).

On the other hand, the insecticidal toxin protein, i.e., the reading frame A of the toxin complex a (tca) called TcaA, produced by the *Photorhabdus* bacteria (mutualistic of heterorhabditid nematodes) showed toxicity against a wide range of agricultural pests, and the phylogenetic tree constructed for TcaA revealed that this toxin did not have any ancestral relationship with *Bt* toxins (Maithri et al., 2013). Furthermore, other Tc toxins from *Photorhabdus* bacteria could be cloned into *Arabidopsis* as a model plant in order to offer protection from arthropod pests. Thus, these toxins are proposed as a potential substitutional to Bt toxin (Clarke, 2008). If so, the prospective transgenic plants will be checked by legislation of health safety organizations to confirm that such plants with genes of *Photorhabdus* bacteria may be generated without health or environmental hazards (Piedra-Buena et al., 2015). Moreover, the growing dissatisfaction with *Bt* as a biopesticide, due to related pest resistance, has encouraged scientists to investigate more safe and efficient alternatives. They (e.g., Jallouli et al., 2013; Gerdes et al., 2015) could define such alternative bioinsecticides as *P. luminescens* or *P. temperata* bacteria, either alone or with their mutualistic heterorhabditid nematodes. Moreover, such plant-incorporated protectants (PIPs) require continuous research to investigate the ongoing environmental fate of these related molecules especially the RNAi-based PIPs (Kumar et al., 2021).

5. Other Challenges in Field Application of Biopesticides

Although the above-mentioned demerits of bioinsecticides are regarded as basics and common wariness in their applications, other issues are less recognized among masses applying these bioinsecticides. These particular disadvantages may pertain to specific group(s) or segments in biopesticide markets. Facing challenges of definite cases are discussed herein:

5.1. High Costs

Several bionematicides seem to be cheaper than their corresponding chemical nematicides (Table 1). On the contrary, EPNs used as bioinsecticides are often more expensive than their corresponding chemical insecticides (Georgis, 2002; Abd-Elgawad, 2017b, 2019b; Shapiro-Ilan et al., 2020). In order to face this challenge, cost-effective scenario should be sought. Consequently, the EPN producer could also act as the distributor. Furthermore, the producer would better nominate some of its employees with experience in EPN application. Those applicators are quite fit to go everywhere for successful crop pest management especially by using BCAs. Surely, those experts can recognize any confronting EPN issues, e.g., the viability of IJs, contamination, relevant biotic/abiotic factors in the field, and nematode fate/persistence. In such a manner, i.e., being responsible for producing, distributing, applying and following up EPNs all by the same company will compensate for EPN formulation and production costs to realize profit margins. This multiple service could not only be less costly but also more attractive to the growers and stakeholders. Furthermore, it may establish 'feedback loop' that properly joins production, formulation and application techniques with a program of field trials to upgrade the biocontrol processes.

Moreover, in order to perfectly meet this challenge, awareness-raising of more farmers, and agricultural extensions of the nematodes and other biopesticides for effective crop pest management should be earnestly attempted. Several approaches should be followed to achieve this goal. On-farm trials and workshops should be held to teach growers and stakeholders how biopesticides-related practices, equipment, and products can adequately work in their cropping systems. Relevant biopesticide companies and governmental sectors should offer broad and deep training for perfection of their standardization and usage. Notably, a suitable EPN for the target pest should properly interact with favorable economics. Other potential factors to be included in training are adequate use of EPNs; e.g., compatibility with typical application technology, consideration of existing environmental settings, shelf-life duration, performance of the EPNs relative to other ready-made control methods, and the reliable supply of an EPN product with high quality. Admittedly, if advanced techniques for synergistically or additively incorporate these EPNs with chemicals or other non-chemical control materialize and are broadly disseminated, it may reduce costs and open up further markets and products that currently remain underutilized. During these extension and counseling sessions, end-users should be enlightened of not only the direct but also the indirect benefits gained by using EPNs against insect pests especially their superior efficacy in protected environments or soil habitats. The indirect gains are materialized in the safety of humans and the environment seems imperative to widen the application of these EPNs or other costly biopesticides market as they recompense for the relatively high price. Taking into account the chemical insecticide damage/pollution to the environment generated from soil, plants, animals and water, it is apparent that such adverse effects and their impact on health and social services exceed many times the original price of a biopesticide.

Eventually, the economic feasibility of biopesticides in a given market should be considered within the context of their cost compared to other control options, the value of the commodity that affects the potential profit margin, and the growers' perceived need to control the pest (Koppenhöfer et al., 2020b). To overcome or alleviate the cost issue, finding new BCA isolate/strain or genetically engineered nematodes for specific desirable attributes, e.g., persistence and resistance to environmental extremes such as high temperature and desiccation should be tried. Also, utilizing economized application strategies should be followed. For example, for EPNs, spraying them round the base of the intended plants, putting them in 'tea bags' as a slow-release technique, or dipping root systems into EPN suspension should be followed. The consequences would mostly mean fewer EPNs applied. These approaches can offer a good level of crop protection against pests but at a lower cost, an explicit solution is to utilize less by applying more efficiently. Clearly, efforts to optimize application methods and upgrade 'biorational' control techniques and agents should continue to occupy priority areas for both scientists and industry alike.

Before field application of biopesticides for successful crop pest management, it is substantial to accurately characterize all related factors especially the previously-known reaction of the targeted pest species, growth stage of the plant, region and environmental factors, time of the problem, solutions available, and acceptable costs. Advised steps to get a good microbial pest control product for promising application strategy would be:

(i) Isolation and characterization of perfect BCA species/strain,
(ii) Laboratory bioassays for efficacy,
(iii) Assessment of mass-culture efficiency,
(iv) Mode-of-action and toxicological properties,
(v) Glasshouse tests, and
(vi) Evaluation of efficacy under field or commercial conditions (Koul, 2011).

To aid with the grasping of the potential of biopesticides practical use in successful crop pest management, remarkable examples for related cost-effective processes concerning isolation, extraction, mass production, formulation, and application technology of biocontrol agents are provided hereafter. Addressing related factors and their likes in different types of other biopesticides can lead to overcoming obstacles and challenges facing their successful application for crop pest management.

5.2. EPN Isolation and Extraction

Although the prevalence of EPNs in soil samples was generally found to range from 0.7-50 per cent, recent sampling techniques with repeated extraction process exceeded this range (Abd-Elgawad, 2020e). This rational sampling could both detect more EPN isolates and allow the application of different indices of dispersion to study their spatial distribution pattern. The technique focused on using both stratified random and systematic sampling methods from weed-infested soil under tree canopy during the season of abundant insect pests to maximize EPN positive samples. Moreover, multiple cycles of *Galleria*-baiting technique were followed to optimize nematode extraction. Each sample contained in a plastic cup was examined twice weekly for the first 4 weeks and once weekly thereafter in consecutive cycles. Thus, a cycle consists of about 3-4 days in the first 4 weeks or 7 days thereafter. A total of 11 cycles in 7 weeks were conducted during the extraction process.

Each cycle ended by examining the cups in order: *i*) to isolate insect cadavers showing symptoms of EPN infection (termed 'suspicious cadavers'), which were transferred to White traps to check for agreement with Koch's postulates (Pelczar and Reid, 1972); suspicious cadavers that failed to produce infectives were considered negative; and/or *ii*) to discard the other dead insects. A new following cycle begins with replacing the infected cadavers/dead insects by new living *G. mellonella* larvae. Soil samples that did not produce any suspect cadavers during the first 4 weeks were not assayed thereafter. The remaining samples were assayed until all of the samples failed to produce any suspicious cadavers for at least 1 week.

This technique of sampling and extraction resulted in recovering EPNs from all the surveyed groves (100%) and from 37 of 60 (61.7%) soil samples. Such functional sampling aimed at timely centering on isolating biocontrol agents of certain pests should be expanded in parallel to large-scale surveys for coordination in search of such reliable BCAs. This technique may avoid hurdles and costs of strain improvement, which means to modify genetics of the biocontrol agent in order to carry out a process more effectively (Abd-Elgawad, 2021a).

5.3. Cost-Effective *Photorhabdus* Mass Culture with Boosting Insecticidal Activity

Photorhabdus bacteria are another BCA group. These insect pathogenic bacteria are found exclusively in mutualistic association with heterorhabditid nematodes. Stand-alone *Photorhabdus* bacteria are reported to be selling well in markets for insect management (Kushwah and Somvanshi, 2015). *Photorhabdus* toxins are considered next to *Bt* toxins in their potential for use in insect-pest management in agriculture. In the recent past, many difficulties in trait deterioration of *Photorhabdus* spp. were apparent during their *in vitro* mass culture. Furthermore, researchers are challenged by a cost and benefit tradeoff. The necessity of high bacterial yield in terms of reduced culture time and scale-up of the mass production are opposed by the requirement to preserve *Photorhabdus* traits during the process of culturing (Wang et al., 2007). Such bacterial attributes are significant for biocontrol programs whether via *Photorhabdus* alone or EPN-*Photorhabdus* complex. Recently, the bacterial traits could be characterized from the standpoint of harnessing them during their growth and metabolic phases. Consequently, the full *Photorhabdus* capacity could be employed for insect pathogenicity. Orozco-Hidalgo et al., (2019) found that the bacterial inoculation at 36 h of culture could offer the highest *H. indica*-IJ yield. The authors stressed the merit of glucose as a substrate to increase/sustain the bacteria so that higher nematode recovery than *in vivo* IJ yield could be obtained.

Very recently, an important discovery could offer the best conditions for enhancing the mass culture and the insecticidal efficacy of *P. temperata* multiplied on wastewater of food industry (Keskes et al., 2021). They used both special lay out (Box-Behnken design) and response surface methodology (RSM) to optimize the bacterial mass-production. This lay out was based on three factors, Carbon/Nitrogen (C/N) ratio, sodium chloride concentration, and bacterial inoculum size. Both insecticidal efficacy and bacterial mass culture were deemed to represent the system in terms of the assessed responses. Furthermore, when wastewater from the food industry was optimized for utilization as an inexpensive raw material for *P. temperate* culturing, its production costed only $ 35/ kg (Keskes et al., 2021), compared to $ 679/ kg medium that used yeast extract and glucose as main components (Jallouli et al., 2008).

About 95 per cent trimming in the total production cost was achieved while the produced bacteria had high insecticidal activity. Thus, this technique for optimizing the important combined factors via RSM to yield superior bioinsecticide with high bacterial mass production in a short time (48 h incubation) should be expanded to other species of *Photorhabdus*. This cheap technique should be considered for large-scale practice as it also gives a share in getting low-cost formulations able to compete with conventional chemical compounds.

5.4. Developing Biopesticide Formulations

The formulation of microbial control agents has been extensively reviewed by Burges (1998). However, there is a real challenge to continuously improve biopesticide formulations in many (especially developing) countries. For adequate application accounts, biopesticidal formulations can be divided into three main groups. The first one contains biopesticides in the form of fermentation products and botanical extracts that can be used functionally as chemical pesticides dissolved in the carrier liquid (Bateman, 2014). Thus, the application aspects of control agents should consider whether the active ingredient is essentially particulate in nature or can be dissolved in the carrier liquid as a basis for determining successful product development.

Additionally, BCAs are usually particulates in nature and may represent the second group that comprises fungi, bacteria, entomopathogenic nematodes and viruses. Their formulations are efficacious against arthropod pests, plant diseases, or weeds. Therefore, if a BCA is to be delivered as a spray, the living organism; such as colony-forming units (CFU), propagules, spores, nematode-infective juveniles (IJs) must be suspended and adequately distributed so that they may have a reasonable chance of attaining the target spot alive. These are the two basic groups where firm quality control is decisive. The third group includes a wide variety of formulations. Among them, the EPN products as EPN suspension (in water) on polyether-polyurethane sponge as an inert carrier are broadly utilized in the cottage industry (Grewal, 2002). However, other EPN formulations have comprised polyacrylamide and alginate gels, diatomaceous earth, clay activated charcoal, paste, peat, water dispersible granules, polyurethane sponge, and vermiculite (Koppenhöfer et al., 2020b). Shelf-life in most EPN commercial products has been boosted by decreasing nematode activity and metabolism via metabolic inhibition, cold storage, physical trapping, or by the induction of anhydrobiosis.

General assessment of the probable success of biopesticide formulation application systems (other than the BCA attributes) should consider the following components: 1) the size spectra of suspended BCA particles in the formulations. They should fit the known dimensions of spores, propagules, IJs in order to give a good score of both formulation quality and potential for optimization, 2) in-flight droplet size spectra of nozzle output, ideally pursued by spray deposit investigations on intended plants, grasses, 3) the volume application rate should be standardized to properly disperse the suspension for each specific biopesticidal group, and 4) considering the features of droplets and the dilution rate in order to optimize the biopesticidal contact with the target pest (Bateman, 2014). On the other hand, optimization of droplet size has been fostered substantially for biopesticide applications in forests to minimize impact in streams within the forest ecosystem. It is focused on increasing deposition within the tree canopy and avoiding large droplets which can fall to ground level (Matthews et al., 2014).

Formulations that are user-friendly and can timely be added to the tank with irrigation water with or without fertilizers are desirable for not only pest control but also cost-effectiveness, e.g., in terms of boosting seed germination rate (Shehata et al., 2021). Additionally, the above-mentioned variety of EPN formulations has facilitated EPN applications, provided product stability especially for reasonable shelf-life, and improved efficacy. Formulations that permit aboveground applications by offering protection against desiccation and ultraviolet radiation can also extend EPN practical use. A sprayable gel, Barricade®, can be utilized aboveground to kill target insects such as the wood boring pest *Synanthedon pictipes*. Thus, numerous additives can further enhance aboveground EPN-applications like wood, flour, foam, and chitosan. Some of these formulations are recommended for applications in organic farming (Shapiro-Ilan et al., 2017). On the contrary, developing other commercial biopesticides, with room temperature shelf-life, have proven more challenging for a variety of temperature-sensitive BCAs. While spores of many entomopathogenic fungi cannot tolerate high temperatures, and consequently do not continue on the foliage for long, *Bacillus* spp.-based products possess long shelf-life at room temperature as they produce endospores (Kesho, 2020).

Bacillus subtilis and *B. pumilus* may be applied as seed treatment against PPNs and other plant pathogens too (Abd-Elgawad and Askary, 2018; Abd-El-Kareem et al., 2021). They proved to have ease of mass-culture and good stability with storage. Also, a drench product and seed treatment products (VOTiVO® and PONCHO®/VOTiVO® mix) contain *B. firmus.* VOTiVO is utilized on golf greens against main PPN genera on turfgrass. For effective PPN control, it is advised to apply the product 2-7 days before planting, and post application irrigation with minimum of 7.5 cm (Wilson and Jackson, 2013). The seed treatment method challenges high costs in field application of biopesticides as it permits a much decreased bacterial load to be applied per unit area. It may be in the order of a 1,000-fold dropping, *B. firmus*-seed treatment on the rhizosphere of freshly developing roots. Moreover, *B. firmus* is mostly sold as a seed treatment combined with the insecticide PONCHO (Clothianidin). This mixture (PONCHO/VOTiVO) forms an insecticide and seed treatment material to control both arthropod pests and PPNs on many crops, e.g., soybean, cotton, corn, sorghum, and sugarbeets. It contains 40.3 per cent clothianidin, 8.1per cent *B. firmus*, and 51.6 per cent other ingredients, 2 x 10^9 *B. firmus* colony forming units (cfu)/ ml and 500 g/ l clothianidin (Wilson and Jackson, 2013). Mounting interest by many companies is directed not only to *Bacillus* spp. but others as bacterial nematicides too. The obligate parasitic nature and biodiversity of the bacteria *Pasteuria* spp. can offer safe tools to manage PPN. Their added spores may promptly initiate and deliver season-long activity, operate under relatively harsh ecological conditions and variable soil pH, temperature and moisture. They do not hurt non-target organisms.

Kokalis-Burelle (2015) stated three methods of *P. penetrans* application: seed, transplant, and post-plant treatments for management of *Meloidogyne incognita* on tomato and cucumber and *M. arenaria* on snapdragon. Abd-Elgawad (2022) stressed that their field application for successful crop pest management should preferably be integrated with other approaches to control PPN in organic farms, e.g., nematode-resistant cultivars and crop rotation. Their endospores are resistant to mechanical shearing, drying, and heat. Furthermore, *Pasteuria* isolates should be screened to select the most adequate one(s) for biocontrol in specific areas to comply with the related biotic and abiotic factors.

Within the frame of such formulations, there should be (i) an obvious relationship between numbers of BCA particles and their effectiveness, (ii) a primary necessity to maintain the BCA alive, and (iii) the accidental chance of horizontal transmission (secondary cycling due to multiplication of the BCA). Yet, this latter is practically hard to confirm and should not be counted upon as it relies on environmental conditions (Bateman, 2014). Eventually, rigorous quality control of the final biopesticidal formulation is crucial under all circumstances.

5.5. Application Technology

EPNs can be applied using most standard agricultural equipment including various sprayers or irrigation systems (Shapiro-Ilan et al., 2017). Optimum application depends on a variety of factors including environmental conditions (avoiding drought, temperature extremes, and ultraviolet radiation), appropriate application rates, nozzle type and volume, agitation, pressure and recycling time, and sprayer type and spray distribution pattern.

In this respect, a number of typical tank mixtures designed for selected commercial BCA products were presented by Bateman (2014). Before initiating spray application, the latter author recommended an assessment of the contents of a 124 micrometre (µm) droplet, which could be near the median of many fine sprays and be present in most droplet spectra of biopesticidal sprays. Because many biopesticides have particulate nature, i.e., in the form of minute separate particles, small opening nozzles can frequently be blocked especially with relatively large particulates or even small ones that have tendency to lump or mass. Such cases may promote the use of higher volume application rate than is needed by chemical pesticides.

Clearly, unlike particulate chemical products that can be milled, biologicals like fungus spores are relatively soft, so a device called the 'MycoHarvester' was established to face this challenge. It is designed to separate conidia safely and efficiently from the solid substrate normally used, e.g., grains such as rice (Cherry et al., 1999). It enabled achieving particle size spectra comparable with certain conventional chemical formulations (Bateman, 2014).

The issue of blocking nozzles is also sensitive for EPNs. However, it is usually satisfactory for the EPNs that are mostly soil-applied treatments; e.g., drenches. On the other hand, applications at higher volumes for foliage treatments need cautious approaches and adequate technology as it may imply a decreased concentration of the IJs. Lello et al., (1996) found that higher output hydraulic nozzles deposited the most IJs of *Steinernema carpocapsae* on plant leaves, but more than 89 per cent of droplets contained none of the IJs. In other studies, Mason et al., (1998a, b) applied EPNs to control foliar pest, diamond backmoth (*Plutella xylostella*) on crucifers using a rotary atomiser.

The IJ deposition was enhanced by elevating flow rate and/or concentration, but nematodes could be separated by centrifugal force from the liquid during application. Therefore, the design of the rotary atomiser is significant and needs reformation as relatively flat, rather than saucer or cup form. Piggott et al., (2003) could optimize such application devices for EPNs through redesigning a spinning disc applicator to get better application of the IJs. Thus, the new design enabled enhanced EPN application with greater deposition of IJs per unit area. However, growers are most likely to apply available and common equipment because EPN market is still small.

Figure 1. Effective methods for applying entomopathogenic nematodes using boom sprayer, sprayers linked to tractor, and perforated greenhouse tray are properly matching the target site for foliage- or soil-applied treatments (Adapted from Dr. R. Ehlers).

Needless to say that proper agitation of suspended particulate matter during application is critical to guarantee even distribution of BCAs. Equipment for biopesticidal applications should adequately match the intended site and related settings. For EPNs, handheld systems such as backpack sprayers and water cans are efficient for small plot applications. However, for larger areas or in open fields for commercial production, a suitable spray apparatus is required (Figure 1). Other traditional but still effective approaches for application comprise trunk-sprayers, microjet irrigation systems, subsurface injection or baits equipment (Shapiro-Ilan et al., 2012b, 2020).

The efficacy of nematode-infective juveniles can be enhanced via novel application approaches. For example, relative to traditional applications in aqueous sprays, applying nematodes in their infected-hosts (Dolinski et al., 2015), with phoretic hosts as quite different method of EPN dispersal after application (Shapiro-Ilan and Brown, 2013), or in pre-infected hosts that reach cryptic habitats (Gumus et al., 2015) has yielded enhanced efficacy. Specific plant volatiles could be utilized to boost EPN efficacy by guiding IJ behaviour and increasing infectivity (Willett et al., 2018). Moreover, other "boosters" that directly enhance EPN activity will result in elevated efficacy. Ascaroside-based nematode pheromones or infected host macerate can back EPN infectivity and dispersal and proved useful against soil pests (Koppenhöfer et al., 2020b). Also, previous results indicated that some EPNs such as the Turkish isolate of *Steinernema weiseri* (Bazman et al., 2008) and *S. riobrave* Texas (Gazit et al., 2000) have the potential to kill *Ceratitis capitata* larvae as soon as they fall to the ground for pupation or even after they pupate (Abd-Elgawad et al., 2012). Thus, their field application under the fruit tree canopy (prior to adult emergence) under the appropriate conditions of temperature and moisture is a promising IPM strategy. In accordance with this attract-and-kill approach, EPN application by a trap tree method similar to that used against the plum curculio *Conotrachelus nenuphar* (Herbst.) (Coleoptera: Curculionidae) may be effective against *C. capitata* (Shapiro-Ilan et al., 2013). In this novel approach, branches of several perimeter-row trees are baited with synergistic lures; this results in the aggregation of adult *C. capitata* on the trees, which can then be specifically targeted with insecticide (Leskey et al., 2008). A suitable EPN strain/species is then sprayed onto the soil under the trap tree canopy to control the ground-dwelling stages of the flies (Figure 2). Conditions of temperature and soil moisture that are favorable for the selected EPN strain/species should be ensured (Gazit et al., 2000; Bazman et al., 2008; Rohde et al., 2010, 2020) to optimize EPN efficacy and persistence after application.

Figure 2. A tree baited with synergistic lure to aggregate *Ceratitis capitata* adults in the trap while the proper entomopathogenic nematodes are sprayed on soil under the tree canopy to kill the pupae of the insect and the fallen larvae of the fruits that the insects attacked due to its proximity to the lure.

An iterative approach to predict and improve biopesticide formulation and application systems has been proposed. It comprises the timely release of

EPNs followed by spatial modeling of their activity (Abd-Elgawad, 2017b). Early release models can guide subsequent applications by providing insight into which aspects of the EPN and environment are the strongest drivers of efficacy and persistence. For instance, the window of chance for EPNs to attack *C. capitata* larvae in the soil should be identified and targeted to establish short-term equilibrium of EPN-host dynamics. This is especially important as some Medfly larvae that are susceptible to EPNs (Abd-Elgawad et al., 2012) found in Egypt infecting fallen mango (Abd-Elgawad, 2014) and citrus (Abd-Elgawad, 2020e) fruits in naturally EPN-infested soil. Additional research is needed in order to study the feasibility of this modelling system.

An alternative approach, not based on modelling is to characterize the factors affecting the nematode efficacy, both positively and negatively. Then, researchers and stakeholders focus on maximizing the positive factors and avoiding or eliminating the negative ones (e.g., Duncan et al., 2013; Campos-Herrera et al., 2019; Dritsoulas et al., 2021). Thus, from the above-mentioned biopesticide formulation application systems, effective IPM programs for *C. capitata* or other citrus pests are likely to be developed.

5.6. Unintended Efficacy of a Biopesticide

This may sometimes be expected. It could happen especially with a few but relatively broad spectrum biopesticides. Unfortunately, spinosad as a biopesticide showed lethal and non-lethal risks for non-target native pollinators, e.g., *Melipona quadrifasciata* in Brazil (Tomé et al., 2015). Therefore, bioinsecticides especially broad spectrum ones should be liable to risk assessment analysis to examine their sublethal and lethal components. Moreover, the interaction of biopesticides with closely related biotic factors should be a priority area of research. This will enable us to anticipate and thus avoid any unwanted/side effect given the fact that the biologicals found in the field settings might have undesirable interaction with the biopesticide, the habitat of the pest makes biopesticide intervention has bad impact on beneficial organism(s).

On the contrary, numerous biopesticides possess high levels of selectivity. For instance, biopesticides based on baculoviruses, e.g., the *Cydia pomonella* granulovirus (CpGV), typically are selective for just one/few insect species. In Europe and USA, CpGV is effectively utilized as an inundative biopesticide against codling moth on apples. Thus, it does not hurt other natural enemies and wildlife. However, it means that such biopesticides are relegated to niche markets with humble profits. In other words, the attributes that make most biopesticide products attractive from the standpoint of human and environmental safety may also act to limit their commercial success (Chandler et al., 2011).

5.7. Farmers' Risk Aversion

Required but less known issue of risk aversion to optimize biopesticidal applications in the biocontrol realm should further be addressed. This aversion usually has increased levels in less educated growers especially in developing countries. They believe that risks of producing unmarketable or low-grade crops are high. Consequently, growers are forced to be risk averse concerning new, untested crop protection technologies in terms of the introduced biopesticide application. Growers have achieved scale economies in conventional pesticide application as an output of learning by doing. The notion is that one would turn into more productive at a function the more it is repeated. In contrast, the restricted evidence base and limited experience with biopesticide technologies create uncertainty for growers (Chandler et al., 2011). Growers' risk averse preferences can lead to sub-optimal patterns of adoption of these technologies. Therefore, the above-mentioned extension and training sessions should also stress on all direct and indirect benefits of these new technologies and how to tackle the potential downsides.

5.8. Untapped IPM Economies

Various IPM tactics should work together as a technology package with optimized performance. In this concern, the area of synergism is of increasing interest in the US and elsewhere, as a synergistic interaction has the benefit of reducing the necessary dose of both biopesticide and the chemical pesticide. Such strategies proved more effective against pests and less expensive as control measure with various groups of biopesticides such as EPNs (Stevens and Lewis, 2017) and fungal and bacterial biocontrol agents (Abd-Elgawad and Askary, 2018). Therefore, additive or synergistic interactions can open up new markets and new possibilities. For instance, investigations on the possible impacts of combining EPN with either chemical pesticides or microbial control approaches continue to increase. The list of potentially synergistic chemicals includes carbaryl, chlorantraniliprole, chlorpyrifos, dimethoate, endosulfan, fonofos, tefluthrin, and imidacloprid (Stevens and Lewis, 2017). The extent (or even existence) of the synergism can be very dependent on the specific combination of pest, EPN species, and application. Therefore, further research is necessary to develop system-specific synergisms that are broad based. Other aforementioned issues related to unexploited opportunities for successful applications of biopesticides should be considered from the economical view.

Conclusion

Development in field application of biopesticides for successful crop pest management has been progressing but still needs to be continuously driven by quite motivated scholars and research groups. This is especially important to improve various aspects of practical biopesticidal use. Successful facing of challenges related to cost-effective processes concerning isolation, extraction, mass production, formulation, and application technology of biocontrol agents that achieved remarkable success in several segments should be expanded. Therefore, more collaboration among relevant enterprises/companies and research institutes is desperately needed to improve biopesticides as many of them are frequently less effective, more inconsistent, and/or slower acting than chemical nematicides. Both biopesticides and chemical pesticides are currently used for pest suppression, and therefore, synergistic or additive tactics that use biopesticides with other agricultural inputs in integrated pest management programs should be optimized. Moreover, the potential of BCAs to recycle in host populations bodes well for further applications to exploit long term pest management with a real chance for environmentally friendly economic growth.

Acknowledgments

This article is supported in part by the US-Egypt Project cycle 17 (no. 172) entitled "Preparing and evaluating IPM tactics for increasing strawberry and citrus production." The study was also supported in part by the NRC In-house project No. 12050105 entitled "Pesticide alternatives against soil-borne pathogens and pests attacking economically important solanaceous crops."

References

Abd-Elgawad, M. M. M. (2014). Spatial patterns of *Tuta absoluta* and heterorhabditid nematodes. *Russian Journal of Nematology*, 22: 89-100.

Abd-Elgawad, M. M. M. (2017a). Toxic secretions of *Photorhabdus* and their efficacy against crop insect pests. In: *Biocontrol Agents: Entomopathogenic and Slug Parasitic Nematodes* (Eds. Abd-Elgawad, M. M. M., Askary, T. H. and Coupland, J.). CAB International, Wallingford, UK, pp. 231-260.

Abd-Elgawad, M. M. M. (2017b). Comments on the economic use of entomopathogenic nematodes against insect pests. *Bulletin of The National Research Centre*, 41: 66-84.

Abd-Elgawad, M. M. M. (2019a). Plant-parasitic nematodes of strawberry in Egypt: A Review. *Bulletin of The National Research Centre*, 43: 7. https://doi.org/10.11 86/s42269-019-0049-2.

Abd-Elgawad, M. M. M. (2019b). Towards optimization of entomopathogenic nematodes for more service in the biological control of insect pests. *Egyptian Journal of Biological Pest Control*, 29: 77 https://doi.org/10.1186/s41938-019-0181-1.

Abd-Elgawad, M. M. M. (2020a). Plant-parasitic nematodes and their biocontrol agents: current status and future vistas. In: *Management of Phytonematodes: Recent Advances and Future Challenges* (Eds. Ansari, R. A., Rizvi, R. and Mahmood, I.). Springer, Singapore, pp. 171-204.

Abd-Elgawad, M. M. M. (2020b). Optimizing biological control agents for controlling nematodes of tomato in Egypt. *Egyptian Journal of Biological Pest Control*, 30: 58. https://doi.org/10.1186/s41938-020-00252-x.

Abd-Elgawad, M. M. M. (2020c). Biological control agents in the integrated nematode management of potato in Egypt. *Egyptian Journal of Biological Pest Control*, 30: 121. https://doi.org/10.1186/s41938-020-00325-x.

Abd-Elgawad, M. M. M. (2020d). Biological control agents in the integrated nematode management of pepper in Egypt. *Egyptian Journal of Biological Pest Control*, 30: 70 https://doi.org/10.1186/s41938-020-00273-6.

Abd-Elgawad, M. M. M. (2020e). Can rational sampling maximise isolation and fix distribution measure of entomopathogenic nematodes? *Nematology*, 22: 907-916. doi: https://doi.org/10.1163/15685411-00003350.

Abd-Elgawad, M. M. M. (2021a). Optimizing sampling and extraction methods for plant-parasitic and entomopathogenic nematodes. *Plants*, 10: 629. https://doi.org/10.3390/ plants10040629.

Abd-Elgawad, M. M. M. (2021b). The Mediterranean Fruit Fly (Diptera: Tephritidae), a Key Pest of Citrus in Egypt. *Journal of Integrated Pest Management* 12: 28. https://doi.org/10.1093/jipm/pmab025.

Abd-Elgawad, M. M. M. (2021c). Biological control of nematodes infecting eggplant in Egypt. *Bulletin National Research Centre, Egypt*, 45: 6 https://doi.org/10.1186/s42 269-020-00463-0.

Abd-Elgawad, M. M. M. (2022). *Pasteuria* species for Management of Nematodes in Organic farms. In: *Organic Management of Nematodes in Agriculture* (Eds. Chaudhary, K. K. and Meghvansi, M. K.). Springer, Switzerland (in press).

Abd-Elgawad, M. M. M. and Askary, T. H. (2015). Impact of phytonematodes on agriculture economy. In: *Biocontrol Agents of Phytonematodes* (Eds. Askary, T. H. and Martinelli, P. R. P.). CAB International, Wallingford, UK, pp. 3-49.

Abd-Elgawad, M. M. M. and Askary, T. H. (2018). Fungal and bacterial nematicides in integrated nematode management strategies. *Egyptian Journal of Biological Pest Control*, 28: 74. https://doi.org/10.1186/s41938-018-0080-x.

Abd-Elgawad, M. M. M., Abdel-Razek, A. S. and Abd El-Wahab, A. E. (2012). Efficacy of kaolin, entomopathogenic nematodes and fungi on the Mediterranean fruit fly infecting citrus. *Egyptian Journal of Agronematology*, 11: 178-192.

Abd-El-Kareem, F., Elshahawy, I. E. and Abd-Elgawad, M. M. M. (2021). Application of *Bacillus pumilus* isolates for management of black rot disease in strawberry. *Egyptian Journal of Biological Pest Control*, 31: 25. https://doi.org/10.1186/s41938-021-00371-z.

Anonymous (2018). *Adopted recommendations to combat agricultural pests* (in Arabic). Commercial Al-Ahram Press, Qalioub, Egypt, Agricultural Pesticide Committee, Ministry of Agriculture, Media Support Center Press.

Atkins, S. D., Hidalgo-Diaz, L., Kalisz, H., Mauchline, T. H., Hirsch, P. R. and Kerry, B. R. (2003). Development of a new management strategy for the control of root-knot nematodes (*Meloidogyne* spp.) in organic vegetable production. *Pest Management Science*, 59: 183-189.

Bateman, R. (2014). Application of biopesticides. In: *Pesticide Application Methods, 4th Edition* (Eds. Matthews, G. A., Bateman, R. P. and Miller, P. C. H.). John Wiley & Sons Ltd., UK, pp. 411-428.

Bazman, I., Ozer, N. and Hazir, S. (2008). Bionomics of the entomopathogenic nematode, *Steinernema weiseri* (Rhabditida: Steinernematidae). *Nematology*, 10: 735-742.

Belgacem, I., Li Destri Nicosia, M. G., Pangallo, S., Abdelfattah, A., Benuzzi, M., Agosteo, G. E. and Schena, L. (2021). Pomegranate peel extracts as safe natural treatments to control plant diseases and increase the shelf-life and safety of fresh fruits and vegetables. *Plants*, 10: 453. https://doi.org/10.3390/plants10030453.

Burges, H. D. (1998). *Formulation of Microbial Biopesticides: Beneficial Microorganisms, Nematodes and Seed Treatments*. Springer, Switzerland.

Campos-Herrera, R., Pathak, E., El-Borai, F. E., Schumann, A., Abd-Elgawad, M. M., Duncan, L. W. (2013). New citriculture system suppresses native and augmented entomopathogenic nematodes. *Biological Control*, 66: 183-194.

Campos-Herrera, R., Stuart, R. J., Pathak, E., El-Borai, F. E., Duncan, L. W. (2019). Temporal patterns of entomopathogenic nematodes in Florida citrus orchards: Evidence of natural regulation by microorganisms and nematode competitors. *Soil Biology & Biochemistry*, 128: 193-204.

Chandler, D., Bailey, A. S., Tatchell, G. M., Davidson, G., Greaves, J., Grant, W. P. (2011). The development, regulation and use of biopesticides for integrated pest management. *Philosophical Transactions* of the *Royal Society B*: *Biological Sciences*, *12*, 366 (1573): 1987-1998. https://doi.org/10.1098/rstb.2010.0390.

Chandler, D., Davidson, G., Grant, W. P., Greaves, J. and Tatchell, G. M. (2008). Microbial biopesticides for integrated crop management: an assessment of environmental and regulatory sustainability. *Trends in Food Science and Technology*, 19: 275-283.

Cheng, F., Wang, J., Song, Z., Cheng, J., Zhang, D. and Liu, Y. (2017). Nematicidal effects of 5-Aminolevulinic acid on plant-parasitic nematodes. *Journal of Nematology*, 49: 295-303.

Cherry, A., Jenkins, N., Heviefo, G., Bateman, R. and Lomer, C. (1999). A West African pilot scale production plant for aerial conidia of *Metarhizium* sp for use as a mycoinsecticide against locusts and grasshoppers. *Biocontrol Science and Technology*, 9: 35-51.

Clarke, D. J. (2008). *Photorhabdus*: a model for the analysis of pathogenicity and mutualism. *Cell Microbiology*, 10: 2159-2167.

Dahlin, P., Eder, R., Consoli, E., Krauss, J., Kiewnick, S. (2019). Integrated control of *Meloidogyne incognita* in tomatoes using fluopyram and *Purpureocillium lilacinum* strain 251. *Crop Protection*, 124: 104874. https://doi.org/10.1016/j.cropro.2019.104874.

Daneholt, B. (2006). Advanced information: RNA interference. *The Nobel Prize in Physiology or Medicine 2006*. Retrieved on 2007-01-25.

Dara, S. K. (2019). The new integrated pest management paradigm for the modern age. *Journal of Integrated Pest Management*, 10: 1-9.

d'Errico, G., Marra, R., Crescenzi, A., Davino, S. W., Fanigliulo, A., Woo, S. L., et al., (2019). Integrated management strategies of *Meloidogyne incognita* and *Pseudopyrenochaeta lycopersici* on tomato using a *Bacillus firmus*-based product and two synthetic nematicides in two consecutive crop cycles in greenhouse. *Crop Protection*, 122: 159-164.

Dolinski, C., Choo, H. Y. and Duncan, L. W. (2012). Grower acceptance of entomopathogenic nematodes: case studies on three continents. *Journal of Nematology*, 44: 226-235.

Dolinski, C., Shapiro-Ilan, D. I. and Lewis, E. E. (2015). Insect cadaver applications: pros and cons. In: *Nematode Pathogenesis of Insect and Other Pests* (Ed. Campos-Herrera, R.). Springer, Switzerland, pp. 207-230.

Dritsoulas, A., El-Borai, F. E., Shehata, I. E., Hammam, M. M., El-Ashry, R. M., Mohamed, M. M., Abd-Elgawad, M. M. M. and Duncan, L. W. (2021). Reclaimed desert habitats favor entomopathogenic nematode and microarthropod abundance compared to ancient farmlands in the Nile Basin. *Journal of Nematology*, 53: 1-13.

Duncan, L. W., Stuart, R. J., El-Borai, F. E., Campos-Herrera, R., Pathak, E., Giurcanu, M., Graham, J. H. (2013). Modifying orchard planting sites conserves entomopathogenic nematodes, reduces weevil herbivory and increases citrus tree growth, survival and fruit yield. *Biological Control*, 64: 26-36.

El-Nagdi, W. M. A., Youssefi, M. M. A., Abd-El-Khair, H., Abd-Elgawad, M. M. M., Dawood, M. G. (2019). Effectiveness of *Bacillus subtilis*, *B. pumilus*, *Pseudomonas fluorescens* on *Meloidogyne incognita* infecting cowpea. *Pakistan Journal of Nematology*, 37: 35-43.

EPA (2012). *Regulating Biopesticides*. www.epa.gov/opp00001/biopesticides/. Accessed on 13 June, 2021.

European Commission (2008). *Encouraging Innovation in Biopesticide Development.* http://ec.europa.eu/environment/integration/research/newsalert/pdf/134na5.pdf.

Fire, A., Xu, S., Montgomery, M. K., Kostas, S. A., Driver, S. E., Mello, C. C. (1998). Potent and specific genetic interference by double-stranded RNA in *Caenoehabditis elegans*. *Nature*, 391: 806-811.

Fletcher, S. J., Reeves, P. T., Hoang, B. T., Mitter, N. (2020). A perspective on RNAi-based biopesticides. *Frontiers in Plant Science*, 11:51. doi.org/10.3389/fpls.2020.00051.

Forghani, F. and Hajihassani, A. (2020). Recent advances in the development of environmentally benign treatments to control root-knot nematodes. *Frontiers in Plant Science*, 11, 1125. https://doi.org/10.3389/fpls.2020.01125.

Gazit, Y., Rossler, Y. and Glazer, I. (2000). Evaluation of entomopathogenic nematodes for the control of Mediterranean fruit fly (Diptera: Tephritidae). *Biocontrol Science and Technology*, 10: 157-164.

Geldhof, P., Visser, A., Clark, D., Saunders, G., Britton, C., Gilleard, J., Berriman, M., Knox, D. (2007). RNA interference in parasitic helminths: current situation, potential pitfalls and future prospects. *Parasitology*, 134: 609-619.

Georgis, R. (2002). The Biosys Experiment: an insider's perspective. In: *Entomopathogenic Nematology* (Ed. Gaugler, R.). CAB International, Wallingford, UK, pp. 357-372.

Gerdes, E., Upadhyay, D., Mandjiny, S., Bullard-Dillard, R., Storms, M., Menefee, M. and Holmes L. D. (2015). *Photorhabdus luminescens*: virulent properties and agricultural applications. *American Journal of Agriculture and Forestry*, 3: 171-177.

Grewal, P. (2002). Formulation and application technology. In: *Entomopathogenic Nematology* (Ed. Gaugler, R). CAB International, Wallingford, UK, pp. 265-288.

Gumus, A., Karagoz, M., Shapiro-Ilan, D. I. and Hazir, S. (2015). A novel approach to biocontrol: release of live insect hosts pre-infected with entomopathogenic nematodes. *Journal of Invertebrate Pathology*, 130: 56-60.

Haryuni, H., Harahap, A. F. P., Supartini, Priyatmojo, A., Gozan, M. (2020). The effects of biopesticide and *Fusarium oxysporum* f.sp. *vanillae* on the nutrient content of binucleate *Rhizoctonia*-induced vanilla plant. *International Journal of Agronomy*, 5092893. https://doi.org/10.1155/2020/5092893.

Hussaini, S. S. (2017). Entomopathogenic nematodes: Ecology, diversity and geographical distribution. In: *Biocontrol Agents: Entomopathogenic and Slug Parasitic Nematodes* (Eds. Abd-Elgawad, M. M. M., Askary, T. H. and Coupland, J.). CAB International, Wallingford, UK, pp. 88-142.

Isman, M. B. (2006). Botanical insecticides, deterrents, and repellants in modern agriculture and an increasingly regulated world. *Annual Review of Entomology*, 51: 45-66.

Jaffuel, G., Sbaiti, I., Turlings, T. C. J. (2020). Encapsulated entomopathogenic nematodes can protect maize plants from *Diabrotica balteata* larvae. *Insects*, 11:27. https://doi.org/10.3390/insects11010027.

Jallouli, W., Abdelkefi-Mesrati, L., Tounsi, S., Jaoua, S. and Zouari, N. (2013). Potential of *Photorhabdus temperata* K122 bioinsecticide in protecting wheat flour against *Ephestia kuehniella*. *Journal of Stored Products Research*, 53: 61-66.

Jallouli, W., Hammami, W., Zouari, N., Jaoua, S. (2008). Medium optimization for biomass production and morphology variance overcome of *Photorhabdus temperata* ssp. *temperata* strain K122. *Process Biochemistry*, 43: 1338-1344.

Kesho, A. (2020). Microbial bio-pesticides and their use in integrated pest management. *Chemical and Biomolecular Engineering*, 5: 26-34.

Keskes, S., Jallouli, W., Atitallah, I. B., Driss, F., Sahli, E., Chamkha, M. and Touns, S. (2021). Development of a cost-effective medium for *Photorhabdus temperata* bioinsecticide production from wastewater and exploration of performance kinetic. *Scientific Reports*, 11: 779 https://doi.org/10.1038/s41598-020-80773-5.

Kokalis-Burelle, N. (2015). *Pasteuria penetrans* for control of *Meloidogyne incognita* on tomato and cucumber and *M. arenaria* on snapdragon. *Journal of Nematology*, 47: 207-213.

Koppenhöfer, A. M., Kostromytska, O. S., Shaohui, W. (2020a). Optimizing the use of entomopathogenic nematodes for the management of *Listronotus maculicollis* (Coleoptera: Curculionidae): split applications and combinations with imidacloprid. *Crop Protection*, 137: 105229. https://doi.org/10.1016/j.cropro.2020.105229.

Koppenhöfer, A. M., Shapiro-Ilan, D. I. and Hiltpold, I. (2020b). Entomopathogenic nematodes in sustainable food production. *Frontiers in Sustainable Food Systems*, 4:125. https://doi.org/10.3389/fsufs.2020.00125.

Koul, O. (2011). Microbial biopesticides: Opportunities and challenges. *CAB Reviews: Perspectives in Agriculture, Veterinary Science, Nutrition and Natural Resources*, 6: 1-26.

Kumar, J., Ramlal, A., Mallick, D., Mishra, V. (2021). An overview of some biopesticides and their importance in plant protection for commercial acceptance. *Plants*, 10:1185. https://doi.org/10.3390/plants10061185.

Kushwah, J. and Somvanshi, V. S. (2015). *Photorhabdus*: a microbial factory of insect-killing toxins. In: *Microbial Factories: Biodiversity, Biopolymers, Bioactive Molecules* (Ed. Kalia, V. C.). Springer, New Delhi, India, pp. 235-240.

Lello, E. R., Patel, M. N., Matthews, G. A. and Wright, D. J. (1996). Application technology for entomopathogenic nematodes against foliar pests. *Crop Protection*, 15: 567-574.

Leskey, T. C., Piñero, J. C. and Prokopy, R. J. (2008). Odor-baited trap trees: a novel management tool for the plum curculio, *Conotrachelus nenuphar* (Herbst) (Coleoptera: Curculionidae). *Journal* of *Economic Entomology*, 101: 1302-1309.

Liu, Y. B., Tabashnik, B., Dennehy, T., Patin, A. and Bartlet, A. (1999). Development time and resistance to Bt crops. *Nature*, 400: 519.

Maithri, S. K., Ramesh, K. V. and Mutangana, D. (2013). Theoretical structure prediction of TcaA from *Photorhabdus luminescens* and aminopeptidase N receptor from *Helicoverpa armigera*. *Research Journal of Recent Sciences*, 2: 40-49.

Mason, J. M., Matthews, G. A. and Wright, D. J. (1998a). Appraisal of spinning disc technology for the application of entomopathogenic nematodes. *Crop Protection*, 17: 453-461.

Mason, J. M., Matthews, G. A. and Wright, D. J. (1998b). Screening and selection of adjuvants for the spray application of entomopathogenic nematodes against a foliar pest. *Crop Protection*, 17: 463-470.

Matthews, G. A., Bateman, R. P. and Miller, P. C. H. (2014). *Pesticide Application Methods* (4th Edition). John Wiley & Sons, Ltd., UK.

Nielsen, A. L., Spence, K. O., Nakatani, J. and Lewis, E. E. (2011). Effect of soil salinity on entomopathogenic nematode survival and behaviour. *Nematology*, 3: 859-867.

Orozco-Hidalgo, M. T., Quevedo-Hidalgo, B. and Sáenz-Aponte, A. (2019). Growth kinetics and pathogenicity of *Photorhabdus luminescens* subsp. Akhurstii SL0708. *Egyptian Journal of Biological Pest Control*, 29:71. https://doi.org/10.1186/s41938-019-0172-2.

Pal, G. K. and Kumar, B. (2013). Antifungal activity of some common weed extracts against wilt causing fungi, *Fusarium oxysporum*. *Current Discovery*, 2: 62-67.

Pelczar, M. J. and Reid, R. D. (1972). *Microbiology*. McGraw-Hill, New York, USA.

Piedra-Buena, A., López-Cepero, J. and Campos-Herrera, R. (2015). Entomopathogenic nematode production and application: regulation, ecological impact and non-target effects. In: *Nematodes Pathogenesis of Insects and Other Pests* (Ed. Campos-Herrera, R.). Springer, Switzerland, pp. 255-282.

Piggott, S. J., Clayton, R., Matthews, G. A. and Wright, D. J. (2003). Development of a new application apparatus for entomopathogenic nematodes. *Pest Management Science,* 59: 1344-1348.

Radhakrishnan, S., Shanmugam, S. and Ramasamy, R. (2017). Biocontrol efficacy of entomopathogenic nematodes against black cutworms, *Agrotis ipsilon* (Hufnagel) (Noctuidae: Lepidoptera) in potato. *Chemical Science Review and Letters*, 6: 219-224.

Regalado, A. (2015). With BioDirect, Monsanto hopes RNA sprays can someday deliver drought tolerance and other traits to plants on demand. *MIT Technology Review*. https://www.organicconsumers.org/news/monsanto-hopes-rna-sprays-can-someday-deliver-drought-tolerance-and-other-traits-plants-demand.

Rohde, C., Moino, A. Jr., da Silva, M. A. T., Carvalho, F. D. and Ferreira, C. S. (2010). Influence of soil temperature and moisture on the infectivity of entomopathogenic nematodes (Rhabditida: Heterorhabditidae, Steinernematidae) against larvae of *Ceratitis capitata* (Wiedemann) (Diptera: Tephritidae). *Neotropical Entomology*, 39: 608-611.

Rohde, C., Mertz, N. R. and Junior, A. M. (2020). Entomopathogenic nematodes on control of Mediterranean fruit fly (Diptera: Tephritidae). *Revista Caatinga, Mossoró*, 33: 974-984.

Saurabh, S., Vidyarthi, A. S., Prasad, D. (2014). RNA interference: concept to reality in crop improvement. *Planta*, 239: 543-64.

Shapiro-Ilan, D. I., Gardner, W. A., Wells, L. and Wood, B. W. (2012a). Cumulative impact of a clover cover crop on the persistence and efficacy of *Beauveria bassiana* in suppressing the pecan weevil (Coleoptera: Curculionidae). *Environmental Entomology*, 41: 298-307.

Shapiro-Ilan, D., Arthurs, S. P. and Lacey, L. A. (2017). Microbial control of arthropod pests of orchards in temperate climates. In: *Microbial Control of Insect and Mite Pests* (Ed. Lacey, L. A.). Elsevier, The Netherlands, pp. 253-267.

Shapiro-Ilan, D. I. and Brown, I. (2013). Earthworms as phoretic hosts for *Steinernema carpocapsae* and *Beauveria bassiana*: implications for enhanced biological control. *Biological Control*, 66: 41-48.

Shapiro-Ilan, D. I., Han, R. and Dolinksi, C. (2012b). Entomopathogenic nematode production and application technology. *Journal of Nematology*, 44: 206-217.

Shapiro-Ilan, D., Selcuk, H. and Glazer, I. (2020). Advances in use of entomopathogenic nematodes in integrated pest management. In: *Integrated Management of Insect Pests: Current and Future Developments* (Eds. Kogan, M. and Heinrichs, E. A.). Burleigh Dodds Science Publishing, UK, 1-30.

Shapiro-Ilan, D. I., Wright, S. E., Leskey, T. C., Tuttle, A. F., Cooley, D. R. and Leskey, T. C. (2013). Using entomopathogenic nematodes for biological control of plum curculio, *Conotrachelus nenuphar*: effects of irrigation and species in apple orchards. *Biological Control*, 67: 123-129.

Shehata, I. E., Hammam, M. M. A. and Abd-Elgawad, M. M. M. (2021). Effects of inorganic fertilizers on virulence of the entomopathogenic nematode *Steinernema glaseri* and peanut germination under field conditions. *Agronomy*, 11: 945. https://doi.org/10.3390/agronomy11050945.

Stevens, G., and Lewis, E. (2017). Status of entomopathogenic nematodes in Integrated Pest Management strategies in the USA. In: *Biocontrol Agents: Entomopathogenic and Slug Parasitic Nematodes* (Eds. Abd-Elgawad, M. M. M., Askary, T. H. and Coupland, J.). CAB International, Wallingford, UK, pp. 289-311.

Tembo, Y., Mkindi, A. G., Mkenda, P. A., Mpumi, N., Mwanauta, R., Stevenson, P. C., Ndakidemi, P. A. and Belmain, S. R. (2018). Pesticidal plant extracts improve yield and reduce insect pests on legume crops without harming beneficial arthropods. *Frontiers in Plant Science*, 9: 1425. https://doi.org/10.3389/fpls.2018.01425.

Timper, P. (2014). Conserving and enhancing biological control of nematodes. *Journal of Nematology*, 46: 75-89.

Tomé, H. V. V., Barbosa, W. F., Martins, G. F. and Guedes, R. N. C. (2015). Spinosad in the native stingless bee *Melipona quadrifasciata*: Regrettable non-target toxicity of a bioinsecticide. *Chemosphere*, 124: 103-109.

Valcárcel, F., Olmeda, A. S., González, M. G., Andrés, M. F., Navarro-Rocha, J. and González-Coloma, A. (2021). Acaricidal and insect antifeedant effects of essential oils from selected aromatic plants and their main components. *Frontiers in Agronomy*, 3: 662802. https://doi.org/10.3389/fagro.2021.662802.

Van Valen, L. (1973). A new evolutionary law. *Evolutionary Theory,* 1: 1-30.

Wang, Y., Bilgrami, A. L., Shapiro-Ilan, D., Gaugler, R. (2007). Stability of entomopathogenic bacteria, *Xenorhabdus nematophila* and *Photorhabdus luminescens*, during *in vitro* culture. *Journal of Industrial Microbiology and Biotechnology*, 34: 73-81.

Weiss, B., Davidkova, G. and Zhou, L. W. (1999). Antisense RNA gene therapy for studying and modulating biological processes. *Cellular and Molecular Life Sciences*, 55: 334-58.

Willett, D. S., Alborn, H. T., Stelinski, L. L. and Shapiro-Ilan, D. I. (2018). Risk taking of educated nematodes. *PLoS ONE* 13:e0205804. https://doi.org/0.1371/journal.pone.0205804.

Wilson, M. J. and Jackson, T. A. (2013). Progress in the commercialisation of bionematicides. *BioControl*, 58: 715–722.

About the Editor

Dr. Tarique Hassan Askary

Dr. Tarique Hassan Askary is an Associate Professor in the Division of Entomology at Sher-e-Kashmir University of Agricultural Sciences & Technology of Kashmir (SKUAST-K), Jammu & Kashmir, India. He has a Ph.D. in Plant Protection from Aligarh Muslim University, Aligarh, India in 2015. He has over 70 publications to his credit which includes research papers, book chapters, books and reviews in several national and international Journals. Dr. Askary's three books are *Key Nematode Pests of Pigeon Pea: Systematics, Biology and Disease Management* published by LAP Lambert Academic Publishing, Germany in 2019, *Biocontrol Agents of Phytonematodes*, and *Biocontrol Agents: Entomopathogenic and Slug Parasitic Nematodes* published by CABI, UK in 2015 & 2017, respectively. In 2018, he was honoured for outstanding contribution in the field of animal science by Animal Science Congress, held at University of Kashmir, India.

List of Contributors

Abd-Elgawad, Mahfouz M.M. Plant Pathology Department, National Research Center, El-Tahrir Street, Dokki 12622, Giza, Egypt.
E-mail: mahfouzian2000@yahoo.com.

Ahad, Ishtiyaq. Division of Entomology, Faculty of Agriculture, Sher-e-Kashmir University of Agricultural Sciences and Technology of Kashmir, Wadura, Sopore-193201, Jammu and Kashmir, India.
E-mail: drishtiyaqraina@gmail.com.

Akhil, Nune. Division of Entomology, Faculty of Agriculture, Sher-e-Kashmir University of Agricultural Sciences and Technology of Kashmir, Wadura, Sopore-193201, Jammu and Kashmir, India.
E-mail: nunebachi@gmail.com.

Askary, Tarique Hassan. Division of Entomology, Faculty of Agriculture, Sher-e-Kashmir University of Agricultural Sciences and Technology of Kashmir, Wadura, Sopore-193201, Jammu and Kashmir, India.
E-mail: tariq_askary@rediffmail.com.

Aziz, Danishta. Division of Entomology, Faculty of Agriculture, Sher-e-Kashmir University of Agricultural Sciences and Technology of Kashmir, Wadura, Sopore-193201, Jammu and Kashmir, India.
E-mail: daniaziz9974@gmail.com.

Bali, Kamlesh. Division of Entomology, Faculty of Agriculture, Sher-e-Kashmir University of Agricultural Sciences and Technology of Jammu, Jammu and Kashmir, India.
E-mail: balikamlesh76@gmail.com.

Bhat, Aashaq Hussain. Experimental Biology Research Group, Institute of Biology, Faculty of Science, University of Neuchâtel, Rue Emile-Argand 11, 2000, Neuchâtel, Switzerland.
E-mail: aashiqhussainbhat10@mail.com.

Campos, Vicente P. Department of Plant Pathology, Federal University of Lavras, Brazil.
E-mail: vpcampos@ufla.br.

Dass, Wasim Muzamil. Department of Zoology, University of Kashmir, Srinagar, Jammu & Kashmir, India.
E-mail: wasimmuzamil913@gmail.com.

Devi, Gitanjali. Division of Nematology, Assam Agriculturual University, Jorhat, Assam, India.
E-mail: gitanjali_devi@yahoo.in.

Gani, Mudasir. Division of Entomology, Faculty of Agriculture, Sher-e-Kashmir University of Agricultural Sciences and Technology of Kashmir, Wadura, Sopore-193201, Jammu and Kashmir, India.
E-mail: mudasir32@gmail.com.

Gupta, Rakesh Kumar. Division of Entomology, Faculty of Agriculture, Sher-e-Kashmir University of Agricultural Sciences and Technology of Jammu, Jammu and Kashmir, India.
E-mail: rkguptaentoskuast@gmail.com.

Guroo, Mushtaq Ahmad. Division of Entomology, Faculty of Agriculture, Sher-e-Kashmir University of Agricultural Sciences and Technology of Jammu, Jammu and Kashmir, India.
E-mail: gurookvk@gmail.com.

Hassan, Taskeena. Department of Zoology, Aligarh Muslim University, Uttar Pradesh, India.
E mail: taskeenahassan7@gmail.com.

Haveri, Rachappa V. Department of Agricultural Entomology, University of Agricultural Sciences, Raichur, Karnataka, India.
E-mail: rachis1@rediffmail.com.

Jan, Ulfat. Division of Entomology, Faculty of Agriculture, Sher-e-Kashmir University of Agricultural Sciences and Technology of Kashmir, Wadura, Sopore-193201, Jammu and Kashmir, India.
E-mail: sheirulfatent@gmail.com.

Jaspher, Ewany. Experimental Biology Research Group, Institute of Biology, Faculty of Science, University of Neuchâtel, Rue Emile-Argand 11, 2000, Neuchâtel, Switzerland.
E-mail: ewanyjaspher@yahoo.com.

K., Sankari Meena. ICAR-Indian Institute of Oilseeds Research, Hyderabad, Telangana.
E-mail: meenaa5@rediffmail.com.

Keshari, Nishi. Department of Nematology, Rajendra Prasad Central Agricultural University, Pusa, Samastipur, Bihar, India.
E-mail: nishinema@gmail.com.

Khalil, Aroosa. Division of Fruit Sciences, Faculty of Horticulture, Sher-e-Kashmir University of Agricultural Sciences and Technology of Kashmir, Srinagar, Jammu & Kashmir, India.
E-mail: aroosakhalil11@gmail.com.

Khan, Akhtar Ali. Division of Entomology, Faculty of Horticulture, Sher-e-Kashmir University of Agricultural Sciences and Technology of Kashmir, Srinagar, Jammu and Kashmir, India.
E-mail: akhtaralikhan47@rediffmail.com.

Khanum, Tabassum Ara. National Nematological Research Centre, University of Karachi, Karachi-75270, Pakistan.
E-mail: tabassumak@uok.edu.pk.

Khurshid, Maheen. ICAR-Central Institute of Temperate Horticulture, Srinagar, Jammu & Kashmir, India.
E-mail: shawlmaheen@gmail.com.

Kudagamage, Chandrasiri. Former Director General, Department of Agriculture, Sri Lanka.
E-mail: chandrakudagamage@gmail.com.

M., Parthiban. Division of Entomology, Faculty of Agriculture, Sher-e-Kashmir University of Agricultural Sciences and Technology of Kashmir, Wadura, Sopore-193201, Jammu and Kashmir, India.
E-mail: parthiworld8@gmail.com.

Majeed, Ishfaq. Department of Zoology, University of Kashmir, Srinagar, Jammu & Kashmir, India.
E-mail: malikishfaqmajeed@gmail.com.

Mantoo, Mohammad Ayoub. Division of Entomology, Faculty of Agriculture, Sher-e-Kashmir University of Agricultural Sciences and Technology of Kashmir, Wadura, Sopore-193201, Jammu and Kashmir, India.
E-mail: mantooskuastk@gmail.com.

Mehraj, Sumaya. Department of Zoology, University of Kashmir, Srinagar, Jammu & Kashmir, India.
E-mail: sumayamehraj@gmail.com.

Muniz, Maria de Fátima S. Agricultural Science Center, Federal University of Alagoas, Brazil.
E-mail: f.muniz@ceca.ufal.br.

Mushtaq, Tahmina. Division of Entomology, Faculty of Horticulture, Sher-e-Kashmir University of Agricultural Sciences and Technology of Kashmir, Srinagar, Jammu and Kashmir, India.
E-mail: drtahmina@skuastkashmir.ac.in.

N. M., Sunil Kumar. Department of Agricultural Entomology, University of Agricultural Sciences, Raichur, Karnataka, India.
E-mail: nmsunilkumar@rediffmail.com.

Parvaiz, Baber. Division of Entomology, Faculty of Horticulture, Sher-e-Kashmir University of Agricultural Sciences and Technology of Kashmir, Srinagar, Jammu and Kashmir, India.
E-mail: mirbaber31@gmail.com.

Pathania, S. S. Division of Entomology, Faculty of Horticulture, Sher-e-Kashmir University of Agricultural Sciences and Technology of Kashmir, Srinagar, Jammu and Kashmir, India.
E-mail: pathaniasjs@gmail.com

Poornima, G. Department of Agricultural Entomology, University of Agricultural Sciences, Raichur, Karnataka, India.
E-mail: poornimaghyr96@gmail.com.

Prasad, R. D. ICAR-Indian Institute of Oilseeds Research, Hyderabad, Telangana.
E-mail: ravulapalliprasad@gmail.com.

Rafiq, Summira. Division of Entomology, Faculty of Agriculture, Sher-e-Kashmir University of Agricultural Sciences and Technology of Kashmir, Wadura, Sopore-193201, Jammu and Kashmir, India.
E-mail: summirarafiq5@gmail.com.

Rajapakse, Rohan. Department of Agricultural Biology, Faculty of Agriculture, University of Ruhuna, Sri Lanka.
E-mail: rhsrajapakse@gmail.com.

Ramya, A. R., Division of Entomology, Faculty of Agriculture, Sher-e-Kashmir University of Agricultural Sciences and Technology of Kashmir, Wadura, Sopore-193201, Jammu and Kashmir, India.
E-mail: a.r.ramya1998@gmail.com.

Rana, Aasha. Department of Zoology, Shrimati Kamla Devi Gauri Dutta Mittal PG College, Churu, Rajasthan, India.
E-mail: aasha.aasharana@ymail.com.

Rasool, Jameela. Division of Entomology, Faculty of Agriculture, Sher-e-Kashmir University of Agricultural Sciences and Technology of Kashmir, Wadura, Sopore-193201, Jammu and Kashmir, India.
E-mail: malikjameela13@gmail.com.

Ratnasekera, Disna. Department of Agricultural Biology, Faculty of Agriculture, University of Ruhuna, Sri Lanka.
E-mail: disnaratnasekera@gmail.com.

Rehman, Afreen. Division of Entomology, Faculty of Agriculture, Sher-e-Kashmir University of Agricultural Sciences and Technology of Kashmir, Wadura, Sopore-193201, Jammu and Kashmir, India.
E-mail: aafreenrehman100@gmail.com.

Rocha, Fernando S. Phytopathology Research Laboratory, Federal University of Minas Gerais, Brazil.
E-mail: rochafsplant@yahoo.com.br.

S., Lakshmi Prasanna. ICAR-Indian Institute of Oilseeds Research, Hyderabad, Telangana.
E-mail: prasannalakshmi.sama@gmail.com.

Shafi, Insha. Division of Entomology, Faculty of Agriculture, Sher-e-Kashmir University of Agricultural Sciences and Technology of Kashmir, Wadura, Sopore-193201, Jammu and Kashmir, India.
E-mail: inshashafi47@gmail.com.

Shobarani, Department of Agricultural Entomology, University of Agricultural Sciences, Raichur, Karnataka, India.
E-mail: srani.ent@gmail.com.

Sofi, Muneer. Division of Entomology, Faculty of Horticulture, Sher-e-Kashmir University of Agricultural Sciences and Technology of Kashmir, Srinagar, Jammu and Kashmir, India.
E-mail: muneerento@yahoo.co.in.

Sreenivas, A. G. Department of Agricultural Entomology, University of Agricultural Sciences, Raichur, Karnataka, India.
E-mail: agsreenivas@gmail.com.

Sujatha, M. ICAR-Indian Institute of Oilseeds Research, Hyderabad, Telangana.
E-mail: mulpuri.sujata@icar.gov.in.

Tak, Hidayatullah. Department of Zoology, University of Kashmir, Srinagar, Jammu & Kashmir, India.
E-mail: drhidayattak@yahoo.com.

Tegelli, Raju G. Department of Agricultural Entomology, University of Agricultural Sciences, Raichur, Karnataka, India.
E-mail: raju.teggelli@gmail.com.

Terra, Willian C. Department of Plant Pathology, Federal University of Lavras, Brazil.
E-mail: terranema@gmail.com.

Vijayalakshmi, D. Department of Agricultural Entomology, University of Agricultural Sciences, Raichur, Karnataka, India.
E-mail: vcpatil1362@gmail.com.

Wani, Danish Mushtaq. Division of Entomology, Faculty of Horticulture, Sher-e-Kashmir University of Agricultural Sciences and Technology of Kashmir, Srinagar, Jammu and Kashmir, India.
E-mail: danishmushtaq2030@gmail.com.

War, Waseem Ahmad. Division of Entomology, Faculty of Agriculture, Sher-e-Kashmir University of Agricultural Sciences and Technology of Kashmir, Wadura, Sopore-193201, Jammu and Kashmir, India. E-mail: warwaseem78@gmail.com.

Wijayaratne, L.K.W. Department of Plant Sciences, Faculty of Agriculture, Rajarata University of Sri Lanka, Puliyankulama, Anuradhapura, Sri Lanka.
E-mail: wollylk@yahoo.com.

Index

A

B

D

E

F

G

M

N

O

S

W

Y

Z